EARTHQUAKE ENGINEERING IN BRITAIN

Proceedings of a conference organized by the Institution of Civil Engineers and the Society of Earthquake and Civil Engineering Dynamics, held at the University of East Anglia, 18-19 April 1985.

EARTHQUAKE ENGINEERING IN BRITAIN

Thomas Telford, London

Conference organized by the Institution of Civil Engineers in association with the Society for Earthquake and Civil Engineering Dynamics, held at the University of East Anglia, 18-19 April 1985.

Organizing Committee
Mr. D. E. Key, Chairman, Society for Earthquake and Civil Engineering Dynamics
Dr B. O. Skipp, Soil Mechanics Limited (Chairman)
Dr R. R. Kunar, Structural Dynamics Research Corporation
Dr D. Papastamatiou, Delta Pi Associates

Cataloguing in Publication Data
Earthquake Engineering in Britain: proceedings of a conference organized by the Institution of Civil Engineers and the Society of Earthquake and Civil Engineering Dynamics held at the University of East Anglia, 18–19 April 1985.
1. Earthquake engineering—Great Britain. I. Institution of Civil Engineers. II. Society of Earthquake and Civil Engineering Dynamics.
624.1′762′0941 TA654.6

ISBN 0 7277 0246 7

First published 1985

Published for the Institution of Civil Engineers by Thomas Telford Ltd, 26-34 Old Street, PO Box 101, London EC1P 1JH

Printed in Great Britain by Billing and Sons, Worcester

Contents

KEYNOTE ADDRESS

D. KEY, Chairman, Society for Earthquake and Civil Engineering Dynamics

On behalf of the members and committee of SECED I welcome all those participating in the conference to the University of East Anglia.

To introduce the reasons for having this conference it is first necessary to introduce, to many of you, the Society for Earthquake and Civil Engineering Dynamics, SECED. SECED operates under the aegis of the Institution of Civil Engineers and is the British National Section of the International Association for Earthquake Engineering, IAEE. The function of SECED is to promote co-operation and the advancement of knowledge in its field, in pursuance of which it holds meetings and conferences for discussion and publishes papers and reports. It provides a medium for co-operation between industry and research and a channel through which the many specialists concerned in earthquakes and dynamics can express their views - as was recently done on Eurocode 8 (the European Code for Structures in Seismic Regions). It is also publishing the first directory of services available in the fields of earthquake engineering and dynamics, and has recently undertaken a role in conjunction with the Science and Engineering Research Council to act as a link between industry and research.

Membership of SECED covers a wide range of disciplines, including geologists, seismologists, engineering seismologists, civil, structural, mechanical and soils engineers. Membership is open to all those engaged in relevant disciplines and is not limited to members of the Institution of Civil Engineers. Attendance at meeting is open to members and non members alike.

The conference subject is at the interface between two important areas of activity in Britain. Both British earthquake engineering and British seismology are concerned with their discipline world wide. It is a well known fact that, in Britain, we do not have earthquakes, and like most "well known facts" this is not true. However it is true to say that the level of seismicity in Britain is low. The reason for bringing engineers and seismologists together at this conference is the growing realisation, that, for many classes of civil engineering structures, where the acceptable risk of failure is very small, even the low level of British seismicity is a matter of serious concern to the engineer. Such structures include nuclear installations, manufacturing or storage facilities for dangerous chemicals, dams, long span bridges, power stations and off-shore platforms. Within these fields there is a great deal of engineering to be done.

Having emphasised the world wide nature of British engineering, the international character of British seismology cannot go without mention. Although Chang Heng invented the first seismoscope in AD 132, the origins of the modern seismograph, which made possible modern studies of seismology, are in John Milne's horizontal pendulum seismograph developed in the period 1880 to 1890. During that period Milne, together with Sir James Ewing and Thomas Gray, was one of the co-founders of the Seismological Society of Japan. Since then, British seismologists have played an important role in the expanding study of international seismology.

Due to the random distribution of SECED chairman by profession the selection process has resulted in a structural engineer addressing a gathering whose distribution is strongly biased towards seismology and engineering seismology. To those of you concerned with pure seismology I offer you the hope that any subsequent conference will coincide with a seismologist SECED Chairman. To those concerned with applied seismology I would like to give the viewpoint of, so to speak, the customer, or, at least, one who has to apply data on ground motion to engineering projects.

A seismologist said to me, some years ago, "Engineers always ask me for a ground acceleration, but I don't know how they make use of it". The reason why engineers asked for an acceleration was, that in the early days of earthquake engineering, that was all we could get! As to how engineers made use of it, there appeared to be two answers. One group of engineers thought that, if you multiplied such an acceleration by a mass, you obtained a force. The second group used it as a guide to the damaging potential of an earthquake. The same seismologist then asked, "What information do engineers really need?" to which the answer was simple, "Everything we can get!"

Now, in 1985, we can, in some cases, get a great deal of information on estimated future ground motion, and I am sometimes concerned that so much information can be deduced from so little certain knowledge of what exists beneath the earth's surface, and too brief a historical record.

The degree to which prediction of earthquake motion, based on statistical data, can be considered reliable depends on the amount of data. In areas of low seismicity the amount of data is small so that the reliability of prediction is low. How accurate are ground motion estimates in Britain compared with, say, California? How much conservatism is being built into estimates by the geologist and the seismologist which compound the conservatism of the engineer?

Engineers are, by training, concerned with experience. What they learn from their own practice forms a basis on which theoretical work can be judged. In areas of high seismicity most practising engineers will have seen earthquake damage. In areas with rare earthquakes engineers will generally not have had that experience and will have no instinctive feel for the value of information that they are given on design ground motion. Furthermore, in new areas of construction such as nuclear engineering, there is very little experience of earthquake damage. Without his customary background of practical experience the engineer is more dependant on the quality of information on ground motion with which he is supplied.

In building design practice today, engineers learn new, and sometimes unpleasant lessons from each earthquake that occurs in a populated area. Without these lessons current practice would be substantially different, and certainly less effective. In an area of low seismicity, and especially where new types of structure are being designed we are in danger of developing a whole technology based on theory, without practical experience.

Too much sophistication in the prediction of ground motion may be counter productive. Engineers are given a false sense of security and feel that they are designing for a real event instead of a statistical possibility. Furthermore uncertainties in the engineering design process for earthquakes are many, both in material behaviour and in analysis, so that the refining of ground motion predictions may have no real influence on the final structure.

Perhaps this conference will help engineers and engineering

seismologists to better understand each other. Engineering for earthquakes is still, or should be, strongly based on commonsense and judgement, despite all the magical data turned out by computers, in which we sometimes feel we are getting buried. Is the same true for engineering seismology?

In conclusion, I hope this conference will benefit all the participants. In the course of these two days perhaps we can consider whether this type of forum, bringing together the many disciplines concerned with earthquakes and earthquake engineering, can be extended, repeated, or modified.

EARTHQUAKE ACTIVITY IN BRITAIN

1. The geography and intensity of earthquakes in Britain—the eighteenth century

C. P. MELVILLE, MA, PhD, Earthquake Documentation and Research

SYNOPSIS. This paper presents a catalogue of earthquakes in Britain during the 18th century, and examines to what extent the distribution and Intensity of reported seismicity is influenced by cultural and geographical factors such as source availability and population density. The problems of locating and assessing events in this period are discussed, to establish how confidently such macroseismic data can be applied to seismic hazard analysis or seismo-tectonic understanding of the region. Comparison with the distribution of more recent events shows that this interval is broadly representative of the overall long-term pattern of British earthquakes and reveals pronounced regional variations.

INTRODUCTION

1. Recent reassessments of British seismicity (refs 1-2) have tended to see 1800 as a turning point in the quality and geographical homogeneity of the macroseismic data available. Certainly a number of factors, such as a great increase in sources of information and improvements in communications, particularly after around 1850, make the record of earthquakes in Britain during the 19th century much fuller than at any time previously and allow more confident and positive identification of individual events.

2. Although the period 1750-1850 could perhaps be preferred as a coherent unit in which British earthquakes are first reported prolifically, the 18th century as a whole may be seen as the first period for which it is possible to make a systematic assessment of seismic activity. While difficulties remain in interpretating the accounts of many individual earthquakes, the completeness of the record as a whole can be assessed on the basis of both positive and negative information, at least in the case of England. The quality of the data is sufficient to suggest that the period of observation can be extended back well beyond 1800 to establish a coherent sample of events for seismic hazard analysis, or regional seismo-tectonic studies.

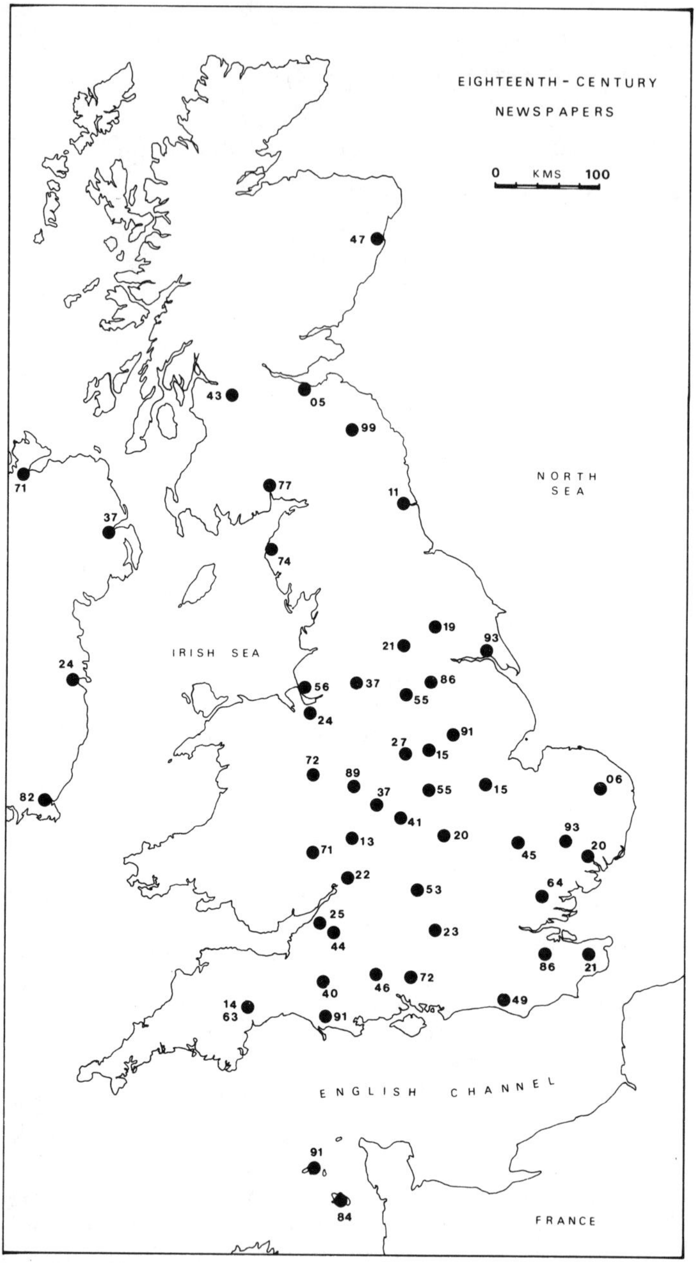

Fig. 1. Places apart from London where newspapers were being published in 1792 [England] and 1793 [Scotland] (ref. 3). A few new places were represented after this date. The numbers indicate the year, from 1700, from which good coverage survives. Several other places had produced papers earlier in the century, which lasted only a few years. Data for the Channel Islands are from local collections; for Ireland see refs 4-5.

SOURCES OF INFORMATION

3. The most important documentary innovation in the 18th century is the appearance and rapid growth of newspapers. The lapse and non-renewal of the Printing Act in 1695 set the scene for full-scale development of the press and London's first daily paper appeared in 1702. Weekly provincial papers were produced around the same time, particularly in Norwich and Bristol, the two largest cities outside London. The significance of this development for the recording and survival of macroseismic data is obvious, since an earthquake is an eminently newsworthy "event". The temporal coverage of events in publications which appear at regularly-spaced and brief intervals is thus close from the beginning of the period.

4. The geographical coverage of events in Britain provided by the provincial press is illustrated on Fig.1, which depicts the places where newspapers were published in 1792, most of them since early in the century. Many provincial papers contained little or no local news in the early period, but earthquakes are generally noted when they occurred. The London press received news from the whole country, reflecting the position of the capital as hub of the dense network of roads that made up the communications system (see ref.6). The survival rate of many of the earliest provincial papers is low, but the London press should provide a record, albeit incomplete, of most significant shocks that occurred. Furthermore, although the papers themselves may no longer be available, the information they contained should have been absorbed into other documentary sources, such as local histories. It will be observed from Fig.1 that the most obvious areas without papers are the Cornish peninsula, the whole of Wales, and the Highlands of Scotland. Northern England though poorly represented is covered by the Newcastle press since an early date. The greatest concentration of publications is in the north and west Midlands, reflecting population trends and industrial developments.

5. Personal memoirs, diaries and correspondence supplement the record derived principally from the press, and in many cases can add useful evidence of the felt area of an earthquake imperfectly identified elsewhere. The survival rate of 18th century private papers is quite high, but the geographical coverage they provide cannot be compared with that of the press. Diaries, whose record is always exactly contemporary and chronologically precise, can nevertheless be used systematically and may contain useful negative evidence of an earthquake, since failure to mention a known event within a given area may be significant for Intensity assessment, depending on the character of the diary.

6. In addition to these sources, technical papers and communications addressed to the Royal Society by the Fellows or their correspondents form a large and important body of disciplined observational and descriptive data about many 18th century earthquakes. Interest was particularly aroused in 1750 by the two London earthquakes and the numerous shocks in

different parts of the country later that year, followed by the catastrophic Lisbon earthquake in 1755. Most of the largest British events are mentioned in some form in the Philosophical Transactions of the Royal Society, which can be supplemented by the Society's archives and the personal papers of the interested parties.

7. Such sources, although less numerous, are similar in kind to the main sources of information during the 19th century, and allow direct comparisons between the two periods.

GEOGRAPHICAL FACTORS

8. The growth in volume of documentary information reflects partly a higher survival rate of the documents themselves, combined with a rise in the number of the reading and writing public, and partly the increased interest in environmental phenomena on the part of the natural philosophers of the Enlightenment (ref.7). It also reflects a growth in the population as a whole (approximately 57% for England and Wales, from under six to over nine millions, see ref.8) and their concentration in buildings and communities where their susceptibility to earthquakes would be increased. Assuming a constant level of earthquake activity, it is clear that more shocks will be reported if more people are affected, particularly if concentrations of people are involved, with the greater scope for news to be communicated and documented that is present in an urban society.

9. The rise in population in England and Wales in the 18th century and their greater concentration in urban environments, are illustrated by Corfield (ref.9). The rate of urban growth exceeded that of the population as a whole, although the towns remained relatively small, and thanks to the development of large provincial centres, England was one of the most densely populated countries in the world by the end of the century. There was also an important geographical dimension to change, with a pronounced shift from the south to the north and the industrial Midlands. Compared with later periods, however, the population was evenly spread and remained predominantly rural (ref.8). The least densely settled areas were the moors of Devon and Cornwall, the whole of Wales, Lincolnshire and northern England: regions characterised by obvious features of physical topography and relatively low agricultural potential. No comparable data are available for Scotland: the total population in 1707 was around 1,048,000, but its regional distribution is unknown. In 1755, more than half the population lived in the north, but the trend of movement to the central lowlands was already well marked by the end of the century (ref.10).

10. As noted by Corfield, the rise of the newspaper industry in 18th century England was a product, as well as a chronicle, of urban life. The growth of towns in this period is accompanied also by the production of local town histories, many of which contain annals of events arranged (in the case of the boroughs) under the year of the mayor in office. Many of the

places plotted on Fig.1, and a few others, produced such histories by the end of the century, thereby preserving much antiquarian material and records of events that might otherwise be lost or difficult to reassemble.

EARTHQUAKES IN 18TH CENTURY BRITAIN

11. Table 1 lists all the apparently genuine earthquakes reported for the period 1700-1799 in the sources discussed above. Not all the events in the catalogue have been studied in the same detail, but an attempt has been made to improve on previous lists, by the retrieval of additional information and the removal of false entries: these are mainly found on closer examination to be associated with meteorite falls (such as the east Midlands event of 28 October 1776). Evidence for others (such as the January 1756 event in the Scottish Highlands) is ambiguous and they have been retained for the present. Aftershocks are not listed, but earthquakes associated with additional shocks are indicated and details will be found in the sources referred to. The full list of shocks in the Comrie swarm from 1789-1799 is not given, but the main events in the early period of this activity are listed, and the most widely felt thereafter. Some confusion surrounds the events before 1792 (largely because of the coincidence of their dates). After 1792, the lists given by Milne and Davison (refs 11-12) are acceptable: but these events have not been confirmed in local news reports.

12. Some small shocks in particular are still based on very few descriptive data, and a few are included which further information may reveal to be spurious. A report of a damaging shock round Methven near Perth on 17 November 1789 was later denied as being without the smallest foundation: "Such accounts are of an alarming nature, and ought not to be circulated, without being well authenticated" (ref.13). With this in mind, the catalogue may be found to err on the "alarmist" side.

13. Reference to sources is kept to a minimum; only those which contribute significantly to the data found by previous studies are noted. It is certain that further research, undertaken on a local level for the whole country, would unearth additional details about some events and perhaps add a few minor shocks, previously unidentified, to the record. Such additions are unlikely to be significant and would need to be numerous (perhaps twice the number of events already retrieved) to have any influence on the completeness of the data sample. As is discussed below, it is considered that the catalogue is complete for the larger events occurring in England.

14. Few if any genuine events have been identified for Ireland during the period: of the shocks listed by Davison (ref.12), most are to be associated with bog-flows or distant events (in Lisbon and Iceland [sic.]).

Table 1. CATALOGUE OF EARTHQUAKES 1700–1799

Date	Time hr/min	Location	Epicentre °N	– °W/E	Q	I-max (Io)	r_4	r_3	Ref.
1703.12.28	1700	N.Lincolnshire	53.5	0.5	P	V	50		2.
1726.10.26*	0700	Dorchester area	50.6	2.4	G	VI–	30+		1.14.
1727.07.19*	0400	Margam	51.6	3.7	G	(VII)	220		15.
1728.03.01	0430	Borders	55.6	2.8	G	V–	30	70?	16.
1731.10.10*	0400	Northamptonshire	52.0	1.3	G	(IV)	15?		1.
1732.07.09	1400	Argyll	56.0	5.3	P	IV+	60		1.
1734.10.25	0350	Channel	50.2	0.7	P	V	90		16.17.
1734.10.28	0300	Derby area	52.8	1.3	P	IV	15		1.
1736.04.30*	1200	Ochil Hills	56.3	3.8	P	V+	15+		16.
1737.05.13	0130	Ludlow area	52.4	3.0	P	IV	12		1.
1738.07.10	0330	Derby area	52.7	1.3	P	V	15?		1.18.
1738.12.30	0700	Halifax area	53.7	1.8	G	(V–)	15		19.
1744.02.05*		Dolgellau	52.7	3.9	P	IV+			1.
1747.07.01	1030	Taunton area	50.9	3.2	P	V	40		1.20.
1748.06.26		N.Derbyshire	53.3	1.9	P	IV	3		1.
1749.02.04*	0830	Leadhills	55.3	3.7	P	(IV)	10?		1.
1749.05.17	1000	Wimborne	50.8	2.0	P	V+	30		1.21.
1750.02.08	1240	London	51.5	0.1	G	(V+)		15+	2.
1750.02.09	0100	Plymouth	50.4	4.2	P	IV			1.
1750.03.08	0530	London	51.5	0.2	G	(V+)		20	2.16.
1750.03.18*	1745	Channel	50.0	1.7	P	IV	95?		1.
1750.04.02	2200	N.Cheshire	53.2	2.8	P	V–	25	55	16.19.
1750.08.23	0645	Lincolnshire	53.2	0.5	G	IV		55+	2.
1750.09.30	1230	E.Midlands	52.6	0.7	G	(V+)	36	60	1.2.
1752.02.23	1430	E.Dartmoor	50.6	3.8	G	V+	8		1.
1752.03.27	1130	Somerset				IV			1.22.
1753.06.08	2330	Skipton area	53.8	2.0	P	V	50		19.
1754.04.19	2300	Offshore Whitby	54.5	1.4	P	V–	50		2.16.
1755.08.01	0655	E.Midlands	53.1	0.5	P	IV		55+	2.
1755.11.17	1900	Irton, Cumbria	54.4	3.4	G	(V+)			19.
1755.12.31	0100	Glasgow area	55.9	4.6	P	IV+	16		1.
1756.01		Rothiemurcus (?)	57.2	3.8	P	IV+			23.24.
1756.01.05	0830	E.Dartmoor	50.6	3.6	G	IV			1.
1756.06.01	2100	Ashford	50.2	0.9E	P	IV+			2.25.
1756.11.17*	2350	Argyll	56.0	5.1	P	IV+	25?		1.
1756.12.26		Redruth area	50.2	5.2	G	IV	6		1.26.
1757.01.10	0230	E.Norfolk	52.0	1.4E	G	IV	20		2.
1757.05.17	1500	Todmorden	53.8	2.1	P	IV	24		19.
1757.07.17	1815	W.Cornwall	50.1	5.3	G	(V)	30		16.
1758.01.24	0200	E.Sussex	51.1	0.0	G	(IV)	6		1.
1759.02.24	2200	Liskeard	50.5	4.5	P	IV	8		1.
1761.02.06	2330	N.Dorset	51.0	2.4	G	(IV)	7		1.
1761.06.09	1145	Shaftesbury	51.0	2.3	P	V	20?		1.27.
1761.08.14	1200	Guernsey	49.5	2.5	P	IV			1.
1764.11.06	0415	S.Oxfordshire	51.7	1.5	P	IV	30		1.
1766.02.10	2330	Mid-Glamorgan	51.7	3.5	P	IV–	10		16.
1766.11.15	1830	Altarnun	50.6	4.5	P	IV			28.
1767.04.20	0900	Stirling area	56.2	3.9	P	IV	8		1.
1768.01.03	0100	Rugby area	52.4	1.2	G	(IV)	5		1.
1768.01.18	0215	Mold (?)	53.1	3.1	P	V–			1.19.
1768.02.15		Llangollen (?)	52.9	3.2	P	IV			1.
1768.05.15	1615	Swaledale	54.5	2.0	P	V		80+	1.19.
1768.10.24		Inverness	57.4	4.2	P	V	12?		1.

Date	Time hr/min	Location	Epicentre °N	- °W/E	Q	I-max (Io)	r 4	r 3	Ref.
1768.12.21.	1700	Worcestershire	52.1	2.2	P	V	25		1.
1769.04.02	0430	N.Devon	51.0	3.5	P	V-	22		29.
1769.11.23	0400	Birmingham area	52.6	1.9	G	IV+	15		30.
1771.04.29	1730	Abingdon	51.6	1.1	G	(IV)	15		31.
1771.07.		Scilly Isles	49.9	6.3	P	IV			1.
1771.08.24	0400	Congleton area	53.2	2.2	P	IV	10		19.
1772.02.17	2355	Shrewsbury	52.7	2.8	P	IV			1.
1772.08.15	1000	Noah Hill	51.1	0.9	P	IV-			32.
1772.11.15	0400	Birmingham area	52.5	1.9	G	IV	6.		1.
1773.04.23*	2330	Channel Isles	49.4	2.2	P	V	110		16.17.33.
1773.07.03*	1630	Shiffnal area (?)	52.6	2.5	P	IV	12		34.
1774.10.25	2200	S.E.Exeter (?)	50.7	3.5	P	(IV-)	1		1.
1775.09.08	2200	Swansea	51.6	4.0	G	(VII-)	120+	215+	15.16.
1776.11.27	0815	Strait of Dover	51.0	1.5E	P	IV	45		2.
1777.09.14	1100	Manchester	53.5	2.4	G	(V+)	40	100	1.19.
1780.08.29	0845	N.Wales	53.1	4.3	P	V+	50		1.
1780.12.09	1630	Kirkby Stephen	54.6	2.1	P	V		90?	1.19.
1781.01.25	0030	Shrewsbury	52.7	2.8	P	V-			1.
1782.10.05	2040	Anglesey ?	53.3	4.0	P	IV	30?		16.
1783.04.23	1315	Coalbrookdale	52.6	2.5	P	IV-			34.
1783.08.09	1600	W.Dartmoor	50.5	4.2	P	V-	35?		35.
1786.08.11	0200	Cumberland	54.5	3.5	G	(VI+)	130+	220	1.16.19.
1787.01.06	1030	N.Glasgow	56.0	4.3	G	IV	10		1.
1787.07.06		Cumberland	54.5	3.0	P	IV	10		1.19.
1789.05.03	0315	Barnstaple	51.1	4.1	G	(V)	20		1.36.
1789.05.16	0615	Robin Hood's Bay	54.4	0.5	P	IV			37.
1789.09.02*	1100	Comrie	56.4	4.0	G	(IV+)			1.38.
1789.09.26		Wenlock, Shrops.	52.5	2.5	P	IV-			39.
1789.11.05*	1755	Comrie	56.4	4.0	G	(V)	25		1.38.
1789.11.10*	1500	Comrie	56.4	4.0	G	(V-)			1.
1789.12.05*		Offshore Sussex	50.8	0.5E	P	IV-		12	1.40.
1789.12.29*	1300	Comrie	56.4	4.0	G	(IV)	5		1.
1790.03.06		N.Wales coast (?)							1.
1791.09.02*	1705	Ochtertyre	56.5	3.9	P	IV-			1.
1791.09.27	2100	Jersey	49.2	2.1	P	IV			1.
1792.01.15	0400	W.Cornwall	50.3	5.0	G	IV-	5		1.
1792.02.25	2040	E.Midlands	52.8	0.6	G	IV+		40	2.
1792.04.05*	2140	Comrie	56.4	4.0	G	(IV-)		12	41.
1792.05.26	2300	Bodmin area	50.5	4.5	P	IV+	c.25		42.
1792.05.28	1100	St. Teath	50.7	4.5	G	(V+)			43.
1792.10.10*		Comrie	56.4	4.0	G	(IV-)			41.
1792.11.10	0530	Loch Rannoch	56.7	4.3	P	IV-			16.2.
1793.09.28*	1600	N.Dorset	51.1	2.0	G	(V)	16		44.45.
1794.07.14		Tavistock	50.5	4.1	P	IV-			1.
1795.03.12*	2300	Lochearnhead	56.4	4.3	P	V-	16		1.
1795.07.17	0330	Manchester area	53.5	2.4	P	IV	12		19.
1795.11.18	2300	Derbyshire	53.0	1.5	G	(VI)	80	140+	1.16.2.
1796.09.20	1420	W.Cornwall	50.1	5.5	P	IV			1.
1797.08.04	2300	W.Argyllshire	56.4	5.5	P	VI-	50+		46.

Note 1) When more than one shock is reported, the event is marked with an asterisk (*). 2) All Longitude co-ordinates given are West of Greenwich unless otherwise indicated. 3) For discussion of the quality assigned to the epicentre (Q), see text. 4) When maximum Intensity observed is considered likely to be the epicentral Intensity, the value is given in brackets.

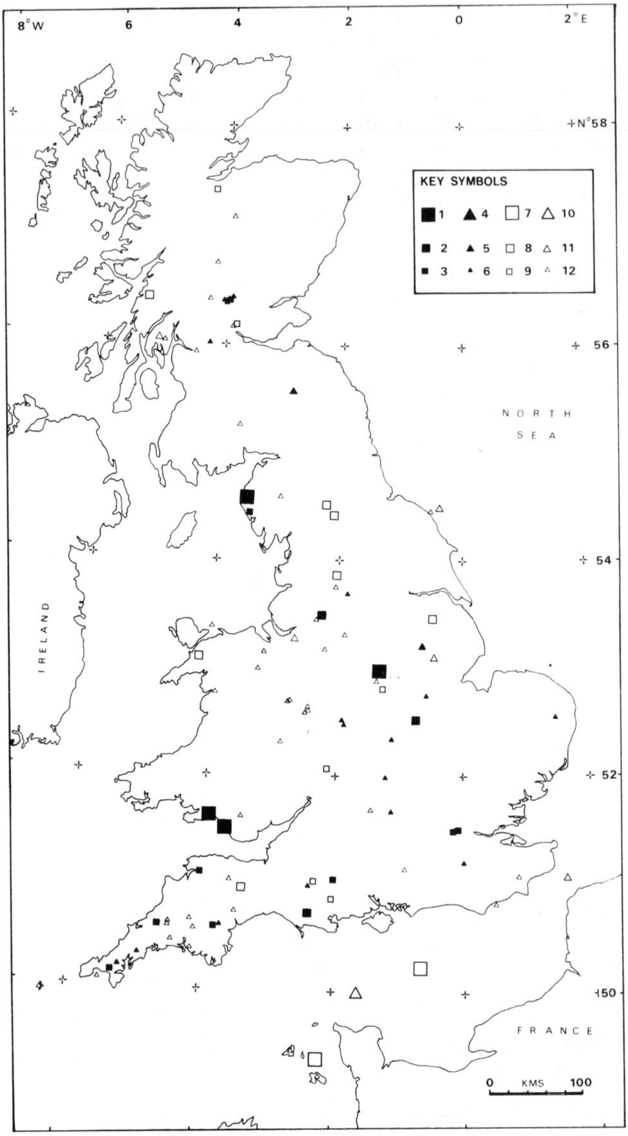

Fig. 2. Distribution and Intensity of British earthquakes in the 18th century. Key: 1,4,7,10: Large events, with a radius (r_4)$\geqslant$75 kms; 2,5,8,11: Medium events, with radius 40 kms$\leqslant r_4 <$75 kms; 3,6,9, 12: Small events, with a radius $r_4 <$40 kms; Squares denote earthquakes associated with high Intensities, $I_{max} \geqslant$ V (MSK); Triangles denote events with low Intensity, $I_{max} <$ V (MSK); 1-6 are well located events; 7-12 are poorly located. E.g., 5 denotes a medium-sized well-located earthquake with a low maximum Intensity.

15. The earthquakes are identified by their location, maximum observed Intensity, and radius of perceptibility, and are plotted on Fig.2. The figure is designed to give prominence to the largest and best-located events with the highest reported Intensities, since these are the most significant. Problems associated with the assessment of these parameters are interdependent and are also influenced by the cultural and geographical factors discussed above.

Epicentral location

16. The earthquakes in Table 1 are located by approximate geographical co-ordinates with a quality factor (Good/Poor). On the scale plotted, and in "global" terms, all the locations are good: that is, the earthquakes occurred nearer the place indicated than anywhere else, so far as it is possible to judge. With very few exceptions, all the earthquakes are associated with particular places. On a local scale, these epicentres have a margin of inaccuracy of anything from 2 to 20 kms, depending on the size of the shock, and are thus less good for the purposes of site-specific hazard assessment or for attempts to associate seismicity with individual faults.

17. On Fig. 2, 60% of the earthquakes have been assigned a Poor location quality, which effectively implies that the information on which the location is based is itself poor. It should be observed that the proportion of Good to Poor locations remains more or less constant throughout the period, despite the greater wealth of data in the latter half of the century. Different aspects of this problem can be identified.

18. Good location implies a detailed account of the earthquake, with evidence both of epicentral Intensities and the extent of the felt area. The best located events are those which caused concentrated damage, or for which the geographical centre of the felt area corresponds with the highest Intensities, allowing I_{max} to be taken as I_o. Several local shocks can also be located confidently on the grounds that the area affected was small and reasonably well defined, or on the basis of the observed Intensity.

19. By contrast, Poor locations are applied to: 1) offshore epicentres; 2) earthquakes apparently affecting a relatively large area, but with no indications of epicentral Intensities; 3) similar earthquakes for which no information is available for the region corresponding to the geographical centre of the apparent felt area; 4) small events for which information is restricted to the effects at only one place, which may or may not be the epicentral area; and 5) dubious events, which may be the result of industrial or mining activity (e.g. in the Coalbrookdale or Leadhills areas) or incorrect original reports.

20. It is clear, therefore, that the problem of locating early events is closely bound up with the problem of Intensity assessment and, more particularly, of identifying the felt area of the earthquake. On the whole, large and damaging shocks are well located; but there is proportionately less

margin for error with small localised events. The distribution of population or source availability may be such that details are lacking for some areas affected. Cases 2) and 3) above represent the frequent situation where the felt area of a shock can be established approximately, but there is no evidence to associate the event with a particular epicentre. Offshore earthquakes clearly defy confident location, but are indicated where predominantly coastal areas are affected and no obvious on-land epicentre presents itself.

Intensity assessment

21. Events are distinguished on Fig. 2 by their maximum Intensity, V (MSK) and greater being considered "high" and under V "low". This distinction is generally easy to apply for any one place. The chief difficulty in the period under survey is distinguishing Intensities at lower levels, away from the epicentral region and in drawing isoseismals to measure Intensity decay. This is particularly critical for the outer isoseismal which defines the felt area of the event (see below).

22. Before the 18th century the fact that an event is reported at all suggests it was felt at a minimum of Intensity IV; little or no negative information is available. The problem is then one of identifying strongest effects of the shock, when reports would only be likely to mention significant damage. In the 18th century, reports are more numerous and minor damage (bricks off chimneys, fissures in plaster) can be mentioned, as well as negative indications. But with this greater range of detail, the simple distinctions become blurred.

23. Frequently only one account is to hand for the effects of the shock at a given place. By picking on the most remarkable features of the shock, an exaggerated impression of the Intensity may result, and data are insufficient to establish the mode value. When no effects on objects or houses are noticed (were people as observant of cracks in their walls as they are now, with insurance companies and potential buyers never far from the back of one's thoughts ?), information is restricted to the effect on people. Problems may arise with conflicting reports or the use of such subjective epithets as "great", "slight", "smart", "sensible" or "amazing" (the latter signifying "frightening" at this period). Slight shocks of brief duration are still capable of causing considerable alarm to susceptible people: in the 18th century, as witnessed by the reaction to the earthquakes of 1750, 1755 and 1777 (in Manchester), a religious dread could be added to the simple physical response. Many reports contain statements that the earthquake was "also felt at" various places, mentioned without details. The result is that an average level of Intensity is suggested for the whole affected area, with elements of Intensity V mingling with elements of Intensity III right out to the edge of the region, and no certain way to discriminate different grades.

24. Depending on the time of day (and even day of the week) of the shock - the quiet of the night or of the sabbath contrasting with the bustle of a market day - and environment in which it was noticed (urban or rural), it is often possible to distinguish Intensity III from Intensity IV reports if the descriptive accounts are adequate, and due consideration is given to possible evidence of the shock not being experienced. This distinction is normally confined to the largest or most damaging earthquakes.

Radius of perceptibility

25. The thresholds at which events are perceived affect the reporting of shocks of low Intensity and thus the identification of the felt area of the earthquake; this in turn classifies its size. Events on Fig.2 are identified as large, medium or small, on the basis of their apparent felt areas. As noted above, the total area can often be determined quite well: the problem is with interpreting the Intensity down to which details are preserved. In practice, r_4 (radius of isoseismal IV) remains the most common limit of (recorded) perception in the 18th century: indications of r_3 are the exception (this situation alters in the 19th century). This level is a function primarily of the limits to the documentary evidence available. While, away from the cities, environmental noise was probably lower than in the present century, there is no reason to suppose that people were more fearful or sensitive than they are now.

26. Table 1 indicates the radius of perceptibility r_4 where known and/or r_3 in a few cases. Despite problems in individual cases, which have to be judged in the context of when and where the earthquake occurred, there is no difficulty in distinguishing the largest events and little difficulty with the medium events. However flexible such classifications, however, there will always be marginal cases.

27. Assessment of such cases can be helped by considering the maximum Intensity value: a shock such as the 26 October 1726 earthquake near Dorchester must be considered a medium-sized event, even though its radius of perceptibility cannot be identified fully beyond about 30 kms. Some of the events occurring in and soon after 1750 are also marginal: their reported felt areas exceed 40 kms, but the threshold of sensitivity in these cases was probably lowered by the impact of the earthquakes in London at the beginning of the year. Even so, the total area in which the shocks should have been experienced at Intensity III was probably considerably larger than the values listed.

28. Similar considerations apply to marginal medium to large earthquakes. For some events in sparsely populated regions such as the north Pennines it is possible to do little more than identify a reported felt area in excess of 75 kms radius, likely to be an exaggeration for r_4 and an underestimate for r_3. The importance of population density and newspaper prominence is clearly illustrated by comparing the 1777 Man-

chester earthquake with an event in Wales three years later. The 1777 earthquake affected a densely populated area and was widely reported in the press: such news stimulates publication of additional reports (indicated by the duration of coverage of the event) and details that may otherwise have gone unreported are thus preserved. The 1780 earthquake in remote Caernarfonshire should have been experienced at Intensity III in places such as Liverpool and Shrewsbury, but the local papers did not even mention the event. Its low impact meant that there was nothing with which to associate otherwise uncoordinated individual observations that may have been made.

29. The largest events (with radii in excess of 75 kms) are clearly identified from the historical record, both by damage associated with them and by the level of information preserved in numerous sources throughout the region affected. Seven such earthquakes occurred in the period, three of them in the English Channel or offshore northern France. The chances of an on-land earthquake of this size escaping attention during the 18th century are negligible.

Fig. 3. Time-distribution of 18th-century earthquakes, by decade. Only those listed on Table 1 are plotted and a distinction is made between Comrie and other events after 1780. In the last decade 1790-99 Davison (ref.12) lists another 45 tremors in addition to those on Table 1. The inclusion of spurious reports would make the change after 1750 even more conspicuous.

COMPLETENESS OF THE RECORD

30. The time distribution of earthquakes catalogued in Table 1 is shown in Fig. 3. It is clear that there is a significant increase in reported shocks after 1750, the catalyst being the London events of that year. After the initial excitement died down, the reporting of earthquakes stabilises from the 1760s at a higher level than before and one that is not far from the 19th century level. It is clear, in view of the factors noted above, that the record of events is incomplete, particularly for the period 1700-1749 but also for earthquakes whose felt areas were small (e.g. less than about 20 kms radius) and for tremors whose maximum Intensity barely reached IV (MSK). Such events are unsung and unlamented (their absence is one of the attractions of 18th century earthquake history); their number is unimportant, but could be estimated from the more recent record of felt reports, and must be nearly twice as many again as the list in Table 1. Medium and large events should be preserved (if not fully identified) more completely: more can be learnt of the sample from its spatial distribution.

31. Comparison of Fig.2 with Fig. 1 and the maps in Corfield (ref.9) illustrates certain points of interest. Working round the country, one sees first that the record of events in Cornwall is relatively good, despite its remoteness, low concentration of population and absence of a local newspaper. This argues for a good data sample. The absence of information from Wales is matched (and explained) by a lack of source material for all but the largest events. Western England and the Midlands are well covered by documentary sources and contain a large number of towns: the information from this area should be relatively complete after 1750. Information from northern England is likely to be less complete for all but the largest events, in view of the physical topography of the region and the low population density; but given the importance of Newcastle and York and their long publication history, the absence of reported seismicity in the northeast is likely to be genuine. The absence of data for Scotland clearly reflects the fact that circumstances were not conducive to the survival of information in the Highlands (even, before 1750, for the largest events), but coverage even of small shocks is available for parts of the the central Lowlands. By contrast, the low level of seismicity reported in East Anglia and the southeast must be representative of the situation, in view of the number of newspaper sources available and the proximity of the region to London. Southern England is also a relatively sensitive area, as indicated by the reporting of shocks of low Intensity originating from offshore in the Channel.

32. It is also clear from Figs. 1 and 2 that epicentres of reported events are not unduly attracted to a main urban centre and that the network of communications is adequate to locate earthquakes in the country. This is the case even where the concentration of towns is greatest, as in the Midlands.

DISTRIBUTION OF BRITISH EARTHQUAKES

33. A final trial of the relevance and representativeness of the sample of earthquakes in the 18th century is to compare it with more recent periods. Principia Mechanica maps (ref.1) plot all reported events from 1800 to c.1960 and the more significant earthquakes thereafter. The size of the events is based on their radius r_3 and is not directly comparable with the 18th century classification on Fig.2, but medium and large 18th century shocks would both be equivalent to the largest events (grades E-F-G) after 1800 and small 18th century shocks would be divided between the two smaller categories (grades A-B and C-D) in ref.1.

34. Our assessment of the completeness of the 18th century record is largely borne out by this comparison, which reveals similar seismicity in Cornwall, south Wales and the Midlands. Central Wales, however, remains free of events, suggesting that absence of data in the 18th century is not simply a function of poor sources of information. The picture for Highland and western Scotland is demonstrated for the first time in the period after 1800, but the lack of information for eastern Scotland, the southern highlands and northeast England is reinforced, as is the situation in the southeast. One area where the 18th century record is not matched more recently is Dorsetshire, where a number of shocks are reported between Salisbury and Dorchester that have not been repeated.

35. More detailed comparisons and analyses must await the unification of the data into a single homogeneous catalogue - we have tried to show how far the 18th century record is suitable for incorporation into such a consistent sample. Many of the problems of precise location and Intensity mapping do not disappear after 1800 and there are good reasons to believe that a homogeneous sample (in terms of data quality for individual events) can be achieved from at least 1750 if not earlier. Geographical coverage may be homogeneous for England and Wales and events of radius larger than about 20 kms from that date, although the same cannot be maintained in the case of Scotland. In terms of the large events, however, the 18th century record is clearly sufficiently detailed to provide confirmation of the regionalisation of seismicity in Britain that is suggested by more recent macroseismic and instrumental data.

ACKNOWLEDGEMENTS

Such an accumulation of data cannot be achieved single-handed and I would like to acknowledge the assistance of many librarians and archivists throughout the country. I would also like to thank Caroline Robertson for her successful foray into the Glasgow newspapers, and Dimitri Papastamatiou for encouraging the production of this paper.

REFERENCES

1. PRINCIPIA MECHANICA LTD. British earthquakes: an assessment for CEGB, BNFL, SSEB, June 1982. Report no.115/82.
2. AMBRASEYS N.N. and MELVILLE C.P. Seismicity of the British Isles and the North Sea, vol.1. Imperial College, London, 1983.
3. CHALMERS G. The life of Thomas Ruddiman. London, 1794.
4. MADDEN R.R. The history of Irish periodical literature, 2 vols. London, 1867.
5. MUNTER R.L. A handlist of Irish newspapers 1685-1750. Cambridge Bibliog.Soc. Monograph no.4, London, 1960.
6. OGILBY J. Brittania, London, 1675.
7. TAYLOR J. Eighteeenth-century earthquake theories: a case-history investigation into the character of the study of the earth in the Enlightenment. Unpub. Ph.D. thesis, University of Oklahoma, 1975.
8. DARBY H.C. (ed.) A new historical geography of England after 1600. Cambridge University Press, 1976.
9. CORFIELD P.J. The impact of English towns 1700-1800. Oxford University Press, 1982.
10. KYD J.G. Scotland's population. Scottish Historical Review, 1949, vol.28, 97-107.
11. MILNE D. Notice of earthquake-shocks felt in Great Britain, and especially Scotland. Edinburgh New Philosophical Journal, 1841, vol.32, reprinted with additions, 1887.
12. DAVISON C. A history of British earthquakes. Cambridge University Press, 1924.
13. CALEDONIAN MERCURY, 30 November 1789.
14. NORRIS'S TAUNTON JOURNAL, 4 November and 2 December 1726.
15. MELVILLE C.P. The seismicity of England: four early earthquakes in western Britain. Bollettino di Geofisica Teorica ed Applicata, 1983, vol.25 no.97, March, 61-79.
16. SOIL MECHANICS LTD. Re-assessment of UK seismicity data, for CEGB, August 1982.Report no.7984.
17. VOGT J. (ed.) Les tremblements de terre en France. BRGM, Orleans, 1979, Mem. no.96.
18. DERBY MERCURY, 13 July 1738.
19. MELVILLE C.P. Seismicity of northwest England before 1800, in preparation.
20. GENERAL EVENING POST, 25-28 July 1747.
21. BATH JOURNAL, 12 June 1749.
22. WESTERN FLYING POST, 30 March 1752.
23. EDINBURGH EVENING COURANT, 7 February 1756.
24. FELIX FARLEY'S BRISTOL JOURNAL, 21 February 1756.
25. KENTISH POST, 5 June 1756.
26. READ'S WEEKLY JOURNAL, 8 January 1757.
27. SALISBURY JOURNAL, 15-22 June 1761.
28. EXETER FLYING POST, 5 December 1766.
29. BATH CHRONICLE, 13 April 1769.
30. ARIS'S BIRMINGHAM GAZETTE, 27 November 1769.
31. JACKSON'S OXFORD JOURNAL, 4 May 1771.
32. JOHNSON W. (ed.) Journals of Gilbert White. Futura, London, 1982.
33. SALISBURY & WINCHESTER JOURNAL, 3 May 1773.

34. SHREWSBURY CHRONICLE, 10 July 1773, 26 April 1783.
35. BATH CHRONICLE, 21-28 August 1783.
36. EXETER FLYING POST, 6 May 1789.
37. YORK COURANT, 26 May 1789.
38. CALEDONIAN MERCURY, 12 November 1789.
39. SHREWSBURY CHRONICLE, 2 October 1789.
40. SUSSEX WEEKLY ADVERTISER, 14 December 1789.
41. EDINBURGH EVENING COURANT, 12 April and 13 October 1792.
42. EXETER FLYING POST, 31 May 1792.
43. WOOLMER'S EXETER GAZETTE, 7 June 1792.
44. SALISBURY & WINCHESTER JOURNAL, 7 October 1793.
45. WARDNER T. to J.Eliot, Ashmore, 29 September 1793, Dorset Record Office, ms. D364/E 10.
46. GLASGOW COURIER, 15 August 1797.

Discussion

MR MELVILLE Introductory remarks
The published paper presents the results of a detailed investigation of the seismicity of a distinct period of earthquake occurrence. It attempts to demonstrate the problems that need to be tackled before a critical catalogue of earthquakes can be established. Comparable work for earlier periods has been undertaken and the recorded seismicity of the interval before 1800 has been compared with the record for the 19th and 20th centuries. The results of this study show not only distinct regional variatons in seismicity in the British Isles, both in terms of 'large' and 'small' events, but also that earthquake hazard is low. Only around 45 reported earthquakes should have exceeded surface-wave magnitudes greater than 4.5 (on the basis of their felt area). Of these, about 30 occurred in England and Wales, the region for which data are available from the 11th century or earlier - i. e. excluding Scotland, Ireland and the Channel Islands. The record of such events is likely to be incomplete for some areas and periods, although other comparable earthquakes may have been noted but not fully identified. Not more than 10 earthquakes recorded in this long period of observation seem to have exceeded magnitude 5.0 and there are good grounds for believing that this data set is complete.

MR J.IRVING, CEGB, Barnwood, Gloucester
Referring to Dr Melville's paper, what specialist techniques can the historian apply to the study of earthquakes to distinguish between real temporal and/or spatial 'gaps' and apparent 'gaps' which may be due to poor or absent historical sources?

DR MELVILLE
In reply to Mr Irving, absence of information about earthquakes occurring in apparent time or space 'gaps' can never be taken categorically to indicate a genuine lack of seismicity; it may, however, indicate that no earthquake larger than a certain size occurred. The 'techniques' available to explore such gaps amount to the collection and assessment of 'negative' information. Where adequate sources are available for information on other events in a given area or period (providing references to floods, storms, comets, local political events, riots and so on), but not on earthquakes, this provides an independent check on the quality of surviving historical data. All contemporary or local sources that do not mention earthquakes, but could be expected to do so in the event of a shock, have to be noted and weighed in the balance. Quality of communications, levels of literacy and the distribution of population can all affect the reporting or non-reporting of earthquakes in different areas and periods. These factors can be investigated systematically. Only when there is a genuine lack of source material can no satisfactory evaluation be made of the extent to which absence of seismicity is genuine rather than the effect of cultural factors. Even here, however, it should still be possible to estimate the size of event that could not fail to have attracted attention, given the conditions of the time.

MR A.M. DAVIE, Geo-Rheological Survey Ltd
The list of 18th Century earthquakes given by Dr Melville (Table 1) has allowed the plotting of the sites to indicate the time-distance-elements cycle to be confirmed. The calculations of time-distance illustrates that an error margin of 1 to less than one tenth of a second can be shown over a period of 205 years, and the nodal measures between epicentres can be measured to 1000 ft error, or by calculations to 0.0001 foot Imperial. These calculations are similar to the sample of over 800 seismic occurrences listed by Dr Marsha Adams, Stanford Research International, in her computer readout, which illustrated that this method of forecasting an earthquake is mathematically possible.

2. ISC records of British earthquakes

R. D. ADAMS, MA, MSc, PhD, International Seismological Centre and University of Reading

SYNOPSIS. The Historical File of the International Seismological Centre contains origin estimates of more than 300,000 events for the period up to the end of 1981. Of these about 200 are located in the British Isles. The smaller events are mainly taken directly from the catalogues of agencies such as the British Geological Survey, but the larger events have had their origin parameters redetermined, often using additional observations from stations throughout Europe and farther afield.

ISC HISTORICAL FILE

1. The International Seismological Centre maintains a continually growing file of past earthquakes, which is available for research studies and to help with the estimation of seismic hazard. The file contains origin estimates of earthquakes dating from 1904, but naturally only major earthquakes are present for early years. The file becomes more complete with time and now contains more than 300,000 events world wide. About 20,000 new events are added each year.

Sources of Data

2. The main sources of data for the ISC Historical File are:

Gutenberg and Richter "Seismicity of the Earth"	1904-1952
International Seismological Summary	1913-1963
Bureau Central International de Séismologie	1935,1950-1963
L R Sykes, Catalogue of oceanic ridge events	1950-1963
US Coast and Geodetic Survey	1928-1960
International Seismological Centre	1964-1981
US National Earthquake Information Service Preliminary Determination of Epicenters	1982-1983
Telex reports from NEIS and other sources	1983 onwards

3. Up to the end of 1963 there may be more than one estimate for an event, but from 1964 to 1981 only the prime estimate from the ISC Bulletin is included. ISC origins for 1982 will be added to the file, replacing those from PDE, early in 1985.

Information Stored

4. The file is now stored on magnetic disc on the ISC's VAX 11/750 computer at Newbury. The information in a single epicentre "record" contains a set number essential fields, which may be of variable length, but must be filled, even if with a null value. The information in these fields includes such details as:

Agency code
Source of information
Date
Time
Latitude
Longitude
Depth

5. There is also provision for optional fields for additional information such as:

Associated phenomena (e.g. tsunami, volcanism)
Flags indicating that the event was felt, damaging etc.
Intensity data
Magnitude data
Number of recording stations
Type of event, e.g. explosion, rock-burst etc.
Comments - any length of text may be added

6. This format is designed to hold the maximum possible information, without taking up storage space unnecessarily with null information.

Retrieval of Information

7. Events can be selected from the file by any combination of criteria relating to time, location, depth, magnitude, intensity, or locating agency. Composite estimates can also be provided, which combine origin parameters from prime estimates with magnitude determinations from other agencies.

8. In addition to giving a listing of events, a line printer plot can also be generated, showing the positions of events superimposed on a representation of coastlines. The earthquakes are represented by a figure which gives the magnitude of the largest event reported at each position. Where no magnitude is assigned, events are shown by an asterisk. Underlining of a symbol indicates multiple events at the same position.

9. Another option lists estimates of expected ground acceleration and felt intensity at a specified location for each event for which a magnitude is given.

10. The information can also be reproduced on magnetic tape, in two formats, one of which contains all the information on the file for the chosen events, but needs some simple programming to interpret, and another which contains only the

basic information in fixed fields in 80-column card images.

FILE OF BRITISH EARTHQUAKES

Sources of Data

11. The files are very incomplete for early years, with information only on earthquakes mentioned in the sources given in the last section. Major British catalogues such as those of Davison and Lilwall (refs 1 and 2) have not been specifically added, and the improvement of files of British earthquakes will await the publication of more definitive catalogues.

12. The major sources of data are the prime estimates given by ISC, and estimates of less well recorded events given by the Institute of Geological Sciences, Edinburgh (IGS), now the British Geological Survey (BGS). The analysis at ISC starts with earthquake origins supplied by other agencies (ref. 3). For Britain, these are IGS and sometimes Dublin Institute of Advanced Studies (DIAS). If enough station readings are supplied with an epicentral estimate, ISC recomputes the origin making use of all available readings. Thus for larger British events ISC redeterminations will also make use of observations from stations farther afield in Europe. In the rare event of enough station readings being submitted to ISC without an accompanying origin estimate, the ISC 'search' procedure will compute an origin from the readings available. Special care has to be taken to prevent unassociated readings from distant earthquakes, and particularly observations of core phase from Pacific earthquakes,being interpreted as new local events (ref. 4). The file also contains some origins determined by the agency LAO, which are determinations made by the large aperture seismic array LASA in Montana. These are extremely unreliable and in many instances arise from misinterpretation of later phases. These origins should be disregarded unless supported by determinations from other agencies.

Earthquake File for Britain

13. As an example of an investigation into the seismic activity in a given region, the area bounded by 49.9^{o} and 60°N, 12°W and 2°E was searched for all events on the file, except those determined by LAO only. This area closely contains the British Isles. The map of events is shown in Fig. 1 with symbols as explained in paragraph 8.

14. A total of 192 events was listed. Of these only 4 occurred before 1950, 5 in the 1950's and 7 in the 1960's. The earliest events given are:

1925	Feb 01	50.0^{o}N	6.5^{o}W	Off Cornwall
1926	Aug 15	52.2^{o}N	2.7^{o}W	Herefordshire
1931	Jun 07	53.8^{o}N	1.2^{o}E	North Sea
1933	Jan 14	54.3^{o}N	2.3^{o}W	Wensleydale

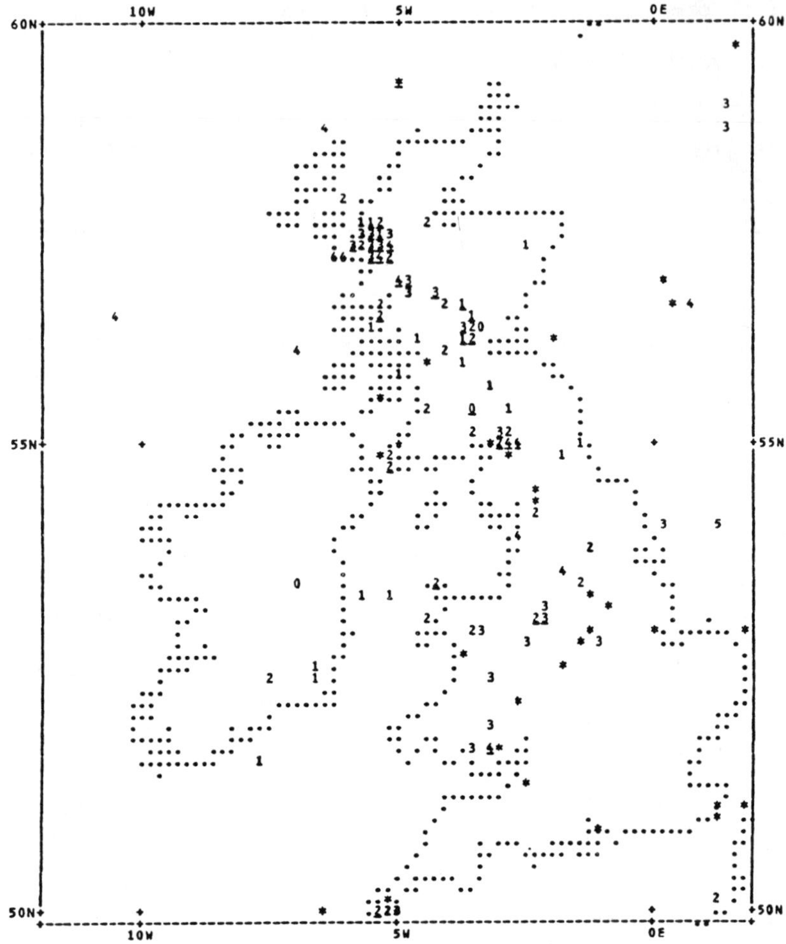

Fig.1. Print-out of earthquakes held in ISC Historical File for area 49.9°-60°N, 12°W-2°E.

15. From 1972 onwards, there is a significant increase in the number of events, corresponding to fuller submission of information from the IGS. Even so, there is a large variation in the yearly number of events submitted, as shown in Table 1.

16. It is seen that most of the variation in annual numbers of events is due to variations in numbers of small shocks submitted by IGS. For the larger events, with enough readings to be redetermined by ISC, the number of events is much more uniform. Table 2 demonstrates that the ISC redeterminations are in general for the larger earthquakes.

Table 1. Annual number of events in file.

Year	Total number of events	No. of events with ISC origins	No. of events with IGS origins only
1972	5	1	4
1973	1	0	1
1974	25	7	18
1975	14	7	7
1976	3	2	1
1977	6	4	2
1978	58	7	51
1979	9	9	0
1980	30	11	19
1981	15	12	1

In 1981 two events had origins supplied by DIAS only.

Table 2. Distribution of events by magnitude.

Magnitude Range	No. of events with ISC origins	No. of events with IGS origins only
$M < 1$	0	6
$1 \leqslant M < 2$	2	27
$2 \leqslant M < 3$	17	41
$3 \leqslant M < 4$	17	20
$4 \leqslant M$	8	5

Only 2 out of 44 events located by ISC have magnitudes less than 2, compared with one third of the events reported by IGS only.

Distribution of earthquakes

17. The distribution of events shown in Fig. 1 arises from a combination of distribution of seismic activity and detecting instruments. As much of the British earthquake activity is in Scotland and the Midlands, there has tended to be a concentration of instruments in these regions, giving a lower detection threshold there. More uniform station coverage has been achieved in recent years, however, as is shown by yearly maps of detection capabilities in catalogues of British earthquakes, published by the British Geological Survey (ref.5).

18. Even allowing for this detection bias, the most active part of the country is Scotland and north west England, with lesser activity extending south west through the Midlands into Wales. Apart from a cluster of activity in Cornwall only a few events of unspecified magnitude are shown to the south east of a line from Bristol to the Wash. This south eastern area is shown as the least active area of Britain, and although it is heavily populated it is also the part of the country that is

least well monitored by seismograph stations.

19. The activity shown in Fig.1 for Ireland is all of extremely low magnitude, and the fact that the earthquakes are shown at all reflects the sensitivity of the recording network operated by DIAS.

20. Several events are plotted in the North Sea, including the largest event in the file, that of 1931 June 07, which is the only one assigned a magnitude above 5 (M_s 5.6 Pasadena). This earthquake was felt at an intensity of VI to VII MM on the Yorkshire coast.

21. Earthquakes of magnitude 4 are scattered widely in the North Sea, Scotland, the Carlisle area, the Midlands and South Wales.

22. Various sequences are included in the lists. In 1974 August, 17 events are located near 57.2°N, 5.4°W at Kintail in western Scotland. The largest on August 10 is allocated a magnitude of 4.4. Another active period was from 1978 March to June, during which there were further events near Kintail and another sequence near Glenalmond in Perthshire.

23. Another prominent sequence is that near Carlisle, associated with the "Boxing Day" earthquake on 1979 December 26 (ref. 6). This event was located by the ISC at 54.93°N, 2.78°W using readings from 83 stations, including many from Europe and a few from Africa and North America. This location is about 15 km south east of that determined by IGS from local readings only. IGS allocated a local magnitude of 5.2 but the body-wave magnitude found by ISC from distant readings was 4.5. Such differences between local and teleseismic magnitudes are not uncommon. One foreshock and seven aftershocks are also listed.

24. The file also contains the earthquakes near Constantine in Cornwall in 1981 February. The largest, of magnitude 3.6 was relocated by 36 stations, including many in northern Europe.

25. At the end of 1984 the file contains definitive origins only for the period up to the end of 1981. Events for 1982 will be added early in 1985. Preliminary determinations of some events of particular interest, such as that in Powys in Wales on 1984 April 15 are already included, however.

CONCLUSION

26. The ISC file of earthquakes is becoming an increasingly complete source of information on the larger British earthquakes. In particular it gives re-determinations of these events, making use of readings from seismograph stations outside the British Isles, and allocating magnitudes derived from teleseismic observations. For smaller events the file is not as complete as that held by the British Geological Survey, but still gives a representative picture of British seismicity.

REFERENCES

1. DAVISON C. A history of British earthquakes. Cambridge University Press, 1924.
2. LILWALL R.C. Seismicity and seismic hazard in Britain. Seismological Bulletin No. 4. Institute of Geological Sciences. HMSO, London, 1976.
3. ADAMS R.D., HUGHES A.A. and McGREGOR D.M. Analysis procedures at the International Seismological Centre. Phys. Earth and Planet Inter., 1982, vol. 30, 85-93.
4. ADAMS R.D. ISC determinations of previously unreported earthquakes in Europe. Proceedings of 2nd international symposium on the analysis of seismicity and on seismic hazard. Czech Academy of Sciences 1981, 132-139.
5. TURBITT T. (Editor) Catalogue of British Earthquakes recorded by the BGS Seismograph Network 1979, 1980 and 1981. Report No. 210. Global Seismology Unit, BGS 1984.
6. KING G.C.P. The Carlisle earthquake. New Scientist, 1980, vol. 85, 575.

3. Investigation of British earthquakes using the national monitoring network of the British Geological Survey

C. W. A. BROWITT, PhD, T. TURBITT, BSc, and S. N. MORGAN,
British Geological Survey

SYNOPSIS. In the past decade, the British Geological Survey (BGS) has expanded its coverage of the UK with modern seismic monitoring stations. In a large part of the Country, standardised instrumentation is permitting objective studies of British earthquakes to be undertaken. The development of this network, its instrumentation, and earthquake activity since 1979 are outlined.

INTRODUCTION

1. The 1884 Colchester earthquake, whilst not the largest earthquake known in the British Isles region, caused considerable damage and has been studied in detail both contemporaneously and in recent years (refs 1 and 2). The time of occurrence, however, pre-dates by several years the development of instruments sufficiently sensitive to measure seismic waves for the accurate location of hypocentres and the objective determination of size (magnitude). As a consequence, studies of such "pre-instrumental" British earthquakes are in marked contrast to those undertaken 100 years after the Colchester earthquake using a dense array of modern instruments (ref. 3).

DEVELOPMENT OF A UK SEISMOGRAPH NETWORK

2. Following the invention of seismographs capable of detecting signals from distant earthquakes in the last decade of the 19th Century, several instruments were established in Britain (ref. 4). For a time in the early 1900's, there was a network of some 10 stations well-distributed in the Country, which have provided some useful information. They were, however, insensitive by modern standards and with long natural periods were unsuitable for the detailed study of small local earthquakes.

3. Interest in British earthquakes and instrumental monitoring fluctuated in the first half of the 20th Century and by the 1950's it was at a low ebb. Modern seismology received a boost in the early 1960's, however, through the requirement for remote monitoring of nuclear tests. This defence seismology stimulated a network of photographic-recording standard seismographs throughout the World including one at the Eskdalemuir Geophysical Observatory in the Scottish Borders. Also in the

early 1960's, the British Ministry of Defence (MOD) installed an array of short-period magnetic-tape recording seismometers near the Eskdalemuir Observatory and a single station at Wolverton in the south of England (ref. 5).

4. In the 1960's, a research group at the Royal Observatory, Edinburgh, (and, later, as part of the Institute of Geological Sciences) was developing portable magnetic tape recording systems specifically for local earthquake monitoring. Some of these were deployed in the Lowlands around Edinburgh principally to test and develop the instruments and techniques for handling the recorded seismograms (ref. 6). By the end of the 1960's, this Lowlands network (LOWNET) had become fairly stable and data was routinely reported on teleseisms and local British earthquakes. However, small earthquakes well outside the 100 km aperture network could not be located with any degree of accuracy although they could be detected and their magnitudes determined owing to the sensitivity of the instruments. This level of instrumental observation continued to the mid-1970's.

5. In the latter part of the 1970's, stimulated partly by public concern about mining-induced seismicity in the Midlands, a network of seismographs was installed in central England and western Wales (ref. 7). This network used a second generation FM analogue magnetic-tape recorder, the Earth Data Geostore, and a new seismometer, the Sensonics Willmore Mk III, which are in use throughout the UK network at present.

BGS UK NETWORK IN THE 1980's

6. The Midlands network of 1978 comprising some 7 seismometer stations, was the first installation of short period instruments specifically intended to monitor background seismicity in the UK. It was not, however, until the 1980's that we achieved a network of sufficient density, sensitivity or uniformity, over a significant part of the Country, to give an objective view of the pattern of earthquake activity, particularly with regard to the depths of occurrence.

7. Figure 1 shows the BGS seismograph network in 1985. Stations have been included where they have existed for 5 years or are likely to be in existence for at least 5 years. Not included, are short-term temporary networks of which two are in operation through 1985; a dense 32-channel, 16-station system in NW Wales, and a 4-station network around the West Midlands. In addition, there are 3 compatible seismographs operated by Leeds University in that part of the Country and the short-period array, EKA, operated by MOD in the Borders. A widespread 8-station MOD network is designed to fulfil their nuclear test monitoring objective and is unsuitable for local earthquake investigations owing to its low pass-band.

8. In Ireland, the Dublin Institute of Advanced Studies (DIAS) operates the same type of instruments (also shown in Figure 1) and data exchange between the two organisations is particularly valuable for earthquakes in Wales and the Irish Sea. BGS stations in Norway and the North Sea have been

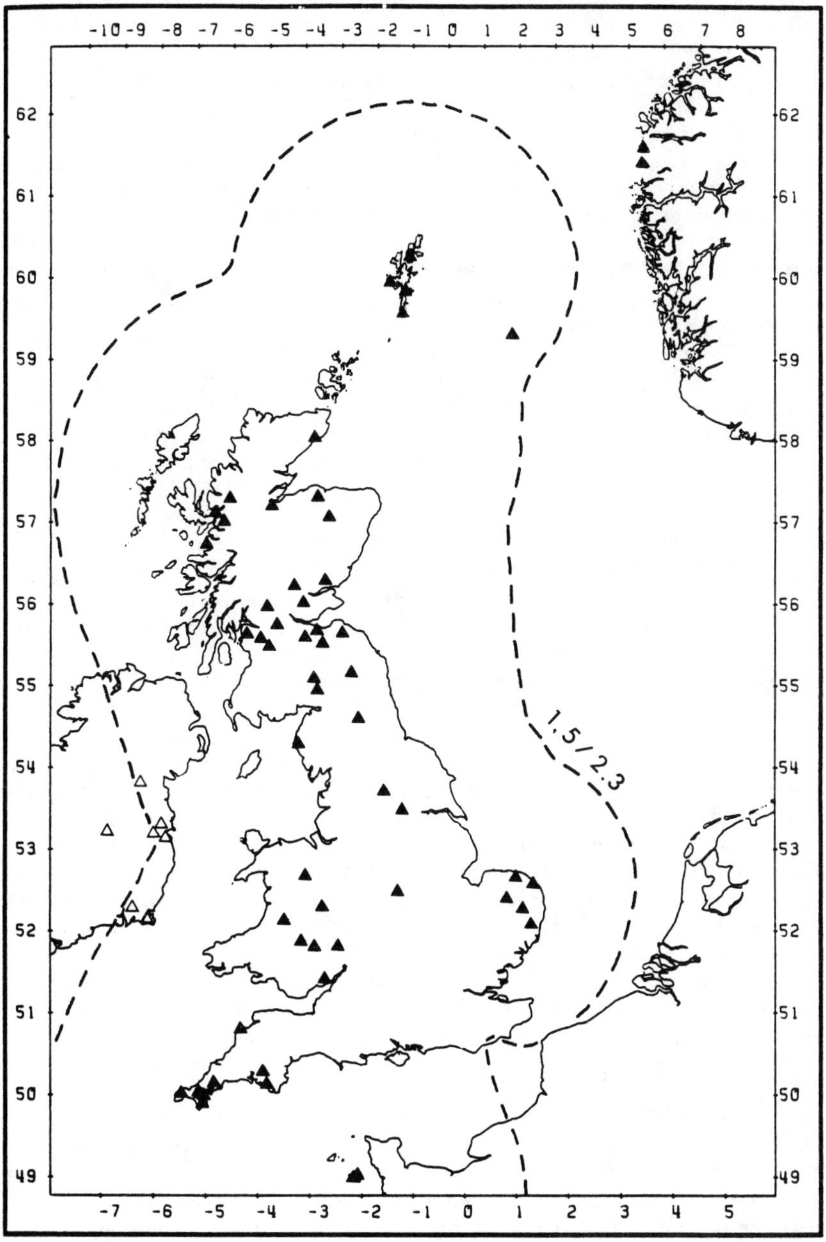

Fig. 1. The BGS UK seismograph network (filled triangles) and the DIAS Irish network (open triangles). The dashed contour gives an indication of the detection capability of the BGS network for 1.5 ML and 2.3 ML earthquakes on quiet and noisy days, respectively.

complemented by 6 Norwegian installations along their west coast with considerable improvement in the joint coverage of the northern North Sea.

9. The non-uniform distribution of stations in the network partly reflects its history and partly the site-specific interests of funding agencies, notably: the Department of Environment in mining tremors in the Midlands, the Department of Energy in policing the Hot Dry Rock geothermal experiments in SW England and in establishing a data base of North Sea seismicity, Renfrew District Council, Jersey New Waterworks Company and Jersey Meteorological Department. The most notable gap in the coverage is in the south and south-east of England which is, unfortunately, matched by a similar gap in NW France.

10. Figure 2 shows the distribution of seismograph stations in low-seismicity, western Europe in 1985. Despite the gaps in the UK, we now have a network with coverage equal to most of our neighbours although many are in the process of upgrading their networks. Germany is notable for its high density of stations stimulated by the Rhine Graben seismicity. With networks operated by several different groups, however, they do not share the UK's standardised instrumentation, operating techniques and centralised coordination.

11. In order to provide an indication of the detection and location capability of the network, contours have been generated to show the size of earthquakes which would be adequately recorded on 3 stations on quiet and noisy days (ref. 8). The magnitude 1.5/2.3 (ML) contour reproduced on Figure 1 shows that, in principle, all earthquakes above magnitude 2.3 are detectable and can have their epicentres located for the land area using the BGS network. In practice, the accuracy of the determination will depend on whether the event is within the network of stations or outside it. Whilst the south-east of England contains the most notable gap, there are other areas where the density of stations is insufficient to determine hypocentres accurately (i.e. including the depth) and where there is no redundancy when an instrument breaks down.

12. Following significant earthquakes, the background strategic network is locally enhanced to improve determinations of hypocentres, seismicity patterns, focal mechanisms and wave-propagation anomalies for the aftershocks (refs 3 and 9).

INSTRUMENTATION

13. Except in the North Sea, the 60 stations are equipped with 1 second period Willmore Mk III seismometers manufactured by Sensonics Ltd. They are grouped into 16 sub-networks of between 1 and 8 stations with signals amplified, frequency modulated, and transmitted by UHF radio to a central magnetic tape recorder (except on Jersey which only has paper recorders). This Earth Data Geostore recorder has a capacity of 10 seismic channels, internal clock and MSF radio, and 2 flutter compensation channels on a 2400 ft, ½ inch tape. It is normally operated at 15/320 inches/sec giving a bandwidth up to 16 Hz and a recording duration of 1 week, although there are options

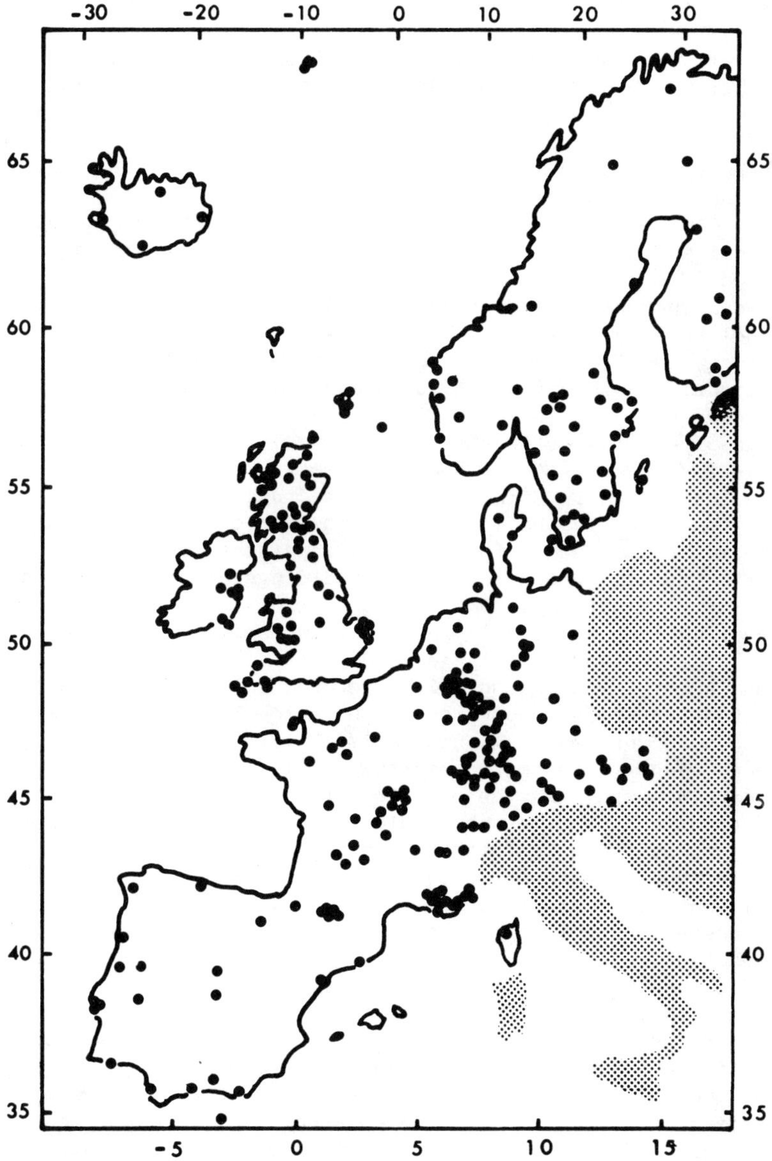

Fig. 2. Distribution of seismograph stations in low seismicity western Europe in 1985. The shaded area is excluded from the station plot.

of twice and half of this speed with equivalent gains and losses in bandwidth and duration. The recorders are operated from a trickle-charged 12 volt battery which can handle power cuts of 3 days. Remote outstations are low-powered (100 ma at 12V) so that a small block of batteries will last 1 year or a small solar panel indefinitely.

14. In the North Sea, a sea-bed seismograph package contains a triaxial set of Geotech S500 piezo-electric sensors and a Graseby hydrophone connected to Earth Data amplifier-modulator units. Signals are transmitted by cable to a 30 m, tension-leg spar buoy and onwards by UHF radio to Mobil's Beryl platform which houses Geostore recorders (refs 10 and 11).

15. The land-based stations all have at least a high-gain vertical seismometer. Fifteen of them have an orthogonal set of high gain seismometers and at 5 of these sites there is a vertical instrument operating with 34 dB less gain. Each channel has a dynamic range of approximately 45 dB. At 4 stations, low-frequency microphones are operated to assist in the discrimination of sonic booms and at 10 sites paper loop recorders are used to permit immediate access to data from events of special interest.

16. The sensitivity of the instruments is high so that signal detection is principally limited by the background vibration noise in the operating bandwidth of 1-16 Hz (flat). Seismometers are, therefore, housed in places remote from traffic, industry, running water, buildings and trees (which move in the wind) and are placed on bedrock to achieve some decoupling from any locally-generated noise.

17. The equipment is largely 10-years old technology but it produces good seismograms which can be expanded in time and amplitude on replay and which can be automatically digitised for computer processing. Drawbacks include the limited dynamic range (compared to digital recording) and a time penalty in extracting the data from the continuously recorded magnetic tape. The latter problem is mitigated by replaying tapes at 160 times real-time which brings the signals into the audible range. An experienced analyst can distinguish between the sounds of local, regional and teleseismic events and, often, quarries and underwater explosions during a 1½ hour pass of a 1 week tape.

18. Microprocessor-based event-triggered recorders which provide a wider dynamic range and compress the data of interest are being used to enhance the network (ref. 12). These devices exist in Shetland and Edinburgh and one is under construction for Jersey. It is not intended that they replace continuous recording which we consider to be essential in a highly populated, highly industrialised Country where there is a need to be able to positively deomonstrate the absence of earthquakes and to discriminate man-made seismic events. In the North Sea, event-triggered systems would not be successful owing to the considerable air-gun activity detectable at ranges of over 100 km.

RECENT SEISMICITY

North Sea

19. Five years of improved seismic monitoring of the North Sea has shown a pattern of seismicity in the North which correlates with the Viking Graben, the Viking Basin and the West Norway coast (ref. 13, Fig. 3). The absence of seismicity in a broad zone around the Shetland Islands during this period is real and the gap between the Viking Graben and the Norwegian coast seismicity is probably real. The possibility of a data shortage in the latter area is not, however, ruled out. Half of the few larger earthquakes in the North Sea over the past century appear to have originated in the Viking seismicity zone which indicates its persistence over a longer period of time that the 5 years of this study.

Britain

20. Figure 4 shows the distribution of earthquakes with magnitudes greater than 1.0 ML in the UK land area for the period 1979–1984 (refs 14 and 8). The distribution of smaller events, particularly in the earlier years, is strongly biased by the distribution of seismograph stations. Figure 5, which shows only those earthquakes with magnitudes ≥ 2.0, is less biased by station geometry. Of the several centres of activity observed only one, in NW Scotland, shows any clear lineation. This approximately N–S trend in the epicentres transects the dominant regional NE–SW oriented Caledonian framework. It is more closely correlated with the orientation of Teriary dykes which were intruded during a period of crustal extension around the time of the opening of the North Atlantic.

21. The maps of recent epicentres shows little activity in North-East Scotland where there is good seismograph coverage, although we know of a significant seismic history in the Inverness region (ref. 15). The absence of events in the South-East of England may be influenced by the absence of seismograph stations.

22. In 1984, several earthquakes were strongly felt: Dumfries in February (ML = 2.6), Powys in April (ML = 2.6), E. Midlands in May (ML = 3.1). On 19 July, the damaging Lleyn earthquake (ML = 5.4) was followed by a series of felt after-shocks, and smaller earthquakes were felt in other areas including Loch Maree, Peebles, and north of Manchester.

Cornwall

23. On 25 and 26 February, 1981, four earthquakes with magnitudes up to 3.5 ML were felt in Cornwall (ref. 16). The local network of seismograph stations, established to monitor activity both external to and associated with Hot Dry Rock hydrofracture experiments conducted by Camborne School of Mines, located these larger earthquakes and 170 smaller ones at depths of between 5 and 6 km near the village of Constantine. Historical records show that there have been 5 felt events within a few km of this locality in the past 150 years.

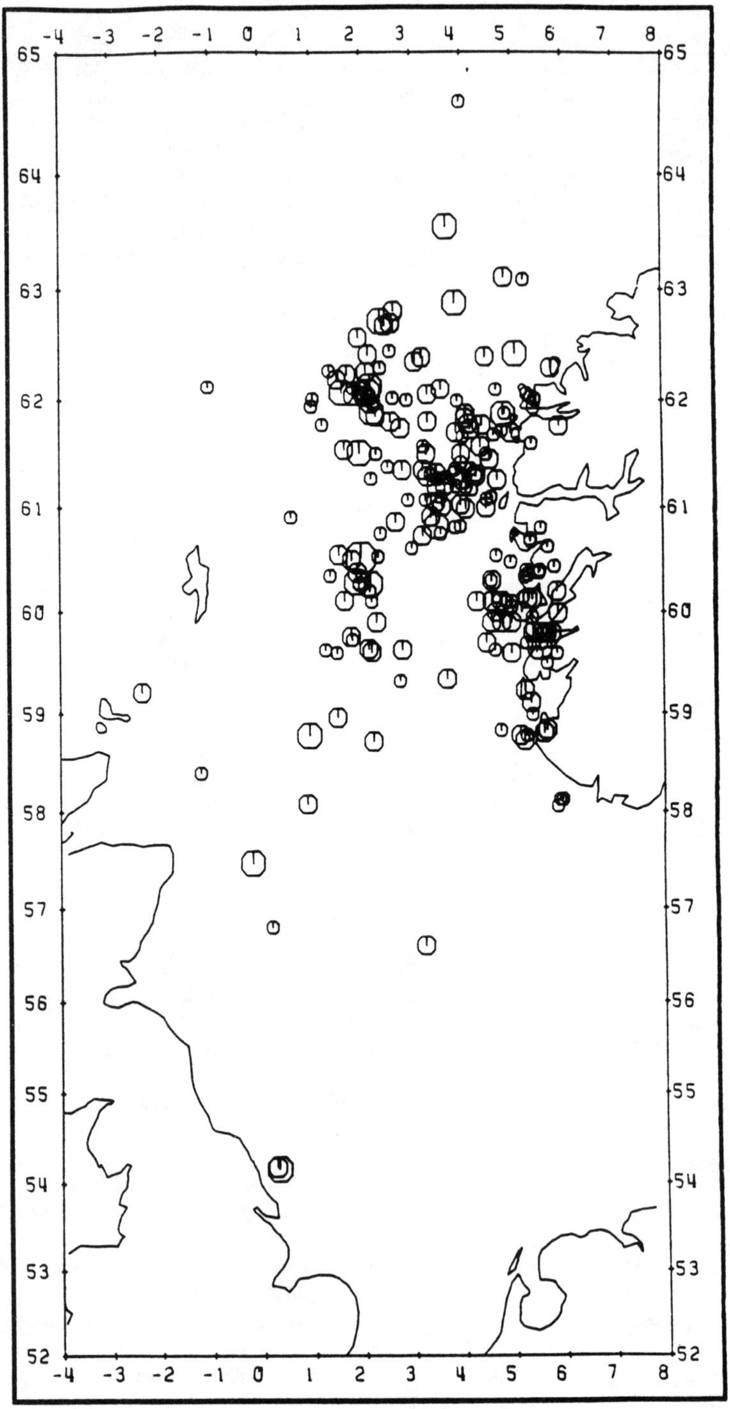

Fig. 3. Seismicity of the North Sea, 1979 to 1984 (after Newmark and Turbitt, ref. 13).

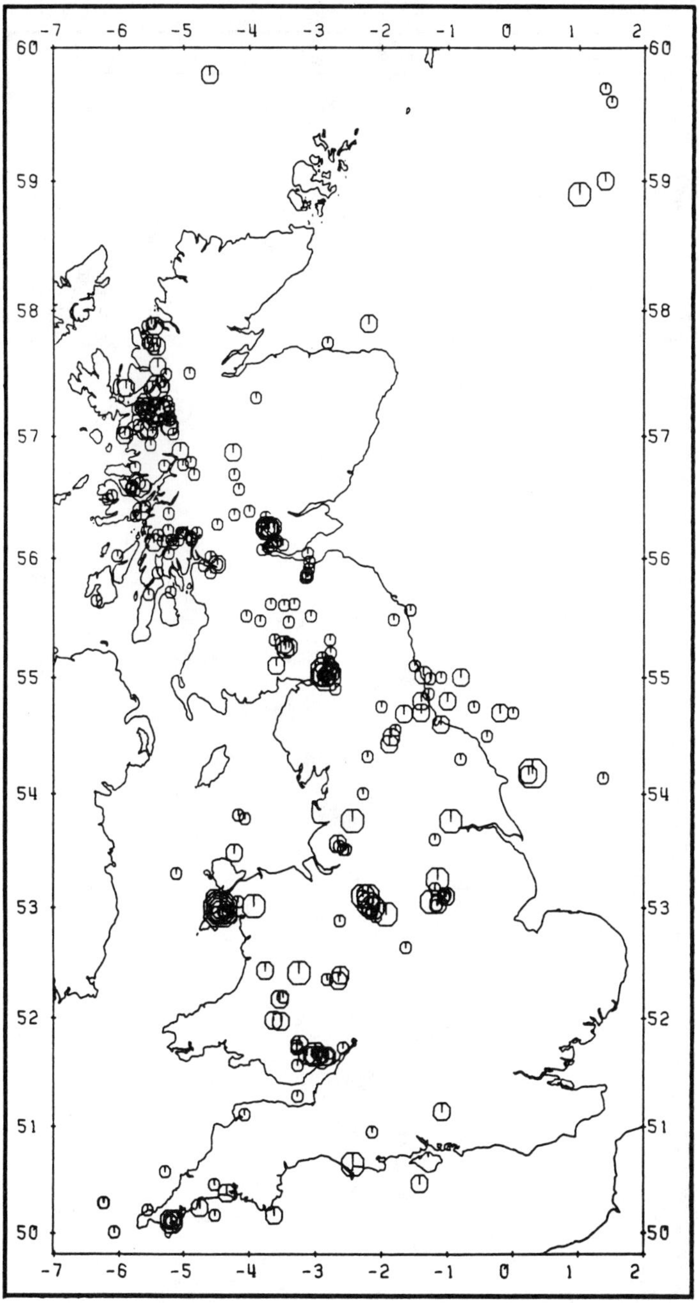

Fig. 4. Epicentres of earthquakes with magnitudes ⩾ 1.0 ML for the period 1979–1984 (after Turbitt, ref. 14).

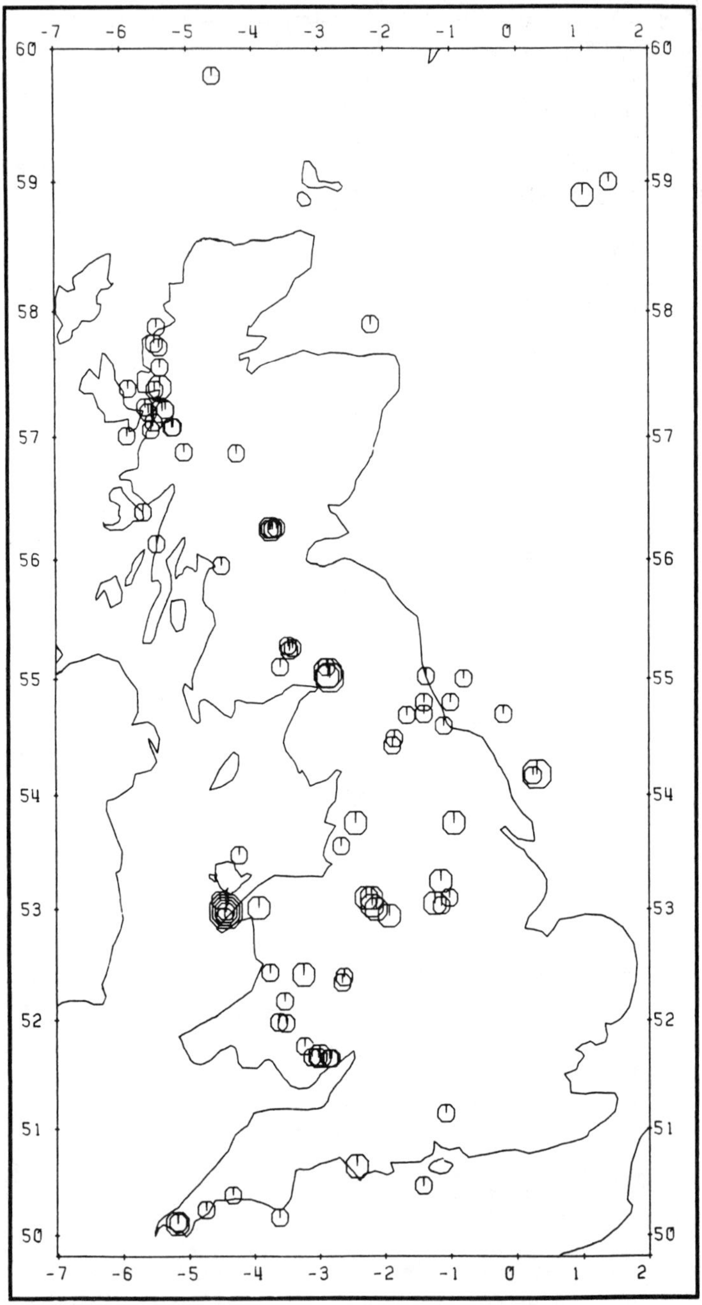

Fig. 5. Epicentres of earthquakes with magnitudes ≥ 2.0 ML for the period 1979-1984 (after Turbitt, ref. 14).

North Wales

24. The Lleyn earthquake of 19 July, 1984, with a magnitude of 5.4 ± 0.3 ML, was one of the largest in Britain during the past 100 years (ref. 3). It caused some damage to chimneys, roofs and plaster walls in the epicentral area, and some, in patches, elsewhere. It was widely felt throughout Wales, in most of England, in SW Scotland and in eastern Ireland, and the provisional isoseismal III (MSK) encloses an area of approximately ¼ million square km. Outside that area isolated felt reports have been obtained mainly from people in tall buildings.

25. The limited damage from an earthquake of this magnitude can be attributed to its depth of occurrence in a remote area. The national networks of BGS, in Britain, and DIAS, in Ireland, placed the main shock on-shore on the Lleyn Peninsula at a depth of 18 km. Using first motion data from 45 stations a focal mechanism was determined indicating a predominantly strike-slip motion on an E-W or N-S striking fault with a normal component, downthrow to the West or North. The maximum compressive stress direction implied by this mechanism is NW-SE. Using data collected in the shear-wave window on triaxial instruments, we expect to obtain an independent determination of stress directions from the analysis of S-wave polarizations (ref. 17).

26. High-gain seismometers throughout most of the network were saturated by the high amplitude wave-train from the main shock. The low-gain vertical seismometers, however, remained on-scale. Figure 6 shows the seismogram from the nearest of these stations to the epicentre, at a range of 140 km, on which the maximum recorded velocity is 0.4 mm/sec at 1.5 Hz.

27. A 16-station network installed in the epicentral region immediately following the main shock has confirmed the unusual depth of the earthquakes. Well-located aftershocks, many of them felt, form a cluster 3 km in diameter between 21 and 24 km in depth. The focal mechanisms are not consistent and include strike-slip, normal and thrust motions. Although there are few well-determined depths for British earthquakes the Lleyn events are thought to be unusually deep. They probably reflect a relatively deep transition between brittle and ductile zones in the North Wales crust which may be associated with a region of reduced heat flow.

Carlisle

28. In contrast to Lleyn, the 46 larger aftershocks of the 4.8 ML Carlisle earthquake of 26 December, 1979, show mainly consistent P-wave polarities on the local network over a period of one year. This has permitted a composite focal mechanism to be determined which shows predominantly strike-slip motion on a NW-SE trending near-vertical fault (ref. 9). The implied NW-SE maximum compressive stress direction is consistent with that obtained for the Lleyn main shock. Further contrasts with the Lleyn aftershocks, however, are the shallower depths of occurrence, 4-6 km, and a clearly defined 6 km long NW-SE trend in the epicentres of the Carlisle events.

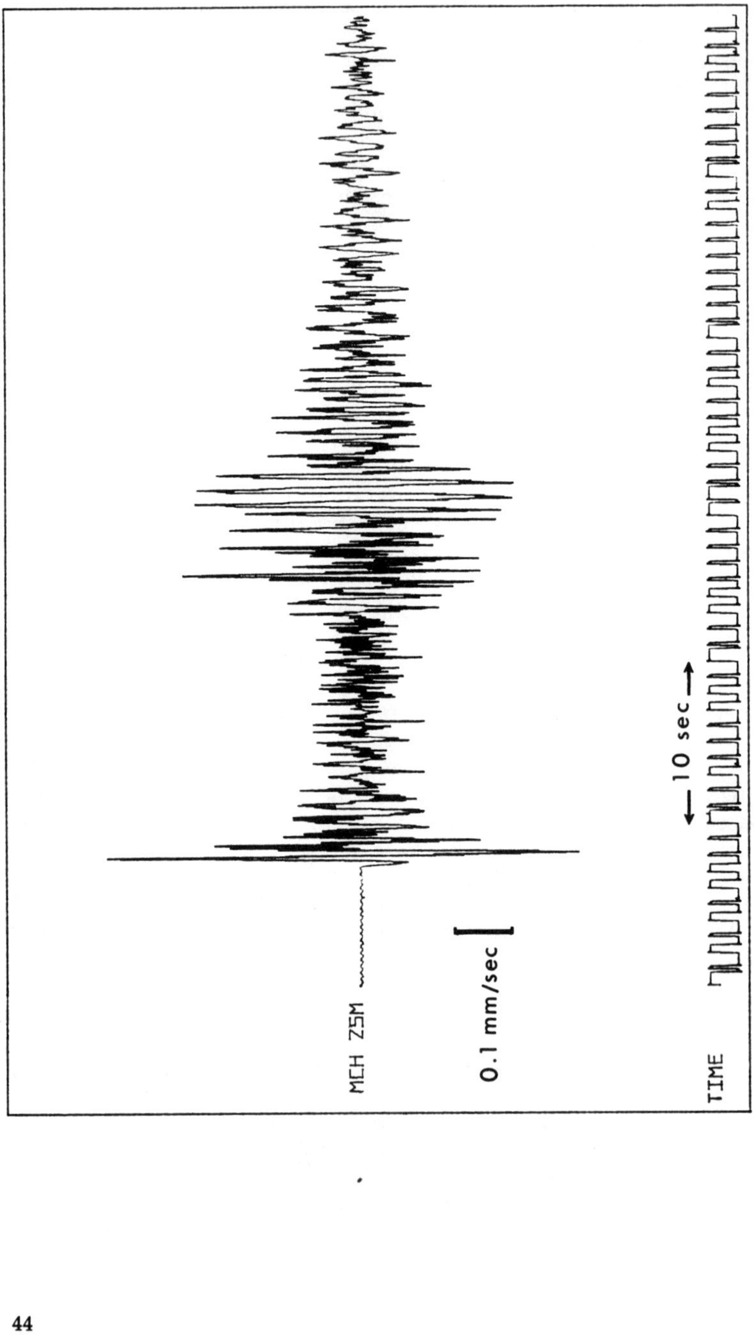

Fig. 6. Seismogram from a vertical component low gain instrument, 140 km from the Lleyn Peninsula earthquake of 19 July 1984.

CONCLUSIONS

29. The 60-station strategic monitoring network of the British Geological Survey is uniform in its instrumentation and the methods applied to the data. The coverage of the UK, however, is not yet uniform and there are some gaps, notably in the south and south-east of England. The network has not been operational long enough to objectively map the distribution of seismicity but, in conjunction with mobile networks, it is yielding new data on specific centres of activity, depths of occurrence, focal mechanisms and stress directions.

30. In addition to its scientific function, the network is used on a day-to-day basis to rapidly provide objective information on natural and man-made vibrations (including sonic booms, underwater explosions, quarry blasts, unscheduled explosions, and mining tremors).

ACKNOWLEDGEMENTS

This work was supported by the Natural Environment Research Council, Department of Energy, Department of the Environment, Jersey New Waterworks Company, Jersey Meteorological Department, Renfrew District Council and North of Scotland Hydro Board. It is published with the approval of the Director of the British Geological Survey (NERC). We thank our many colleagues in BGS who have installed and maintained the network of seismographs and who have analysed and interpreted the data, and we extend that appreciation to station operators, the Universities of Leeds, Leicester, East Anglia and Bergen who house equipment, and to Mobil North Sea who provide facilities on the Beryl oilfield.

REFERENCES

1. MELDOLA R. and White W. Report on the East Anglian earthquake. Macmillan, London, 1885, 223pp.
2. SKIPP B.O. et al. Colchester revisted. Earthquakes and Earthquake Engineering in Britain. Thomas Telford, 1985 (in Press).
3. TURBITT T. et al. The North Wales earthquake of 19 July, 1984. J.geol.Soc.London, 1985, vol. 142, 567-571.
4. NEILSON G. and BURTON P.W. The magnitudes of the larger British earthquakes of the 20th Century. BGS Global Seismology Report No. 206, 1984, 106pp.
5. TRUSCOTT J.R. The Eskdalemuir seismological station. Geophys.J.R.astr.Soc., 1964, vol. 9, 59-68.
6. CRAMPIN S. et al. The LOWNET radio-linked seismometer network in Scotland. Geophys.J.R.astr.Soc., 1970, vol. 21, 207-216.
7. BROWITT C.W.A. Seismograph networks of the Institute of Geological Sciences, UK. Physics of the Earth and Planetary Interiors, 1979, vol. 18, 127-134.
8. TURBITT T. (Editor). Catalogue of British earthquakes recorded by the BGS Seismograph Network 1982, 1983, 1984. BGS Global Seismology Report No. 260, 1985 (in Press).

9. MARROW P.C. and Roberts G. Focal mechanisms of aftershocks of the 1979 Carlisle earthquake. Geophys.J.R.astr.Soc. (in Press).
10. TURBITT T. et al. Instrumentation for North Sea seismic data acquisition. Proceedings NATO Advanced Research Workshop on Seismicity and Seismic Risk in the Offshore North Sea Area, 1982, pp. 155-171. D Reidel, 1983.
11. BROWITT C.W.A. and Turbitt T. Seismic data acquisition spar buoy in the North Sea. Proceedings 1983 Symposium on Buoy Technology, New Orleans, 1983, 224-230.
12. HOULISTON D.J. et al. Event triggered seismic detection systems developed by the British Geological Survey. Earthquakes and Earthquake Engineering in Britain. Thomas Telford, 1985 (in Press).
13. NEWMARK R.H. and Turbitt T. Catalogue of North Sea earthquakes 1980 to 1984 and brief project summary. BGS Global Seismology Report No. 246, 1985, 23pp.
14. TURBITT T. (Editor). Catalogue of British earthquakes recorded by the BGS Seismograph Network 1979, 1980, 1981. BGS Global Seismology Unit Report No. 210, 1984, 53pp.
15. BROWITT C.W.A. et al. Seismicity of the Inverness region. IGS Global Seismology Unit Report No. 76, 1976, 210pp.
16. TURBITT T. et al. Monitoring the background and induced seismicity for the Hot Dry Rock programme in Cornwall. BGS Global Seismology Report No. 195, 1984, 66pp.
17. PEACOCK S. Aftershocks of the North wales earthquake of July 1984. Geophys.J.R.astr.Soc., 1985, vol. 81, 341 (abstract).

Discussion

PROFESSOR R.T. SEVERN, University of Bristol
What developments would Dr Browitt like to see in British earthquake research? What are the gaps?

MR E.D. BOOTH, Ove Arup & Partners, London
Could Dr Browitt comment on areas of the North Sea where deficiencies in the earthquake detection network exist, particularly for the areas of the southern gas fields and the deeper waters to the extreme north.

PROFESSOR W.D. LIAM FINN, University of British Columbia
The Norwegian Large Aperture Seismic Array (NORSAR) 1978 report on North Sea seismicity assumed the North Sea to be a homogeneous seismic zone. Do the preliminary data from Dr Browitt's network suggest a different type of model? In particular does his data suggest that earthquakes should be associated with particular North Sea tectonic structures?

DR BROWITT

In answer to Professor Severn, the most important requirements on the observational side are to fill the gaps in seismograph coverage particularly in the south and south-east of England, and to provide stable funding for the long-term operation of this uniform network so that it is independent of short-duration site-specific projects. Only in this way will be obtained an objective view of the pattern of seismicity, its correlation with geological structures and, in particular, a knowledge of the depths of occurrence in addition to the epicentres.

Replying to Mr Booth, by comparison with the density of seismograph stations considered to be necessary on land in low seismicity western Europe (Fig. 2), the whole of the North Sea is deficient in its earthquake detection network. It is less so in the north between Shetland and the Norwegian coast where recent efforts by the UK Department of Energy and a Norwegian consortium have resulted in the installation of onshore and offshore stations. Even in this area, however, correlations between seismicity and structures can only be shown on a broad basis owing to the inaccuracies in the locations. There is negligible information on the depths of occurrence and on correlation with specific faults.

In reply to Professor Finn's question: Yes. With little objective data available, early risk assessments were forced to assume a homogeneous distribution. Our 5 years of data show a concentration of events in the Viking Graben, Viking Basin and West Norwegian Coast areas. Inaccuracies in locations only permit correlations to be made in this broad sense. Of the few larger events which have been located in the North Sea over the past century or so approximately half have originated in these northern areas. This suggests that the 5-year 'snapshot' which we now have reflects at least 100 years of earthquake activity.

4. Long-term seismicity of Britain

N. N. AMBRASEYS, DSc, PhD, FICE, FGS, Imperial College of Science and Technology and J. A. JACKSON, PhD, Bullard Laboratory, University of Cambridge

INTRODUCTION

One of the greater uncertainties in the engineering assessment of risk and of design ground motions arises from the indiscriminate use of Intensity and magnitude data derived from heterogeneous scales or from magnitude scales based purely on macroseismic data. This heterogeneity of earthquake catalogues introduces great difficulties and occasionally controversy, into assessing seismicity and the design ground motions for engineering projects.

This study presents the results of a uniform assessment of magnitudes and intensities for British earthquakes, Figure 1; it shows that a comparatively simple attenuation model is adequate to predict the fall-off of intensity with distance and that magnitudes can be predicted by use of macroseismic data calibrated against re-assessed instrumental data.[1,4]

To appreciate that such an undertaking is necessary requires recognition of the fact that existing earthquake catalogues of British earthquakes (for instance 8 & 9) often employed for the evaluation of earthquake risk and for the determination of ground motions for application in earthquake engineering practice, are not uniformly compiled in terms of magnitude and intensity. With few exceptions these catalogues use magnitudes, some of which are empirical, derived

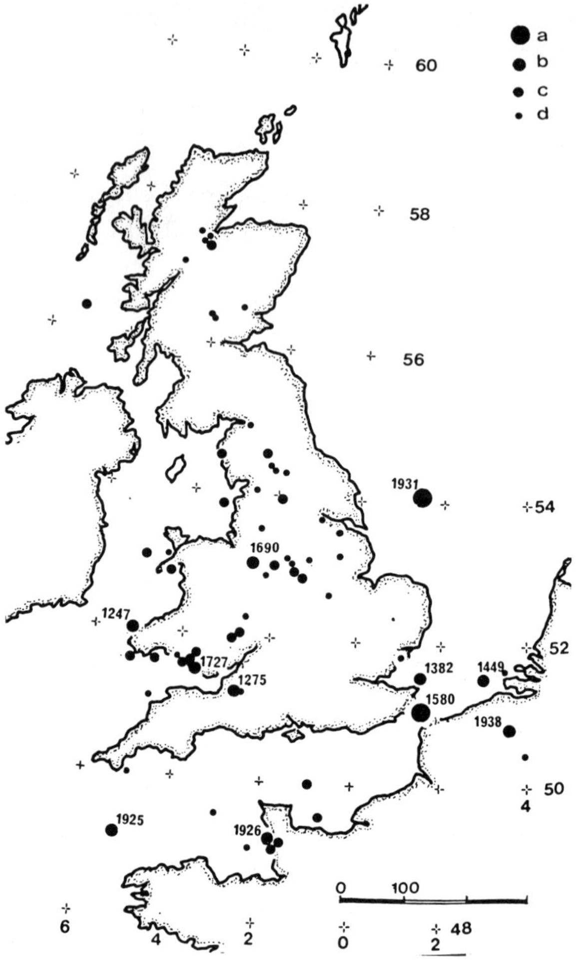

Figure 1. Distribution of British earthquakes during the last seven centuries. a: M = 5.5, b: 5.5 > M ≥ 5.0, c: 5.0 > M ≥ 4.5, d: 4.5 > M ≥ 4.0.

solely from macroseismic data of other parts of the world, while others are not clearly specified.

MAGNITUDE DETERMINATION

It can be shown (1) that, for the larger events in Northwest Europe, surface-wave magnitudes can be estimated uniformly and that neither a period constraint at 20 seconds nor distance effects, for the regional conditions considered, appear to play an important role in the use of the Prague formula $M = \log(A/T)_{max} + 1.66\log(D^{o}) + 3.3 + dM$..........(1) in which $(A/T)_{max}$ is the maximum value of all the ratios of the surface-wave groups on a record, D^{o} is the epicentral distance and dM is a station correction. For smaller events, crustal phases at short distances may be used to derive calibration curves for a magnitude which, at larger distances and longer periods would fit the magnitude scale of surface waves.

The data available for Northwest European earthquakes, i.e. events in the U.K., the North and the Norwegian Seas, Western Scandinavia, Holland, Belgium, the northern part of the Rhine region and north France (6, 7), suggest that on the average, for $D^{o} \geqq 4^{o}$, values of M calculated from equation (1), from medium-period instruments, do not exhibit an equivocal distance dependence. The orthogonal regression of about 50 station magnitudes in the range 3.5 to 6.0, for which both M and M^{M}, as defined by Marshall and Basham (2), can be estimated with distance and path corrections, shows that M^{M} determinations are merely lower by 0.3, a value well within the limits corresponding to the standard deviation of the major axis solution: $M = -0.01 + 1.06(M^{M})$ .... (2)

The coherence between M and body-wave magnitudes m for Northwestern European and U.K. events is rather low. The orthogonal regression of about 60 event magnitudes gives: (1): $M = 1.05(m) - 0.8$ ....(3) with a coefficient of determination of 0.5.

For more recent events, the relation between M and the local magnitude M_L of 140 earthquakes of $M \leq 5.5$, was found to be; $M = 0.22 + 0.89(M_L)$ .... (4)
with a coefficient of determination of 0.9. In this relation M_L values, excluding estimates made by the Institute of Geological Sciences (IGS) for the U.K., come from a variety of sources that involve different types of seismographs and methods of calculation. Treating separately magnitudes reported by IGS as M_L up to 1979, we have:

$$M_L = -0.78 + 1.41(M) \quad (5)$$

with a coefficient of determination of 0.8 ($M \leq 4.5$).

There is in general a decrease in magnitude values for the recalculated events from those given in other sources. However, it is not always possible to make a comparison with magnitudes estimated in other sources unless the scale used is given explicitly. Thus, with Kárnik's estimates, known to be surface-wave determinations, there is a perfect correlation, i.e. M = -0.2 + M(Kárnik), with a calculated 0.2 standard deviation. However, the same is not true for the magnitudes listed in reference (8) for which it is not clear what scale or scales have been used for the determination of U.K. events. If we call M_i the magnitudes reported in that reference for the period 1900 to 1972, for events with $M \geq 3.0$ we have, $M = 3.31 + 0.21(M_i)$, with a coefficient of determination of only 0.1.

FOCAL DEPTH AND ATTENUATION CHARACTERISTICS

In the absence of reliable instrumental source depths in the U.K., one way of estimating focal depths and at the same time assessing the attenuation characteristics of Intensity using macroseismic data, is to employ a modified Kövesligethy technique[3]. It can be shown[4] that the fall-off of Intensity with epicentral distance D_i may be written

$$dI_i = I_o - I_i = b.\log\left[(R_i/h)^n \exp.a(R_i - h)\right] \quad (6)$$

where h is the focal depth and R_i is the focal distance corresponding to an intensity I_i. The focal depth as well as the intensity factor (bn) and the near-field attenuation coefficient (ah/n) may be determined from equation (6) for events for which their epicentral intensity I_o as well as at least three isoseismals are known. For events for which there is sufficient macroseismic information, focal depths and the values of the constants in equation (6) have been estimated using this technique[4,7].

Figures 2a and 2b show the distribution with macroseismic depth of earthquakes of different magnitude for the U.K. and for the whole of Northwest Europe respectively. We notice that depths are confined in the upper 25 kilometres and that in general larger magnitude events tend to nucleate at greater depths. There is some evidence that events with $M \geq 4.5$ in Western Scandinavia and the North and the Norwegian Seas are somewhat deeper (18 ± 4km) than in Northern France (15 ± 4km) or in the U.K. (11 ± 3 km).

For events with epicentres on land in the U.K., the absorption coefficient a in equation (6) was found to be inversely proportional to the focal depth, varying as:

$$a = 0.09(h)^{-1.3} km^{-1} \quad (7)$$

between values as high as $900 \times 10^{-3} km^{-1}$ for rockbursts and $3 \times 10^{-3} km^{-1}$ for the larger earthquakes.

Also the coefficient of geometric spreading n was found to have a definite trend, i.e. to increase with focal depth as:

$$n = 1.24 + 0.03(h) \quad (8)$$

a behaviour of n consistent with the fact that body-waves from deeper sources travel greater distances.

Finally, the coefficient b in equation (6), which is always considered to be a constant and equal to 3.0, was found to vary as:

$$b = 0.75(ah)^{-0.27} \quad (9)$$

This dependence of b on (ah) means that the proportion b of the energy density E in Kövesligethy's basic equation, i.e. $I = C + b.\log(E)$, associated with intensity I, increases with decreasing energy absorption (ah) at the epicentre.

For British earthquakes with epicentres on land or shore, for the period 1800 to 1983, the least-squares fit of equation (6) to 149 data points from 41 events in the range $5.5 \geq dI_i \geq 0$, $310km \geq D_i \geq 5km$, and $15km \geq h \geq 0.3km$, leads to:

$$dI_i = 0.32 + 5.00x10^{-3}(D_i) + 2.16\log(D_i/h) - 0.17\log(h) \quad \text{...........(10)}$$

with a coefficient of determination of 0.94. Figure 3 shows a plot of equation (10) for different values of the focal depth, together with the attenuation law derived for North-west Europe[4] using 351 data-points from 105 events, i.e.

$$dI_i = 0.27 + 3.60x10^{-3}(D_i) + 2.43\log(D_i/h) - 0.21\log(h) \quad \text{............(11)}$$

with a coefficient of determination of 0.91.

So far we have only indicated that the absorption coefficient in equation (6), which is in fact a near-field parameter, was found to be nearly inversely proportional to the macroseismic focal depth, equation (7). This dependence of a on source depth makes it impossible to discriminate any regional variations of the absorption coefficient which in most cases was found to be larger for near-surface than deeper layers, which is reasonable. It is very probable, therefore, that the regional variations of this coefficient calculated from macroseismic data for other parts of the world[5] may reflect differences in focal depths rather than genuine regional variations in a. However, what equation(7) shows is that it would be difficult to use explosions to evaluate the near-field attenuation of ground motions in terms of intensity.

With regard to the attenuation of intensity, equations (10) and (11) show that for rockbursts and for very shallow shocks the fall-off of intensity with distance is very rapid for the higher intensities. This explains why the epicentral intensities of shallow shocks may easily be missed out.

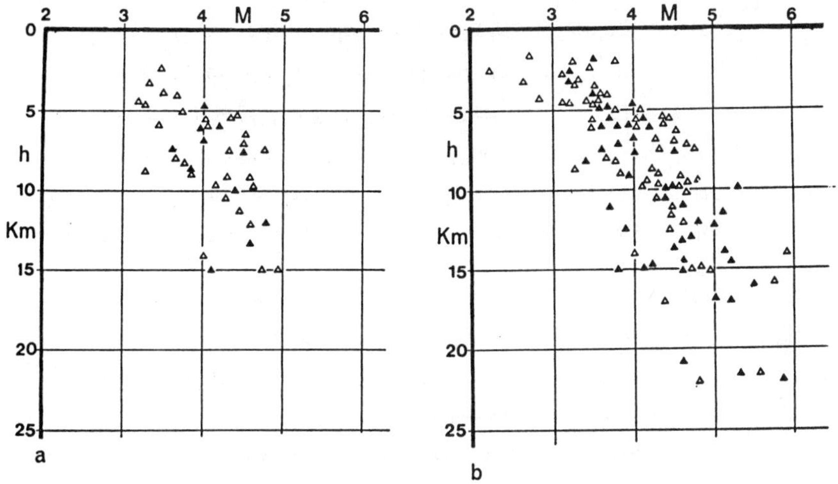

Figures 2a and 2b. Distribution of macroseismic focal depths with magnitude M for U.K. and Northwest European earthquakes respectively.

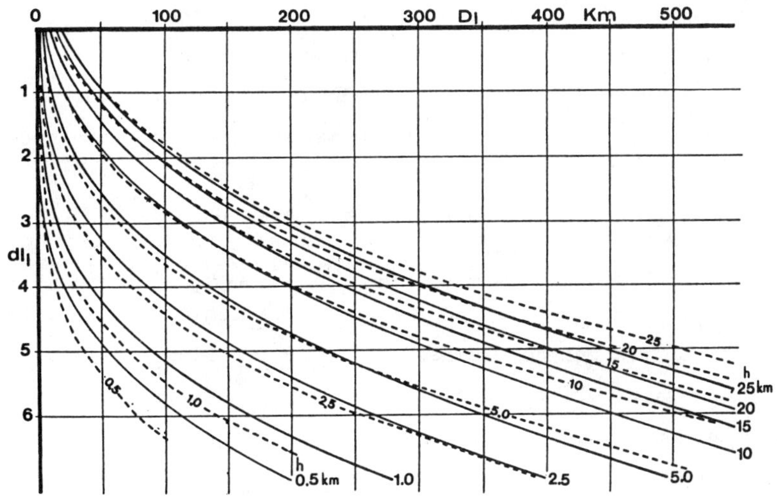

Figure 3. Least-squares solution of equation (6). Solid curves correspond to equation (10) for U.K. events, and dashed curves illustrate equation (11) for Northwest European earthquakes, plotted for different source depths h.

THE ASSESSMENT OF MAGNITUDE FROM MACROSEISMIC DATA

This is done[4] by combining $I = C + b.\log(E)$ with equations $E = E^*(4\pi R^n)^{-1}\exp(-aR)$ and $\log(E)^* = \alpha + \beta M$, where E* is the seismic energy at the source. Assuming $R \approx D$, from 53 sets of values of I_i and D_i, belonging to 17 U.K. earthquakes the magnitudes of which have been estimated from instrumental data, we find that:

$$M_{sc} = -1.41 + 0.73j^{-1}\sum(I_i) + 4.44\times10^{-3}j^{-1}\sum(D_i) + 1.62j^{-1}\cdot\sum\log(D_i) \quad ....(12)$$

with a coefficient of determination of 0.88. In equation(12) j is the number of isoseismals available for the determination of the average value of the surface-wave magnitude M_{sc} of a particular event. Figure 4 shows a plot of equation (12) together with the data-points used in the least-squares solution. From this figure we notice that the U.K. data cover relatively narrow magnitude and intensity ranges and consequently extrapolation of equation (12) outside the limits $3\frac{1}{4} \leq M \leq 4\frac{1}{4}$ and $I_i < 5$, is not recommended. An alternative, therefore, would be to enlarge the area of study to include the whole of Northwest Europe with 151 events and 300 data-points[4] that would widen the magnitude range to $2.5 \leq M \leq 6.0$ and increase the upper bound of intensities to VII (MSK), with the following results:

$$M_{sc} = -1.10 + 0.62j^{-1}\sum(I_i) + 1.30\times10^{-3}j^{-1}\sum(D_i) + 1.62j^{-1}\cdot\sum\log(D_i) \quad ....(13)$$

with a coefficient determination of 0.91. Figure 4 illustrates equation (13) which also fits, with the same coefficient of determination, the data used to derive equation (12).

When only one isoseismal distance is known, a small improvement of the estimate of M_{sc} can be achieved by including in its derivation the epicentral intensity I_o. For 14 earthquakes which have occurred on the land mass of Britain we have: $M_{sc} = 0.06 + 0.10I_o + 1.82\log(D_{III})$ .....(14) with a coefficient of determination of 0.86. A very similar, one-way regression applies to the whole Northwest European region, derived from 63 events, i.e.:

$$M_{sc} = -0.14 + 0.10I_o + 1.93\log(D_{III}) \quad (15)$$

with $r^2 = 0.95$.

THE LONG TERM SEISMICITY OF THE U.K.

This was done using our historical file of British earthquakes. Each event has been assigned a magnitude determined either from instrumental data or from macroseismic observations if none were available from the instrumental, using equations (13) and (15).

Figure 1 shows the location of all events since 1247 with a calculated magnitude equal to or greater than 4.0, within an area of $80x10^4 km^2$. From this figure we notice that during the last seven centuries the largest event on land ($50x10^4 km^2$) did not exceed a magnitude of 5.5 and that during that period, epicentral intensities exceeding a value of VII but not VIII (MSK), occurred only in four earthquakes. Thus, the number of yearly occurrences of events with epicentres on the land mass of Britain of damaging epicentral intensities between VII and VIII is only $1.2x10^{-4} year^{-1} 10^{-4} km^{-2}$, which is a very low rate indeed.

Equally low is the overall moment-rate associated with U.K. earthquakes. This was calculated using the magnitude-moment relationship $\log(M_o) = 17.4 + M$, derived for small magnitude events. Thus, the long-term (1247 - 1984) moment-rate for the whole region is $M_o = 1.3x10^{21}$ dyne-cm/year, while that for the land mass is $M_o = 0.9x10^{21}$ dyne-cm/year. The short-term (1900 - 1984) moment-rate for the whole region and for the land mass of Britain is $3.2x10^{21}$ and $1.0x10^{21}$ dyne-cm/year respectively, comparable to the long-term estimates. The small difference between long- and short-term regional rates by a factor of three suggests that as the historical record for the North Sea area prior to about 1700 is deficient, a number of offshore events of the historical period may have been missed out.

THE SEISMICITY OF THE U.K. IN PERSPECTIVE

In an attempt to place the seismicity of the U.K. in a global context a number of points can be made:

1. The focal depths estimated from macroseismic data, Figure 2a, all lie in the bottom of this range. This result is entirely compatible with continental seismicity in more active regions[15], and encourages us to place some confidence in the relations used in this paper.

2. The lack of earthquakes larger than M = 5.5 on land has two obvious consequences. Firstly, our experience with more active continental regions suggests that it is most unlikely that the bulk of the U.K. seismicity can be associated with particular surface features or faults. Small earthquakes have small source dimensions, and attempts to correlate U.K. seismicity with large, old, surface faults or "lines of weakness" which may now be experiencing "reactivation" is, in our opinion, optimistic at best. Secondly, and as a corollary, faulting during earthquakes of this size very rarely breaks the surface. This reduces the engineering problem to one of coping with transient ground motions or shaking rather than permanent offsets of foundations and abutments by seismic faulting. This does not, of course, remove the risk of slope and foundation failures which may, particularly for large engineering offshore structures, remain significant. This is particularly so in the graben of the North Sea and the Norwegian Sea, where there is some evidence for slightly larger events[7]. In contrast to permanent seismic offsets, which require detailed local geological knowledge, transient ground motions will have to be estimated from strong-motion data.

3. The seismic activity of the U.K. area in global terms, is entirely trivial.

Comparison of seismicity levels in different areas cannot be done in terms of displacement rates (either mm/yr or M_o/yr), which are misleading, as they take no account of the size of each area. Table 1 compares the U.K. activity with that of of other areas in the Alpine-Himalayan belt in terms of M_o/unit area/yr, which gives a true idea of the extremely low activity in the U.K.

4. The geological nature of the activity in the U.K. is enigmatic. Figure 5 shows the activity in Europe and North Africa reported by the USGS in the period Jan. 1963 to Aug 1983 and is probably complete at the $M = 4\frac{1}{2}$ level. Although this map shows how much more active the Mediterranean is than N. Europe, twenty years is not long enough to reveal some important tectonic features such as the Anatolian Fault zone in Turkey. More importantly, in the apparently "aseismic" region of west Africa several destructive earthquakes of magnitude greater than 6.0 occurred in the last 100 years (1906 and 1939 in Ghana, 1945 Cameroon, 1974 Gabon, 1983 Guinea), some of which have evidently been associated with surface faulting. It is worth stressing that, with a historical record almost ten times longer, no earthquake in the U.K. has been found of a size comparable to these in W. Africa: any one of which would account for 10,000 years worth of moment release in the U.K. The same remarks apply equally well in comparison with the eastern U.S.A., where we have no reason to think that events of $M_o = 10^{26}$ dyne-cm or greater are uncommon up the entire eastern seaboard.

5. The seismicity of the U.K. is clearly different from that of eastern U.S.A. or W. Africa in that either (i) no earthquakes of $M \geq 6.0$ occur, or (ii) 700 years is not long enough to reveal such events in the U.K., whereas 100 years is more than adequate in the eastern U.S.A. and W. Africa. In the light of this observation two interesting geological questions suggest themselves:

(a) is there anywhere on the continents seismically quieter than the U.K.? The considerable effort of the last few years as well as the conveniently long and complete historical record available, have revealed a constant low level of small events in the U.K. Is this the norm for "inactive" continental areas, and would com-

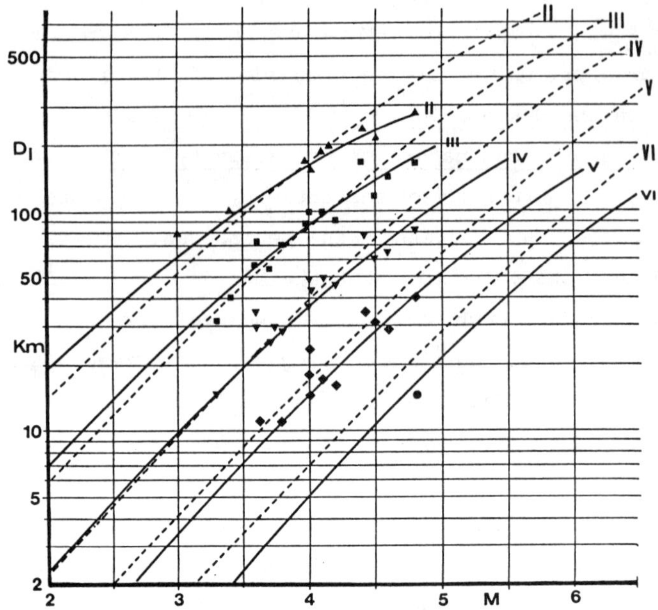

Figure 4. Data-points and least-squares solution of equation (12) for U.K. events and of equation (13) for Northwest European earthquakes (dashed curves).

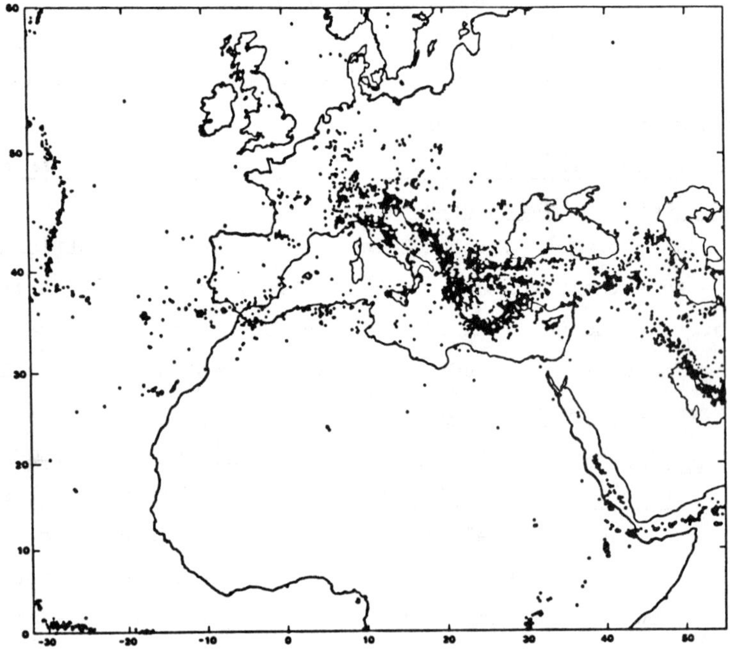

Figure 5

parable effort and data on say, Archaean shields, reveal similar low seismicity?

(b) <u>why is the seismicity of the Atlantic margin of the U.K. different from that of E. U.S.A. and W. Africa?</u> The evidence presented by Sykes[14] suggests that the location of transform faults in a new ocean is controlled by the presence of major pre-existing "zones of weakness" oriented sub-parallel to the direction of continental separation. After separation, such zones may still remain "weak", and many of the larger seismic events on the margin of E. U.S.A. and W. Africa do indeed appear to concentrate on the inland extrapolation of major oceanic transform faults. Such a correlation does not appear to work in Northwest Europe, where the rifting associated with continental separation in the Mesozoic and Tertiary was spread over a series of grabens from the W. Ireland to Poland; a distance of more than 1,000 km which is evidently much wider than that in most other regions of the Atlantic. It is easy to suggest that this fragmentation of NW Europe in the Mesozoic and Tertiary is responsible for its rather different seismicity today; but such a correlation does not imply understanding of why this should be so.

CONCLUSIONS

In summary, the acquisition and analysis of documentary information and instrumental observations of earthquakes in Britain since the middle of the 13th century has resulted in over 2,000 seismic events being identified.

Intensities were re-assessed in terms of the MSK scale and isoseismal maps constructed for almost all earthquakes of magnitude greater than about 3.5, in most cases from primary sources.

Macroseismic epicentres and radii of isoseismals were calculated and a considerable number of offshore events of the instrumental period were relocated.

Surface-wave magnitudes were re-assessed from instrumental data and a calibration function, equation (13), derived from the combination of macroseismic and instrumental data was used to assign magnitudes to all events for which there is macroseismic information available, thus creating a homogeneous body of data covering a period of over 700 years.

Focal depths and absorption coefficients derived from macroseismic data were found to be between 2 and 15 kilometres and $900x10^{-3}km^{-1}$ and $3x10^{-3}km^{-1}$ respectively, while the attenuation of intensity with distance was approximated by equations (10) and (11).

It is shown that the largest earthquake on land during the last seven centuries did not exceed a magnitude 5.5 and that the maximum intensity recurrence rate for VII (MSK), is only $10^{-4}year^{-1}10^{-4}km^{-2}$. This low level of long-term seismic activity is also reflected in the moment-rate of 10^{21}dyne-cm/year.

In our opinion it is unlikely that the small magnitude U.K. events are likely to be reliably associated with faults. We conclude that the seismicity of the U.K. is clearly different from that of eastern USA or W. Africa in that either no earthquakes of $M \geq 6.0$ occur, or 700 years is not long enough to reveal such events in the U.K., whereas 100 years is more than adequate in the eastern U.S.A. or W. Africa.

Summarising evidence from all sources of information, it appears that the damage caused by earthquakes in Britain has been very small. Earthquakes have triggered minor slides and rockfalls, in most cases from steep and unstable cliffs, and in a few cases they have caused the liquefaction of loose deposits. There is no evidence of turbidity currents and submarine slides being associated with U.K. seismicity.

ACKNOWLEDGEMENTS

We would like to thank all those who kindly loaned seismograms and early station bulletins. Also, we would like to thank Prof. M. Båth, Mrs. V. Kárnik, R. Adams, C. Melville, Mr. J. Menu, Dr. S. Sarma, Mrs. H. Anderson, Miss M. Kielmas, for assistance. This research was performed as part of the study of the Seismicity of the British Isles and the North Sea, supported by the London Centre for Marine Technology, S.E.R.C. and N.E.R.C.

Table 1.

Region

TABLE 1

Region	Based on average of	Mo-rate dyne-cm/yr	Area 10^4 xkm^2	Mo-rate/ unit area dyne-cm/ yr/km^2	Ref.
Ordos, China	700 years	$17.0x10^{25}$	100	$17.0x10^{19}$	(11)
Mongolia & Baikal	80 "	$49.3x10^{25}$	360	$13.6x10^{19}$	(12)
Tien Shan	80 "	$22.5x10^{25}$	75	$30x0x10^{19}$	"
Tibetan Plateau	80 "	$4.7x10^{25}$	128	$3x7x10^{19}$	"
E. Tibet, Gansu & Ningxia	80 "	$48.8x10^{25}$	143	$34.1x10^{19}$	"
Xianshiuhe Fault & W. Sichuan	80 "	$5.2x10^{25}$	44	$11.8x10^{19}$	"
Yunnan & Burma E.	80 "	$2.5x10^{25}$	48	$5.2x10^{19}$	"
N. China	80 "	$3.8x10^{25}$	99	$3.8x10^{19}$	"
Greece & W. Turkey	60 "	$27.3x10^{25}$	121	$22.6x10^{19}$	(13)
Italy & Yugoslavia	60 "	$11.9x10^{25}$	96	$12.4x10^{19}$	"
Iran	1200 "	$6.0x10^{25}$	112	$5.3x10^{19}$	(16)
U.K. (Land area)	700 "	$9.0x10^{20}$	50	$1.8x10^{15}$	
"	80 "	$10.0x10^{20}$	50	$2.0x10^{15}$	
U.K. (Whole area)	700 "	$13.0x10^{20}$	80	$1.6x10^{15}$	
"	80 "	$32.0x10^{20}$	80	$4.0x10^{15}$	

REFERENCES

1 Ambraseys, N. (1985). "Magnitude assessment of Northwestern European earthquakes", Journ. Earthq. Eng. & Struct. Dynam., vol. 13, no. 3.

2 Marshall, P., Basham, P. (1972). "Discrimination between earthquakes and underground explosions", Geoph. Journ. R. astr. Soc., vol. 28, p. 431.

3 Sponheuer, W. (1960). "Methoden zur Herdtiefenbestimmung in der Makroseismik", Freib. Forschubgshft., C.88, Berlin.

4 Ambraseys, N. (1985). "Intensity-attenuation and magnitude intensity relationships for NW European earthquakes". Journ. Earthq. Eng. & Struct. Dynam., vol. 13, no. 4.

5 Howell, B., Schultz, T. (1975). "Attenuation of MM Intensity with distance", Bull. Seism. Soc. Am., vol. 65, p. 651.

6 Ambraseys, N., Melville, C. (1983). "The seismicity of the British Isles and the North Sea", vol. 1, Marine Tech. Inst. Report, London.

7 Ambraseys, N. (1985). "The seismicity of western Scandinavia". Journ. Earthq. Eng. & Struct. Dynam., vol. 13, no. 3.

8 Lilwall, R. (1975). "British earthquake data file", Global Seism. Unit, IGS, Edinburgh.

9 Davison, C. (1924). "A history of British earthquakes", Cambridge UP.

10 Wesnousky, S. et al (1984). "Historical seismicity and rates of crustal deformation, North China", Bull. Seism. Soc. Am., vol. 74, p. 1767.

11 Molnar, P., Deng, Q. (1984). "Faulting associated with large earthquakes & the average rate etc", J. Gephys. Res., vol. 89, p. 6203.

12 North, R. (1974). "Seismic slip rates in the Mediterranean" Nature, vol. 252, p. 560.

13 Sykes, L. (1978). "Intraplate seismicity, reactivation of pre-existing zones etc", Rev. Geophys. Space Phys.,

vol. 16, p. 621.

14 Chen, W., Molnar, P. (1983). "Focal depths of intra-continental and intraplate earthquakes etc." J. Geophys. Res., vol. 88, p. 4183.

15 Ambraseys, N., Melville, C. (1982). "A history of Persian earthquakes". Cambridge Univ. Press.

Discussion

MR G. NEILSON, British Geological Survey
Could Professor Ambraseys say whether his instrumental data is available for inspection? What account was taken of the response characteristics of the instruments in determining magnitudes?

DR R.D.ADAMS, International Seismological Centre, Newbury
Although the seismicity of Britain is admittedly low, does Professor Ambraseys not agree that there are many continental regions of comparable area, particularly in Africa and Australia where there is no evidence for seismicity even as low as that in Britain? Of course, the continent of Antarctica is almost completely aseismic.

PROFESSOR FINN
Dr Ambraseys stated that the seismicity of the UK is trivial as demonstrated by the fact that nothing over magnitude M = 5.5 has occurred in over 700 years. Yet in the Norwegian sector of the North Sea ductility events with magnitude M = 7+ have been used on design. Can he reconcile these different viewpoints?

PROFESSOR AMBRASEYS
In reply to Mr Neilson, a list of loaned seismograms and early station bulletins from European stations is available. Also true-to-scale copies of a limited number of seismograms is available for inspection. You may also borrow through our library station bulletins for the period prior to 1955. The response characteristics of seismographs were used only when records were read or when station bulletins reported trace amplitudes.

As Dr Adams stated, there are other continental regions of comparable or even larger area than NW Europe of low, short-term seismicity. However, the question is, is it likely that any 'aseismic region', if scrutinized as thoroughly as we have done the UK, will show a similar behaviour? Or, can you find another part of the world of low seismicity for which seven centuries of data is readily available for comparison? In places like Tibet and East China there is dispersed seismicity with low strain rates compared to the rest of Asia, and occasionally large magnitude 7 earthquakes which do break the surface and are correlated with surface geomorphology. In the UK strain rates are minute even compared to Tibet and Eastern China. The real questions are, is the difference between NW Europe and the eastern USA significant in a geological sense; and are the long-term strain-rates comparable, but released differently in the short term?

In reply to Professor Finn, no, I cannot reconcile the viewpoints he raised. I am well aware of the impact that recent studies of seismicity have had on the design of offshore structures in the North Sea and of the difficulty of finding a sober account of what appears to have been blown up out of proportion. The argument,as we see it, is as follows: because the 150 year record in the eastern USA shows infrequent large events it is sensible to investigate apparently aseismic areas like NW Europe. But, the much longer record of NW Europe shows no large events and that in reality, the single large event of 1819 tells us nothing in itself, except that earthquakes of the same magnitude of about 6.0 must be considered credible events throughout the entire region. Until we understand why the larger earthquakes occurred where they did and how other parts of the region are genuinely different, the subdivision of NW Europe into seismotectonic provinces is not a realistic proposition, particularly when using short-term seismicity data of 20 to 50 years, derived from an heterogeneous body of information. The fact is that the long-term (7 centuries) moment-rate for the region is trivial; it is dominated by a very few events which are in themselves insignificant in hazard terms, and therefore the contribution from events of magnitude 4 or less is absolutely nothing at all. The attempted correlation of magnitude 4 to 5 events with offshore faults is questionable. In areas of normal or thrust faulting small seismic events and aftershocks appear almost never to define planes or be correlated with particular surface faults. It is very likely that a large amount of the relative motions in the North Sea is taken up aseismically, in processes such as creep. The other approach, which may lead to design magnitudes in the North Sea of M = 7.0+, that of extrapolating from magnitudes of instrumentally recorded small events to maximum magnitudes that might occur within a longer period than that sampled, has considerable perils; how can one be assured that the short-term record is statistically representative? In any event, what we have shown here is that the short-term seismicity is not representative.

5. Event triggered seismic detection systems developed by the British Geological Survey

D. J. HOULISTON, J. LAUGHLIN, G. WAUGH, British Geological Survey

SYNOPSIS. The British Geological Survey has developed seismic event triggering systems over the past three years for a range of requirements. Three types of multichannel triggering recorders have been designed, developed and deployed over this period including the system for the observatory type of environment, the version developed to monitor microseismic activity in old and unstable limestone workings and a battery powered system designed for operator flexibility to act as a remote recorder. The latter system downloads the detected event data to a cartridge recorder, or is an intelligent data acquisition module, on-line to a local data processing system.

SYSTEM DESCRIPTIONS

Observatory Triggering System

1. The general specification of the original system, on which the seismic triggering algorithms were developed, is given in Table 1. As illustrated, time channel information is available from the local network Geostore recorder in the Vela coded format and also from an MSF Rugby standard time transmission receiver. Both timing formats are recorded as separate channels.

2. This unit is designed to analyse only three of the ten seismic channels. The remaining seven seismic and two time channels are not analysed but are included in the delay buffer, which provides a 20-second delay for all channels. This is sufficient time for even the slowest of the seismic waves to propagate across the network. The delay of 20 seconds ensures that the event onsets are recorded. By spacing the triggering channels around the network, uniform detection of events is achieved.

3. A dynamic range of 60db (10-bit accuracy) was chosen since this is the typical range available from the UHF transmitted FM carrier data from the remote outstations.

4. The sampling rate of the system is 100 samples/second/channel and anti-aliasing filters provide an upper bandwidth limit of 40Hz for the data.

Table 1. Original seismic event detection system

(OPERATIONAL EARLY 1982)

No: of time channels :	2 (Internal time encoder and standard time transmissions from MSF Rugby).
No. of seismic channels :	10 (Transmitted from radio-linked outstations).
Dynamic range :	60 db (10 bits)
Sampling rate :	100 samples/second/channel
Event buffer :	20 seconds/channel
Recording medium :	1. 9-track IBM compatible reel-to-reel for direct computer analysis (approx 40 Mbytes) 2. Triggered ink-jet recorder through D-A facility
Power source :	Mains operated

Algorithm to detect local, regional and teleseismic events.

Digital-to-analogue converters are provided for the delayed data of all channels for monitoring purposes.

5. The system, designed for observatory applications, records the detected events on a standard 9-track computer compatible reel-to-reel tape recorder to facilitate data analysis since there is no need for transcription or data re-formatting. Digital-to-analogue converter outputs for each channel permit the display of detected events on a triggered ink-jet recorder, providing an instant hard copy for the network analyst. This has proved a useful feature since the type of event detected, and initial event parameters, are measured from this record. It also aids the analyst to pick events of interest, on the recorded tape, which require further analysis by computer. The D-to-A facility also provides an instant check on the system performance.

6. In 1982 the complete system was transferred to Cornwall to be operational during the initial pumping operations at the "Hot Dry Rock" experimental site operated by the Camborne School of Mines on behalf of the Department of Energy and the Commission of the European Communities. It proved to be a useful monitor, providing immediate feedback to the scientists controlling the pumping. After this two month period, the system was reinstalled in Edinburgh and has continued to be totally reliable in its present role.

7. More detailed comment on the system performance and scope is given later.

Cartridge Triggering System

1. The triggering system for the Shetland Islands, funded by the Department of Energy, is designed with operator flexibility in mind, with all data channels being capable of analysis as triggers. It is also designed to operate reliably throughout power failures lasting for several hours. The trickle-charged batteries with the present small 9Ah battery unit will support the system for 4 hours during power failures. This could be easily extended using larger capacity batteries if required.

2. The time channel information, as indicated in the specification (Table 2), is identical to the previous system since the Geostore time code is available and MSF time standard transmissions are easily received in the Shetland Islands.

3. The maximum number of seismic channels is ten, and the operator selects which of these are full trigger channels. The seismic channels which are excluded from the detection analysis are still included in the data stream and are delayed and recorded with the triggering channels. The operator also selects the number of coincident triggers, within 15 seconds, required to start the cartridge recorder.

4. The dynamic range of this system is 72 db (12-bit accuracy) and the sampling rate is 100 samples/second/channel. Anti-aliasing filters provide an upper bandwidth limit of 40Hz. The packing density on the 4-track tape is 8000 bits/inch with up to 26 M bytes stored on one cartridge which permits storage of up to 80 seismic events on one cartridge.

Table 2. Shetland Islands seismic event detection system

(OPERATIONAL JULY 1984)

No. of time channels :	2 (Internal time encoder and standard time transmissions from MSF Rugby).
No. of seismic channels :	Up to 10
Dynamic range :	72 db (12 bits)
Sampling rate :	100 samples/second/channel
Event buffer :	15 seconds/channel
Recording medium :	Cartridge at 8000 bits/inch with up to 26 Mbytes storage.
Power source :	Batteries trickle charged.

Algorithm to detect local, regional and teleseismic events.

Digital-to-analogue converters are provided for the delayed data of all channels for monitoring purposes.

5. The data recorder is a Tandberg TDC 3224 streaming cartridge recorder designed to stream data at 86.66k bytes/second for computer disk back-up. The cartridge recorder has been modified to operate in a start/stop mode with a microprocessor-based dual buffer and controller downloading the detected event data to the recorder. The data are transferred in 24k byte blocks instead of the standard 4k byte blocks, reducing the number of times the deck has to start and stop for each recording. The controller is required to operate at high speed and direct memory access (DMA) techniques are employed with the processor to achieve the fastest possible access time to the 48k byte data buffer. Read-after-write facilities check the recorded data, rewriting blocks with errors to ensure high data integrity.

6. Digital-to-analogue converters are provided for system monitoring or to obtain an instant paper record with a triggered ink-jet recorder. Hardware has been developed and software written by C.J. Fyfe to permit downloading of the data onto an LSI 11 with a Winchester disk at the BGS laboratories in Edinburgh.

Microseismic Monitoring System

1. The erosion of old limestone workings in the Dudley area of the Midlands is causing concern with local authorities and the Department of the Environment. BGS was asked to monitor the microseismicity in the mines to help gather information.

2. The triggering system developed to monitor microseismic activity in old limestone workings, (Browitt et al, 1983) includes some of the features of the previous system. Some of the original triggering software is relevant but apart from the processor, memory and I/O boards,the hardware is totally different (Table 3).

3. MSF Rugby was used as the timing standard throughout an experiment in the West Midlands.

4. The high data sampling rate of 2000 samples/second/channel is necessary because of the high frequency energy from the microseismic activity. Again, 60db dynamic range is adequate for this particular application.

5. Since the transducers for this type of application are closely-spaced (tens of metres), an event buffer for 4 seconds of data is adequate, allowing the various recording systems to switch on and record the detected event onsets.

6. For this type of experiment, a Racal Store 14, 14-Channel tape recorder and a Siemens 16-channel ink-jet recorder have been chosen to operate in a triggered mode and record detected events. They were selected because of the high frequency content of the events and because of the ease of data analysis. Both recorders are capable of continuous recording during simulated roof-fall experiments.

Table 3. Trigger system for microseismicity investigations in old limestone workings

No. of time channels :	1 (MSF Rugby standard time transmissions)
No. of channels : (from underground mine)	6 (One processor for each channel because of high sampling rate)
Dynamic range :	60 db (10 bits)
Sampling rate :	2,000 samples/second/channel
Event buffer :	4 seconds/channel
Recording medium :	Triggered Store 14 (14-track FM recorder) Triggered ink-jet recorder (16 channels)
Power source :	Mains operated

Algorithm to detect local events and local microseimic activity.

Digital-to-analogue converters are provided for the delayed data of all channels.

PERFORMANCE ANALYSIS

1. Since this paper is mainly concerned with the performance of the earthquake detection systems and not the specialised experiments concerning microseismicity of underground limestone workings, the following performance assessments relate to the hardware and detection algorithms of the systems currently analysing the network data in the Lowlands of Scotland and in the Shetland Islands.

2. The detection algorithms (Houliston et al, 1984) were initially developed whilst operating the original system on-line to the radio-linked seismometer stations of LOWNET based around Edinburgh. Two separate algorithms were required for (a) local and regional events, and (b) teleseismic events. Although this sounds cumbersome, it proves to be advantageous since the type of event has already been categorised at the time of detection. This information could be useful in future developments, when flagging events on continuous recorders.

3. The algorithm for detecting local and regional events operates on the first difference between successive data samples and calculates a modified differential function (MDF). This function has a resetting feature as the period of the seismic waveforms increase. The resetting feature is further used to dynamically increase the threshold level required for triggering for longer period waveforms. These combined features provide high sensitivity for the high frequency seismic phases and extremely low sensitivity for the long period microseisms, eliminating the possibility of false triggers from large amplitude microseisms during adverse weather conditions. This algorithm is particularly successful at detecting small local and regional events superimposed on large amplitude, long period, microseisms and provides better detection reliability than STA/LTA (short term average/long term average) algorithms for this type of event.

4. Figure 1 demonstrates the sensitivity of the detection technique. The upper event is relatively easily detected but the centre and lower events provide a greater challenge. All three channels detected the three events though the waveform is barely noticeable on the Broadlaw trace of the lower event. The examples in this figure were obtained on the LOWNET network.

5. Figure 2 illustrates another event less easily detected by a triggering algorithm. All three triggering channels of the LOWNET system detected the event which was a quarry blast in the Moorfoot Hills (0.6ML). The weakest signal was detected at Aberfoyle (EAB) and again demonstrates the sensitivity of the algorithm though it was not programmed to its highest possible sensitivity.

6. The Shetland network is smaller than LOWNET comprising a three-component station, a separate strong motion channel at Lerwick and two radio-linked outstations with vertical seismeters at Yell and Walls. Figure 3 shows the locations of the instruments.

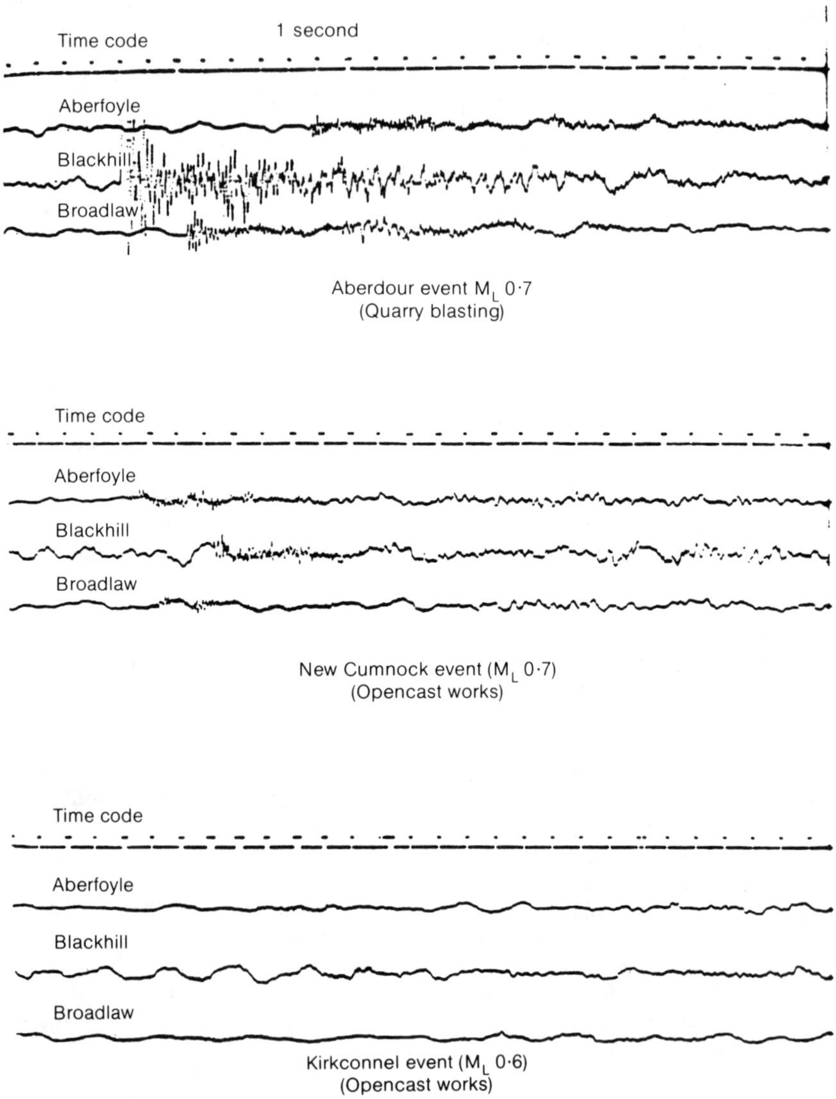

FIGURE 1. Local events detected on the Lowland Network

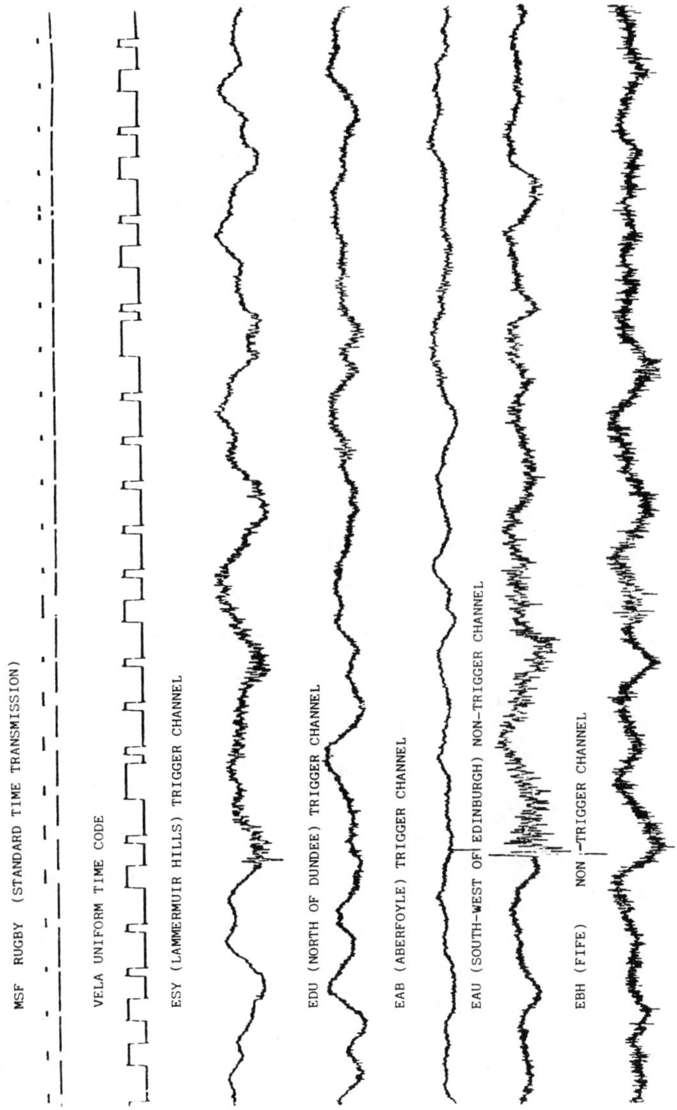

FIGURE 2. A quarry blast in the Moorfoot Hills detected on the Lownet Network

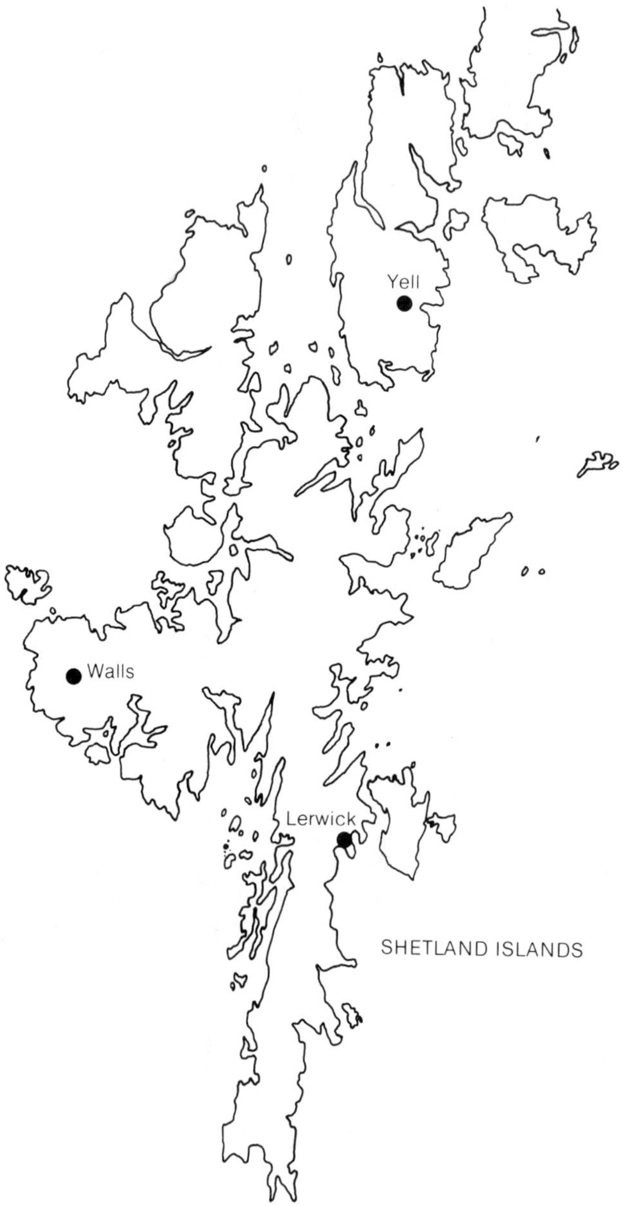

FIGURE 3. Shetland Islands Network

7. An example of an event detected on this network is illustrated in Figure 4. It is the result of a quarry blast 4 km West of Lerwick and is typical of such events in the Shetland Islands, providing no great challenge for the triggering algorithm. The replay gain settings were left identical for all channels to illustrate the relative amplitudes of the waveforms at the different sites. All three triggering channels detected this event.

8. An event missed by the triggering algorithm is illustrated in Figure 5. It is a North Sea earthquake which occurred North-East of the Shetland Islands but since the P waves are not detectable on the waveform and the S waves are not clear on any of the channels it is impossible to locate this event with any accuracy. It is not desirable to increase the sensitivity of the system to detect such events.

9. The teleseismic algorithm, examines samples 0.32 seconds apart to determine if a teleseismic earthquake has been detected. The detection sensitivity takes into account the current microseismic noise levels as with the local and regional algorithm. An example of the sensitivity of this technique, illustrated in Figure 6 is for a magnitude 4.2 (mb) events in the Aegean Sea detected on LOWNET.

10. The Shetland Islands network provides a much greater challenge for teleseismic event detection because of the adverse weather conditions especially in the winter months with wind speeds up to 70 knots. As illustrated in Figure 7 the combination of sea and wind noise results in extremely high microseism levels. The event in this example was recorded during 35 knot winds and occurred in the Tyrrhenian Sea (mb 4.8).

11. In general, large aperture networks with many triggering channels have fewer false triggers than single channel applications, and triggers, due to radio interference and other anomalous spikes on the received data, are reduced. Various approaches are already being considered to reduce false triggers for single channel applications and will be implemented in the portable field system currently being developed for the NERC Seismic Equipment Pool.

ACKNOWLEDGEMENTS
This work was supported by the Department of Energy, the Department of the Environment and the Natural Environment Research Council and is published with the approval of the Director of the British Geological Survey (NERC).

REFERENCES

1. Microseismicity Investigations at Dudley Sports Centre Limestone Workings - 1. by, C.W.A. Browitt, D.J. Houliston, D. McCann, and A. Miller.

2. Automatic real-time event detection for seismic networks. D.J. Houliston, J. Laughlin and G. Waugh. Computers and Geosciences Vol. 10 No. 4, 1984.

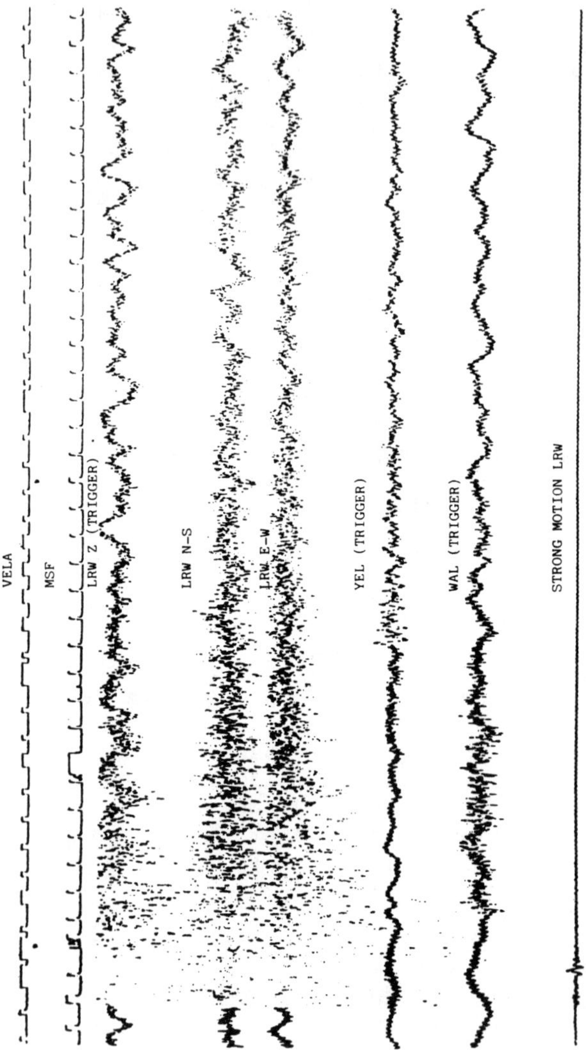

FIGURE 4. Quarry blast in the Shetland Islands detected by the Shetland Islands Network

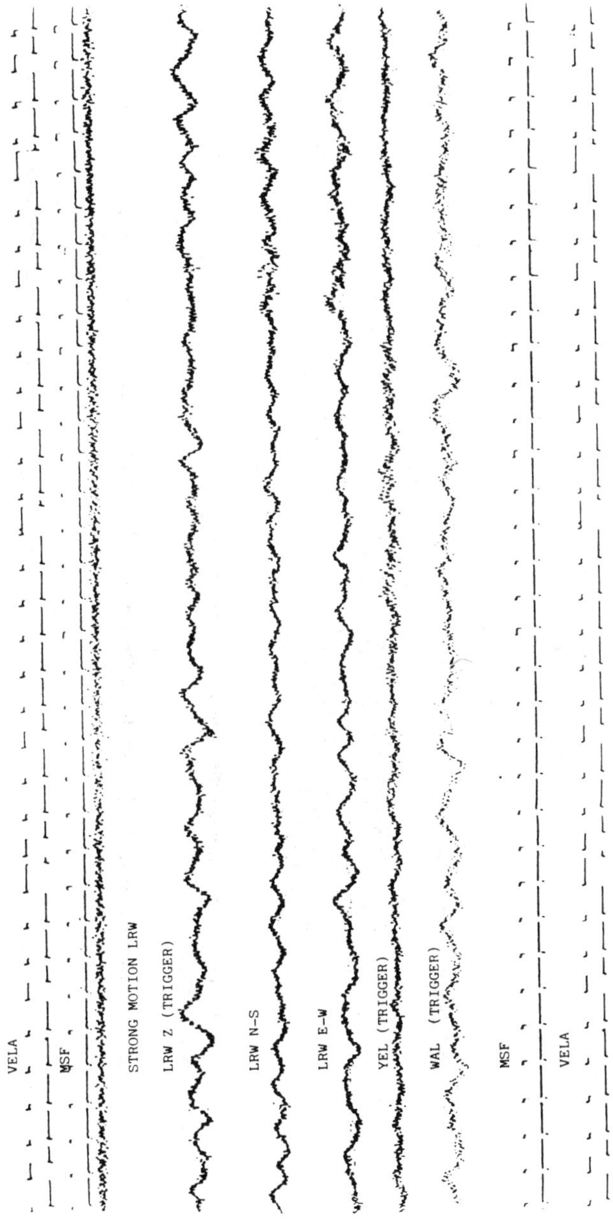

FIGURE 5. North Sea earthquake not detected by the Shetland Islands Network

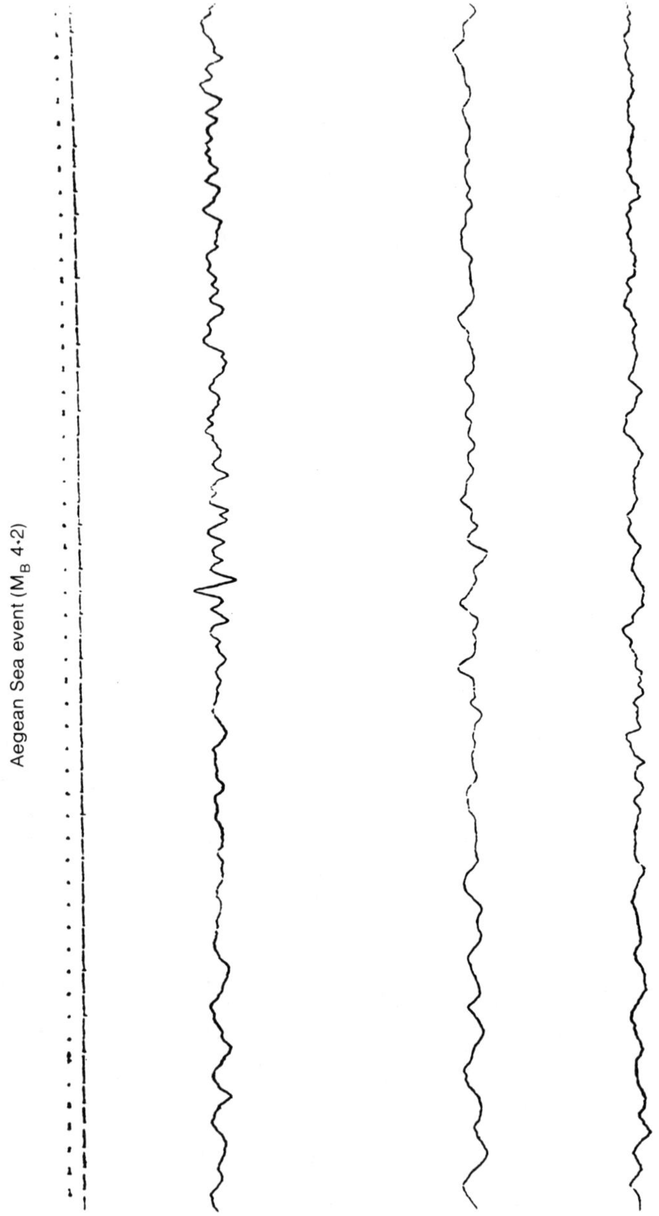

FIGURE 6. Aegean Sea earthquake detected on the Lowland Network

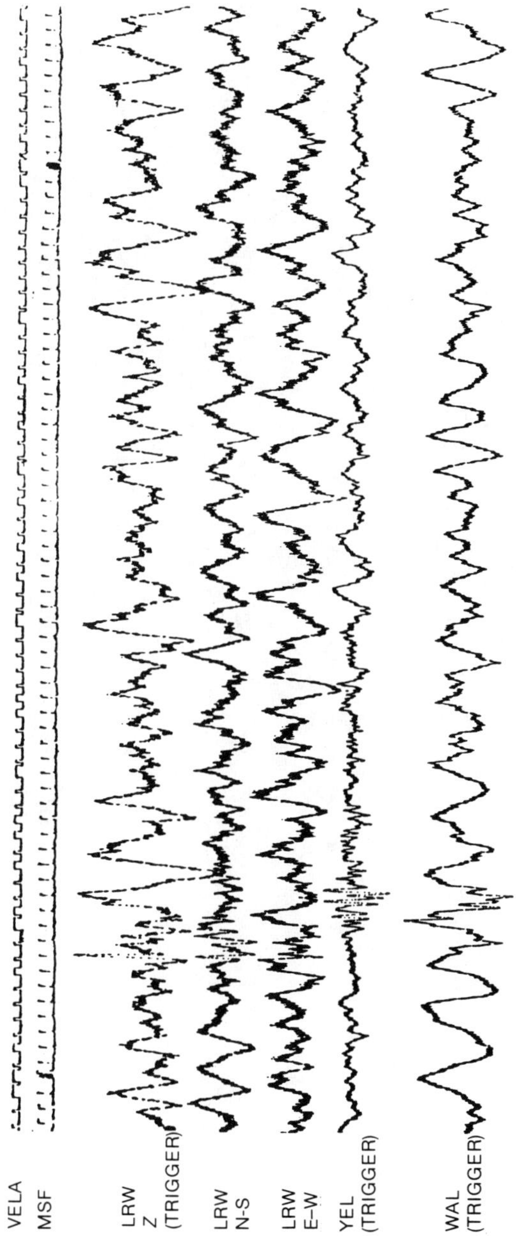

FIGURE 7. Tyrrhenian Sea earthquake detected in the Shetland Islands Network

6. Instrumental magnitudes of British earthquakes

G. NEILSON and P. W. BURTON, BA, MA, PhD, British Geological Survey

THE BRITISH GEOLOGICAL SURVEY INSTRUMENTAL MAGNITUDE STUDY OF BRITISH EARTHQUAKES

1. Quantification of the size of British earthquakes, on an objective basis, is of fundamental importance to the study of seismic risk and seimotectonics in this country. The simplest method of assessment is by means of instrumental magnitudes. The British Geological Survey has an ongoing programme of magnitude studies of British events which has been expanded in scope since 1981 when a programme of inspection of seismogram archives was begun and more than 400 original seismograms from British and European Observatories have been examined in order to calculate the magnitudes, on appropriate scales, of the larger earthquakes which have been felt in Britain since the advent of continuous instrumental recording at the beginning of the 20th Century (ref. 1).

INSTRUMENTAL MAGNITUDE

2. The concept of magnitude as a method of assessing the size of earthquakes was introduced by Richter, in California, in 1935, who based his method on the pioneering work of Wadati in Japan. The observed decrease in amplitude of seismic waves with hypocentral distance, when recorded on a network consisting of identical seismographs, is the physical basis for Richter's original local magnitude scale, M_L. From that time onwards the concept of magnitude has been developed and expanded to include magnitude scales which may be used at teleseismic distances and using various seismic phases. Magnitude values were only calculated for earthquakes on a routine basis by European stations, from the late 1950s onwards. It is therefore necessary to examine original seismograms of the earlier events in order to calculate magnitude values for earthquakes of this period.

DATA

3. Since Britain is an area of low to moderate seismicity, even the largest British earthquakes are not widely recorded by seismographs outside Western Europe and hence the inspection of seismogram archives was restricted to this area. Because

of the small amplitude and, in the case of the older seismographs, the poor time resolution of the records, examination of original seismograms, wherever practicable, was deemed to be the most reliable procedure. Calculation of seismograph transfer functions was also undertaken using instrumental parameters obtained from station records.

CHOICE OF MAGNITUDE SCALES

4. The magnitude scales selected were largely determined by the hypocentral distance of the events, two local scales, Richter's original M_L scale and a modification of the body wave scale - denoted by m_{b*} being the most frequently used. In addition the normal teleseismic scales for both body waves and surface waves, m_b and M_s respectively, were used where appropriate. In cases where one or more of the teleseismic magnitudes could be obtained, in addition to m_{b*} and M_L, this provided a valuable indication of the relationship between the local magnitudes and the teleseismic magnitudes.

RESULTS OF THE STUDY

5. More than 70 earthquakes have been studied so far which have been divided into three categories (a) bigger earthquakes with epicentres on land, (b) bigger earthquakes with epicentres outside the land area of the UK, and (c) smaller earthquakes recorded only on seismographs in Britain and Ireland. In category (a) there are 24 events spanning the years 1903 to 1980. The highest magnitudes in this group belong to the 1926 Hereford event (5.4 m_{b*}, 4.8 M_L) and the 1957 Derby earthquake (4.9 m_{b*}, 5.3 M_L). Category (b) consists mainly of earthquakes in the Channel Islands area and North Sea. The largest magnitude event amongst the latter is the 1931 North Sea earthquake (6.0 m_{b*}, 5.7 M_L). The 36 events in category (c) are confined to those events which occurred before the commissioning of the Lowland seismic network, LOWNET, in 1969. The largest magnitude values are those for the 1903 Caernarvon earthquake (4.8 M_L) and the 1916 Stafford earthquake (4.8 M_L).

6. This study has also shown that, as was to be expected, the magnitude detection threshold on all scales has decreased with time and currently British events with m_{b*} exceeding about 3 will probably be recorded at European stations.

7. The relationship

$$M_L = 0.98\, m_{b*} + 0.31 \tag{1}$$

has been found for the bigger events with epicentres on land. This may be compared with the relationship

$$M_L = 0.72\, m_{b*} + 1 \tag{2}$$

obtained by Jacob and Neilson (ref. 2) in a study of a data set primarily consisting of low magnitude British earthquakes. This tends to suggest that the partition of longitudinal and shear wave energy for higher magnitude British events is not so asymmetric as for smaller magnitude British earthquakes.

ACKNOWLEDGEMENTS
This work was supported by the Natural Environments Research Council and is published with the approval of the Director of the British Geological Survey (NERC).

REFERENCES

1. NEILSON G. and BURTON P.W. The magnitudes of the larger British earthquakes of the 20th Century. British Geological Survey Global Seismology Unit, 1984, Report No 206.
2. JACOB A.W.B. and NEILSON G. Magnitude determination on LOWNET. British Geological Survey Global Seismology Unit, 1977, Report No 86.

7. A comparison of two historical earthquakes in North Wales

R. M. W. MUSSON, BSc, PhD, G. NEILSON and P. W. BURTON, BA, MA, PhD, British Geological Survey

1. Introduction

The recent Lleyn Peninsula earthquake of 1984, July 19, with magnitude of 5.4 +/- 0.3 ML, has focussed considerable attention upon the seismicity and seismic history of the NW Wales area (ref. 1). This paper compares macroseismic data on the two largest previous earthquakes in this area this century. The events of 1903, June 19 and 1940, December 12.

2. In the last four years there has been a considerable revival of interest in the historical seismicity of the UK, a subject which had received relatively little attention since Davison published his History of British Earthquakes in 1924 (ref. 2). This has lead to some debate on appropriate methodology for re-assessing historical earthquakes using macroseismic data, and this subject is at present a developing field (ref. 3).

3. Instrumental and Macroseismic Comparison

The two earthquakes described in this paper not only give an insight into the historical background for the current seismic activity in the Caernarvon area, but also illustrate the way in which the preservation of historical macroseismic data is influenced by considerations other than the characteristics of the earthquake itself. The earthquakes of 1903, June 19 and 1940, December 12 seem to have been very similar if one compares the respective areas under isoseismals 5 and 4 MSK; furthermore, the mean magnitude values of the two earthquakes are 4.8 ML and 4.7 ML respectively (refs. 4-5).

4. The similarity of the two events in magnitude, location and isoseismal pattern, is markedly in contrast with the differences in documentation. The earlier event provides much more data, in much greater detail, to the historical investigator of macroseismics than does the later one. In a recent study conducted by the authors of this paper (ref. 4), investigation of the 1903 Caernarvon earthquake uncovered felt reports from 72 towns, villages and other locations, of which all but 8 were of sufficient detail to allow numerical intensities to be assigned using the MSK scale. In addition, an intensity value was assigned to an ill-defined area (Lleyn Peninsula/S Caernarvonshire) and a further 7 locations were

discerned at which it may be conjectured with reasonable certainty that the earthquake was felt.

5. By contrast, the 1940 earthquake was investigated in the same way and yielded reports from only 16 towns or other locations, and numerical intensities could only be assigned to 6 of these. Additionally, 4 reports referred to ill-defined locations (Anglesey, Caernarvonshire/Merionethshire, the coastal strip from Colwyn Bay to Chester, and Snowdonia).

6. The detail in reporting shows similar variation: the longest account describing the earthquake's effects at any place in 1903 is that for Porthmadoc (about 1,000 words long). The longest account describing the effects at any single place for the later earthquake is only 50 or 60 words long.

7. Nevertheless, the highest intensities are similar. Those of the 1903 event are slightly greater; minor damage to plaster and one or two chimneys was done at a few locations, giving a maximum intensity of 6 MSK. The highest intensities in 1940 seem to have been just below the threshold of damage, although considerable alarm was caused. Maximum intensity for the 1940 event was about 5-6 MSK.

8. In comparing the felt areas it appears that the 1903 event was observed over a wider area than the 1940 earthquake. However, as the patterns of isoseismals 4 and 5 MSK are so similar (ref. 4), and as the earlier event was so much better reported, the conjectural radius of the isoseismal 2 MSK was probably similar in both cases.

9. <u>Discussion of Differences</u>

The explanation of the wide variations in reporting between the two events is to be found in historical and cultural changes, and the individual circumstances of the two events.

(i) In 1903, the newspapers as a rule carried much more detailed reporting of local events. By 1940 standards of reporting had changed towards a much terser style;

(ii) In 1940, the fact that there was a war on meant that a British earthquake was relatively less newsworthy than would otherwise have been the case. Also, wartime restrictions reduced the resources of newspapers.

(iii) The 1903 earthquake occurred in mid-morning, the 1940 earthquake late at night. This evidently affected the number of people who perceived the earthquake; in particular the radius of perceptibility of a nocturnal earthquake tends to be less because of the threshold of waking. Furthermore, one source of data for the 1903 earthquake (school records) is only available because the earthquake took place within school hours and was noted down by some teachers.

(iv) The 1903 earthquake was investigated at the time by a contemporary seismologist (Charles Davison) (ref. 6). Though his original data are no longer extant, and though he quotes little original detail in his report on the earthquake, it is still possible to use his work to trace the outer limits of the felt area at the lowest intensities. This is particularly useful since reports at low intensity, being less newsworthy, are less often printed in newspapers. The 1940 earthquake was not investigated macroseismically at the time so far as is known, and this shows when one compares the the extent of far-field reporting for the two earthquakes.

10. Conclusions
Comparison of these two earthquakes illustrates graphically the importance of being aware of historical and cultural factors when considering the extant macroseismic data of historical earthquakes. In particular, it demonstrates that differences in degree of detail in reporting cannot be simply related to differences in intensity.

11. The evidence shows that the 1903 and 1940 North Wales earthquakes were similar events, and, although smaller in magnitude than the 1984 event in the same area, were strong by British standards. The 1940 earthquake in particular has not previously received full recognition of its significance.

ACKNOWLEDGEMENTS
This work was supported by the Natural Environment Research Council and is published with the approval of the Director of the British Geological Survey (NERC).

REFERENCES

1. TURBITT, T. et al. The earthquake of 19 July, 1984, Lleyn Peninsula, UK (magnitude 5.4). British Geological Survey Global Seismology Unit, Edinburgh, 1984, Report No 239.
2. DAVISON, C. History of British Earthquakes. Cambridge University Press, Cambridge, 1924.
3. BURTON, P.W., MUSSON, R.M.W. and NEILSON, G. Studies of historical British earthquakes. British Geological Survey Global Seismology Unit, Edinburgh, 1984, Report No 237.
4. MUSSON, R.M.W., NEILSON, G. and BURTON, P.W. Macroseismic reports on historical British earthquakes II: North Wales and Irish Sea. British Geological Survey Global Seismology Unit, Edinburgh, 1984, Report No 208.
5. NEILSON, G. and BURTON, P.W. The magnitudes of the larger British earthquakes of the 20th Century. British Geological Survey Global Seismology Unit, Edinburgh, 1984, Report No 206.
6. DAVISON, C. The Caernarvon earthquake of June 19th 1903. Quarterly Journal of the Geological Society, 1904, vol. 60, no. 239, 233-242.

8. Macroseismic focal depth and intensity attenuation for British earthquakes

P. W. BURTON, MA, PhD, R. McGONIGLE, BA, G. NEILSON and R. M. W. MUSSON, BSc, PhD, British Geological Survey

SYNOPSIS. The decay of earthquake intensity (MSK scale) with distance from the macroseismic focus is considered for a selection of British earthquakes. Two formalisms describing intensity attenuation are considered: that of Kovesligethy generates the shallower focal depths, nearer to instrumental determinations, and is preferred. Optimal values of macroseismic attenuation coefficients are obtained and focal depths in the range 5-36 km, that is, throughout the crust.

INTRODUCTION

1. The felt and damaging effects of earthquakes are typically quantified on a macroseismic intensity scale (as opposed to the absolute energy or size of an earthquake, quantified on a magnitude scale or by seismic moment). The intensity or damage potential of an earthquake attenuates with distance from the macroseismic hypocentre or focus.

2. The estimation of focal depth and intensity attenuation is relevant to seismic hazard (and risk) in that it influences hazard assessment in several ways, for example: estimates of maximum ground motion, understanding of seismotectonic correlations and calibration of historical earthquakes when instrumentally determined earthquake parameters are subsequently invoked.

DATA

3. The data used here are extracted from a set of 101 earthquakes occurring in the period 1727-1979 which have been macroseismically evaluated in detail. An impression of the overall range of earthquakes studied, their epicentres and epochs, may be had from Fig. 1. The general methodology used to evaluate these earthquakes is detailed in ref. 1, which also lists the individual earthquakes evaluated in a series of ensuing reports (variously Musson et al and Neilson et al), including presentation of much of the primary data used in assessing felt

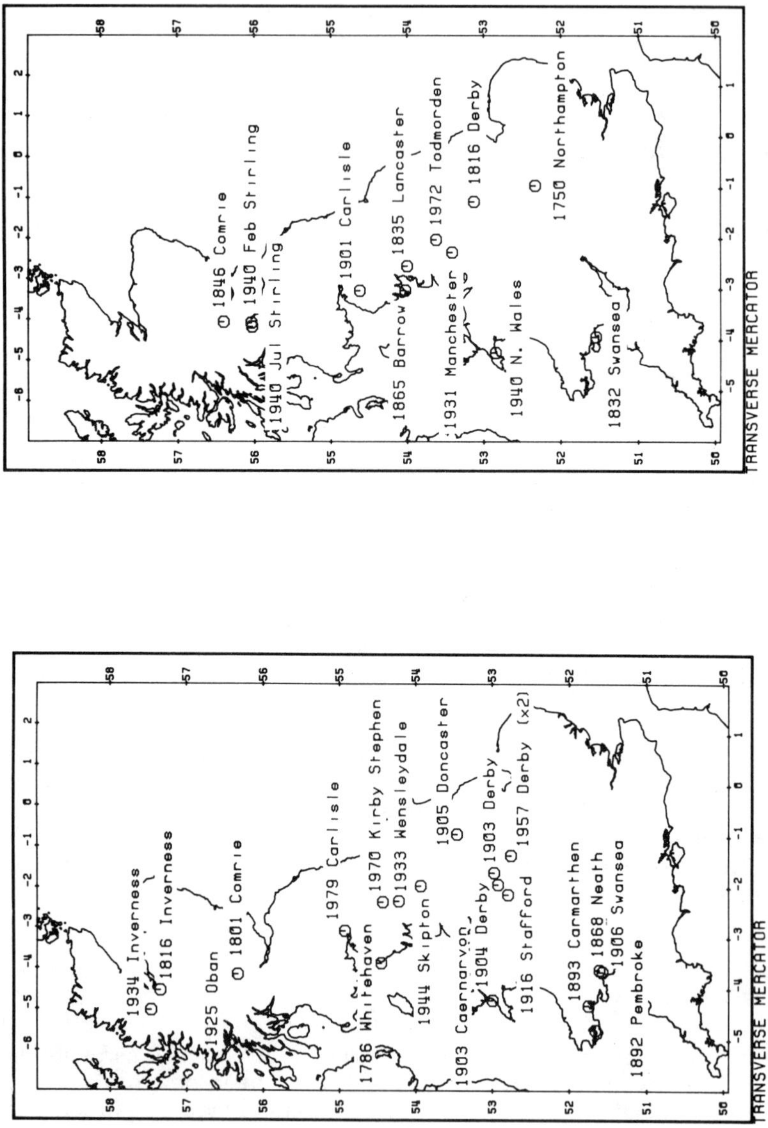

Fig. 1. Representative selection of earthquakes studied in ref. 1.

effects of the earthquakes.

4. For individual earthquakes there is a basic file of intensity data. The intensity scale used throughout is that of Medvedev, Sponheuer and Karnik, commonly referred to as the MSK scale. This scale was devised for European use and is recommended by the European Seismological Commission to facilitate comparison with other European studies.

5. Where possible the map of intensity data arising from an earthquake was contoured using smooth isoseismals. The purpose of each isoseismal is to delimit the "felt" area within which each intensity level was experienced or exceeded.

6. The original isoseismal maps were then digitised and areas within particular isoseismals calculated. Two specimen isoseismal maps are shown in Fig. 2, illustrating the earthquakes of 1903 March 24 Derby and 1934 August 16 Inverness. It should be noted that in the original isoseismal maps that uncertain isoseismals are often indicated by "dashed" lines. The areas within particular isoseismals form the basic data set from which intensity attenuation laws may be derived. An additional item of useful information is the intensity at the earthquake epicentre (I_o). The highest intensities located within the highest value isoseismal, taking into account the location of any known foreshocks or aftershocks, generally gives an observational assessment of I_o.

INTENSITY ATTENUATION AND FOCAL DEPTH

7. The data obtained from the two illustrative earthquakes of Fig. 2 are seen in Fig. 3 rendered into an intensity-epicentral distance plot. It is the relationship between intensity and distance which we wish to analyse here - for British earthquakes. It is immediately obvious that for the same size or energy of earthquake (defined typically in terms of magnitude or seismic moment), the epicentral intensity I_o depends on the focal depth of an earthquake. Also, the rate at which intensity decays with distance is a fundamentally important parameter for seismic hazard and risk studies.

8. There are two general descriptions of intensity attenuation which have arisen this century: Kovesligethy in 1907 (ref. 2, unseen) reviewed by Sponheuer in 1960 (ref. 3), and Blake in 1941 (ref. 4). Both formalisms draw on the observation that peak ground acceleration (a) and macroseismic intensity (I) in an earthquake are related by a power law (see eg ref. 5), typically

$$I = \log(a)^3 + \text{constant}$$

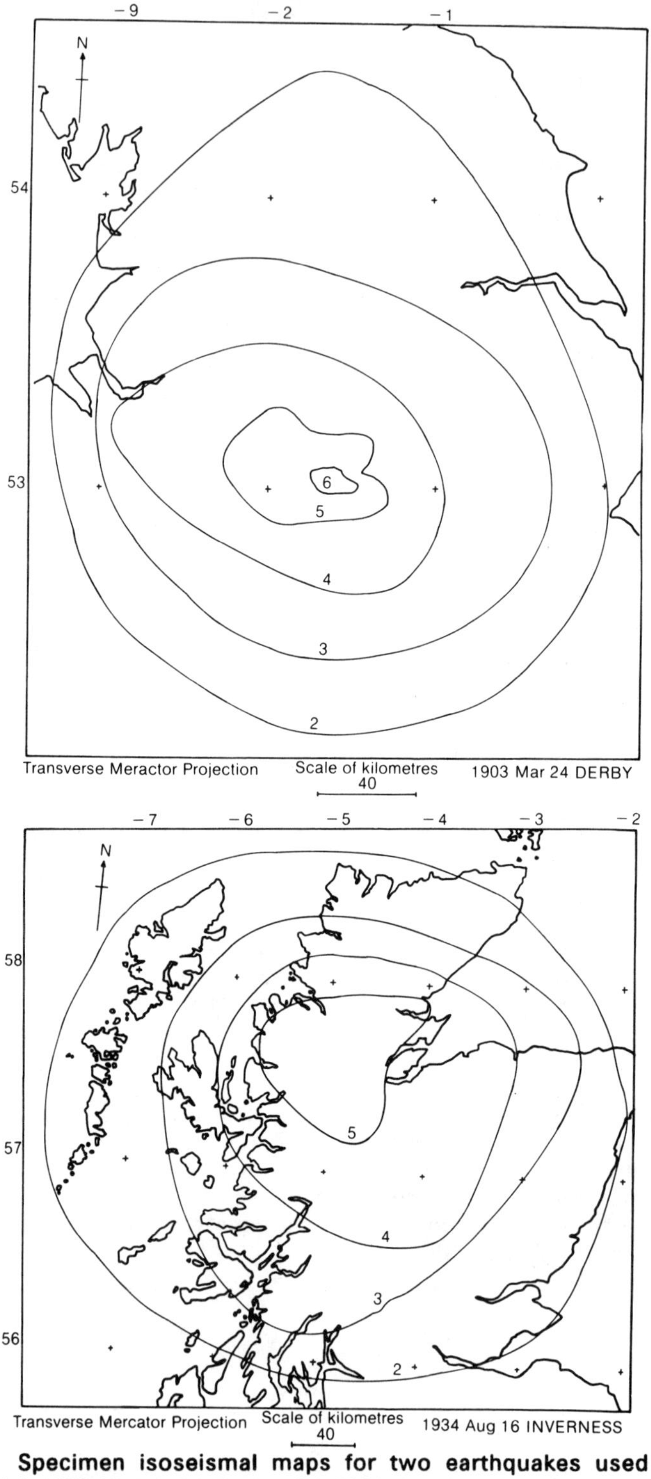

Fig. 2. Specimen isoseismal maps for two earthquakes used in this study: (a) 1903 March 24 Derby, (b) 1934 August 16 Inverness.

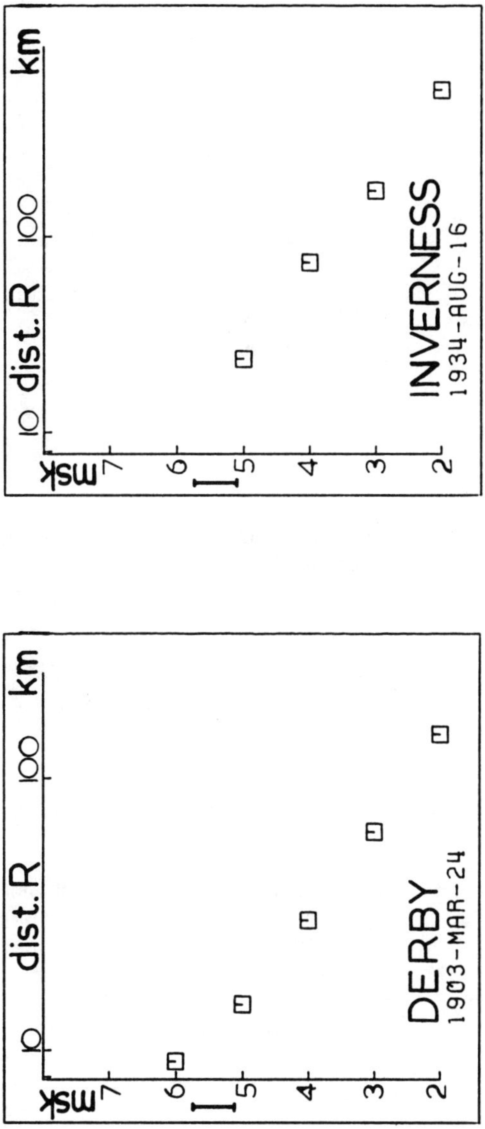

Fig. 3. Intensity–epicentral distance plots for two earthquakes shown in Fig.2.

Kovesligethy

9. The earlier formalism, that of Kovesligethy, models the attenuation using both a logarithmic term (derived from the power law relationship to a) and a factor accounting for the absorption of energy with distance. The result is

$$I_o - I_i = 3\log(r_i/h) + 3\alpha M(r_i - h) \quad (2)$$

where:

I_o : notional intensity at epicentre (on MSK scale)
I_i : intensity of ith isoseismal (on MSK scale)
r_i : hypocentral distance in km to ith isoseismal
(R_i is corresponding epicentral distance)
α : Kovesligethy's "attenuation"
M : a constant = 0.4343 = loge

Blake

10. Blake produced a subsequent modification of Kovesligethy which effectively simplifies the formula but at the expense of it being less readily interpretable. His result is

$$I_o - I_i = \nu\log(r_i/h) \quad (3)$$

where ν is Blake's attenuation.

KOVESLIGETHY AND BLAKE FITTED TO OBSERVATIONAL DATA

11. There is a clear link between h and I_o. High I_o will readily correlate with small h etc. In practice I_o may be observational (effectively I_{max} of the individual acceptable observations within the maximimum isoseismal) or correspond to the theoretical I_o at the notional epicentre required by equations (2) and (3). It is consistent with equations (2) and (3) to take I_o as an unknown to be solved for (rather than an observational datum) along with h and α or h and ν in the respective formalisms. Sponheuer presents graphical means to solve for the parameters of (2), here computer optimisation is preferred.

12. The function chosen for optimisation for both Kovesligethy and Blake is the RMS difference between observed and computed intensity at distances R_i for each of the N isoseismals of the earthquake being considered, that is

$$RMS = \sqrt{(\sum (I_{calc,i} - I_{obs,i})^2/N)} \qquad (4)$$

This function will effectively form a surface determined by h, I_o,α or ν. For a given α or ν, the values of RMS at a given h and I_o pair may be obtained and the RMS values contoured over a range of h and I_o.

13. For each earthquake analysed the range of I_o investigated is from the intensity of the innermost determined isoseismal to that intensity plus two, for instance, the highest isoseismal in Fig. 2(a) is 6 and so a range of 6-8 for I_o is investigated. The depth range searched is 0-60 km (although the diagrams extend only through 0-40 km).

14. The contoured h and I_o surface obtained using Koveslígethy on 1903 March 24 Derby is shown in Fig. 4. The optimal solution, marked in Fig. 4 by the black dot, has the value h = 11.2 km, I_o = 6.2 MSK. The observational estimate of I_o was 6 MSK (ref. 6). Values of α (and ν) will be discussed later.

15. The Blake optimal solution, for 1934 August 16 Inverness, is similarly represented in Fig. 5 and has the value h = 30.1 km, I_o = 6.5 MSK. It is interesting to note for this earthquake that the observational estimate of I_o is "at least 5 MSK" (ref. 7). This is because the innermost isoseismal (intensity 5 MSK) had large area (see Figs 2 and 3) within the Highlands in Scotland, from within which area there was a lack of data. This higher value of the computed I_o was to be expected and is consistent with the type of macroseismic data available on this earthquake.

16. Each earthquake has its own optimum pair of h and I_o. However, the attenuation value is probably a property of the geological structure of the propagation path followed by the energy. Although α may differ regionally it is a reasonable assumption to assume a constant value for it here (the situation will be different for very shallow earthquakes). To this end a constrained optimisation was performed using a "best" data set consisting of ten earthquakes. The "best" data set was defined to be those earthquakes which gave a clear optimal solution for h and I_o when optimised independently of each other. This criterion does not reflect on the quality of the macroseismic data of the "other" earthquakes. The observational variation of intensity with distance for these earthquakes is shown in Fig. 6. The constrained optimal value of the attenuation coefficient was α = .0068 and for ν for the same earthquakes, was ν = 5.63. The contoured h and I_o diagrams shown in Figs 4 and 5 are actually obtained using these constrained attenuation coefficients.

17. A selection of optimal results for these earthquakes is listed in Table 1, along with results obtained by additional means. The I_o obtained from these formalisms is notional and may or may not accord directly with observed effects near the epicentre. Equally, we may have little idea of what intensities actually occurred near the epicentral area (recall 1934 Inverness). Nevertheless, a useable

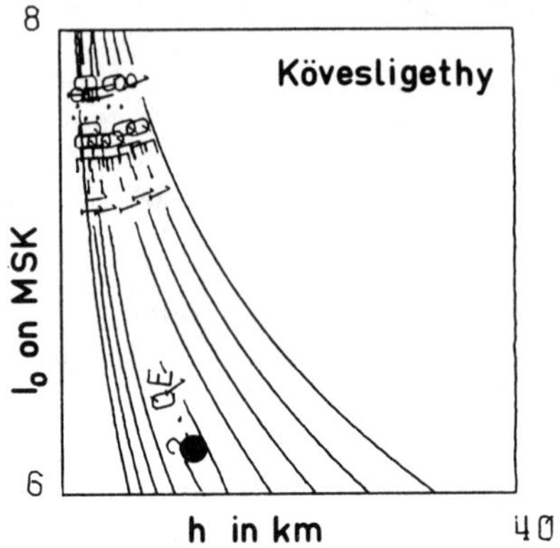

Fig. 4. Contoured h and I_o surface from Koveslígethy for the 1903 March 24 Derby earthquake. The computed optimal solution is shown by the solid black dot.

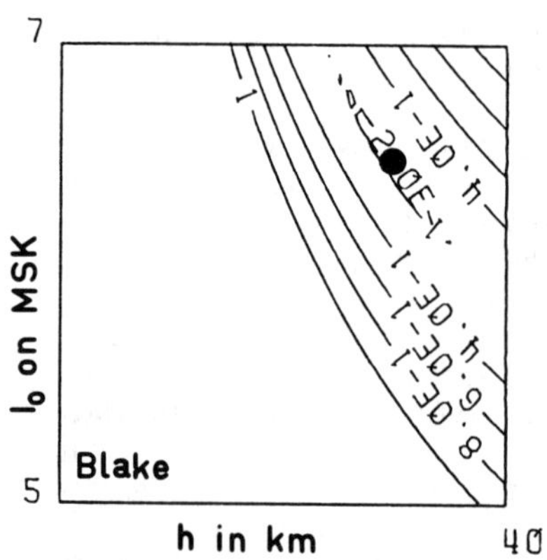

Fig. 5. Contoured h and I_o surface from Blake for the 1934 August 16 Inverness earthquake. The computed optimal solution is shown by the solid black dot.

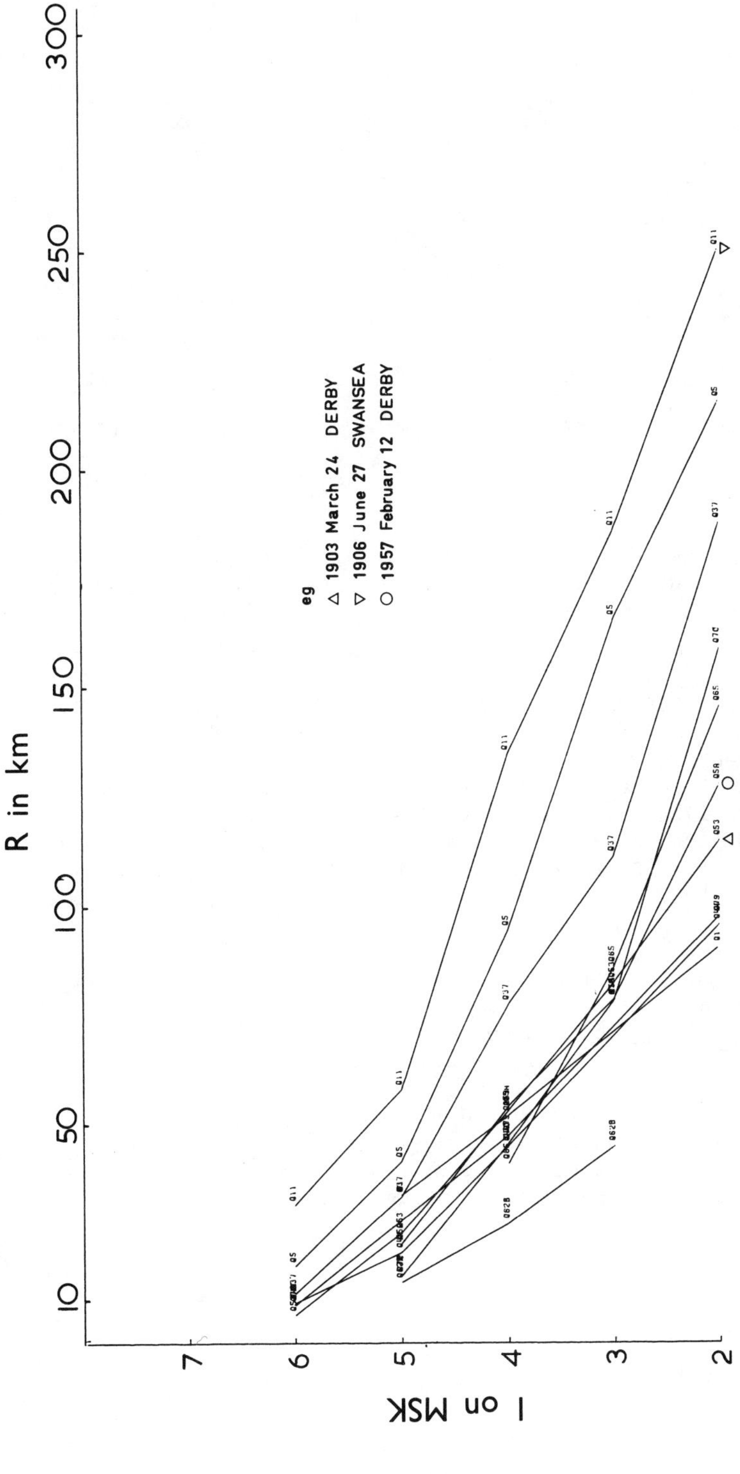

Fig. 6. Intensity–epicentral distance plots for a selection of earthquakes.

Table 1. Paired values of h and I_o.

	KOVESLIGETHY									BLAKE	
	(1)		(2)			(3)		(4)		(5)	
Event	I_o	h	I_o	h	$\bar{h}$	I_o	h	I_o	h	I_o	h
A	7.0	4-7	6.8	5-8	7	7.5	3.3	7.0	6	6.2	17.9
B	6.0	9-14	6.4	7-11	9	6.9	5.3	6.3	11	6.3	18.5
C	6.0	12-15	6.3	10-12	11	6.2	11.2	6.3	11	6.1	23.7
D	6.0	12-18	5.5	17-30	24	5.1	29.0	5.5	23	5.1	42.6
E	7.0	20-34	7.2	18-28	23	6.2	43.0	6.8	23	6.4	47.2
F	7.0	9-14	6.5	14-23	19	6.1	21.9	6.5	16	6.0	35.7
G	5.0	28-37	4.9	30-41	36	4.5	44.1	4.8	34	4.5	54.6
H	5.0	12-21	6.0	6-8	7	7.0	3.1	6.1	6	5.7	15.7
I	6.5	19-34	6.8	16-25	21	6.0	34.8	6.3	24	6.0	45.0
J	6.0	13-16	5.7	16-20	18	6.1	13.0	6.1	12	5.6	30.6

Events:
A 1816 March 17 Mansfield, B 1832 December 30 Swansea,
C 1903 March 24 Derby, D 1904 July 3 Derby,
E 1906 June 27 Swansea, F 1916 January 14 Stafford,
G 1925 December 23 Oban, H 1940 July 16 Stirling,
I 1957 February 11 Derby, J 1957 February 12 Derby.

Notes:
I_o on MSK scale, h in km.
columns headed (1) etc.:
(1) is h using observed I_o,
(2) is h using extrapolated I_o,
mean value $\bar{h}$ also given,
(3) is computed h and I_o,
(4) is "eyeball" h and I_o,
(1)-(4) use Kovesligethy,
(5) is computed h and I_o using Blake.

estimate of observed I_0 is frequently available, and this may be used to obtain a corresponding value of h with an uncertainty. Similarly, an I_0 obtained by extrapolating the two innermost isoseismals to R = 0 km may be used as shown in Fig. 8 to infer a corresponding range of h. These values are entered into Table 1, along with further h and I_0 pairs obtained simply by "eyeball" picking from such contour figures.

18. The isoseismals obtained from the 1970 August 9 Kirkby Stephen event are shown in Fig. 7 (also see ref. 8) and the ensuing constrained optimisations on h and I_0 are shown in Fig. 8, using both Kovesligethy and Blake formalisms. The black dots again mark the optimum minima, which are equally good in both cases. It is immediately apparent that Blake gives greater focal depth than Kovesligethy. Inspection of Table 1 shows this to be consistently true for all the earthquakes. It is seen to be generally the case at these values of attenuation that the Blake formalism will generate greater values of h than does the Kovesligethy formalism. A full and detailed analysis, and contrast of Kovesligethy with Blake, using considerably more earthquakes, will be presented elsewhere.

19. Fig. 9 shows the variation of intensity with distance, using both formalisms, for a hypothetical earthquake of 15 km focal depth, using the constrained values of observed attenuation obtained from British earthquakes. It is evident that for these two curves to converge then the focal depth in the Blake formalism will have to increase, whereas it would decrease for Kovesligethy.

20. When dealing with felt earthquakes it is a moot point whether instrumental or macroseismic hypocentres are to be preferred, or indeed, if they should be the same. However, the few earthquakes for which instrumentally determined focal depths are available (eg 1957 February 11 Derby h = 12 km with aftershock h = 8 km, 1970 August 9 Kirkby Stephen h = 15 km) show focal depths which are consistently less than that determined using the Kovesligethy formalism. h from Kovesligethy is to be preferred to that from Blake (very shallow earthquakes are not considered here).

21. Further resolution of variability in the problem will only be considered briefly. Attenuation will be variable. It is apparent from Fig. 6 that intensity-distance data derived from the earthquake of 1906 June 27 Swansea may be more appropriately fitted by low attenuation values. Fig. 10 shows solutions constrained at α = .005, .007 and .009 respectively, a satisfactory minimum being obtained using α = .005, from which revised h and I_0 values might be obtained (see the black triangle) with respect to the overall constrained optimum using α = .007.

The quality of individual isoseismals may be variable for an individual earthquake (even when this is not apparent visually, or from prior knowledge of low confidence isoseismals). The 1957 February 12 Derby aftershock suffices as an example here. Four isoseismals were drawn to the macroseismic data of this earthquake at intensities 2, 3,

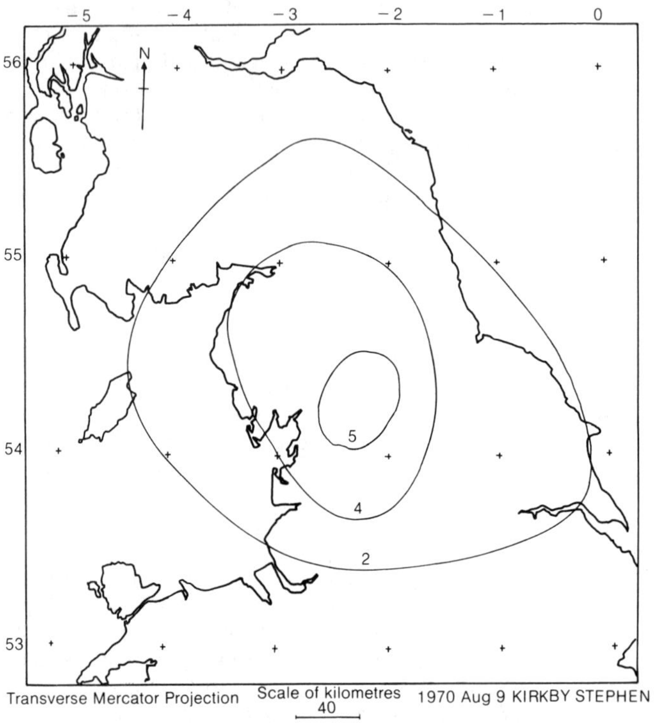

Fig. 7. Isoseismal map of the 1970 August 9 Kirkby Stephen earthquake.

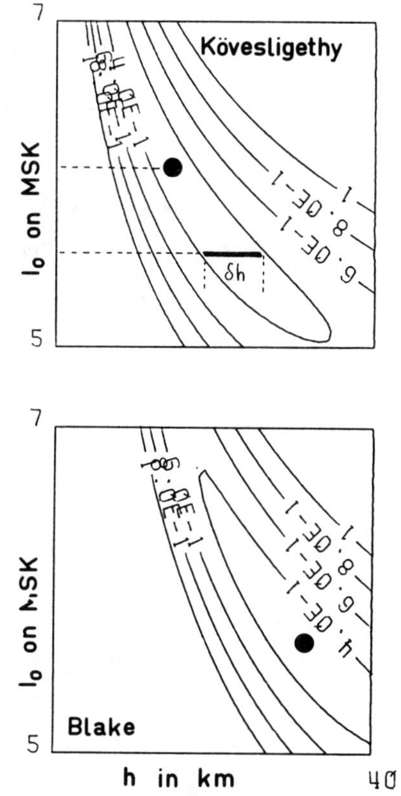

Fig. 8. Contoured h and I_o surface from both Koveslighethy and Blake for the 1970 August 9 Kirkby Stephen earthquake. Computed optimal solutions are shown by solid black dots. An alternative focal depth and its uncertainty, obtained using the extrapolated I_o, is shown in the Koveslighethy diagram.

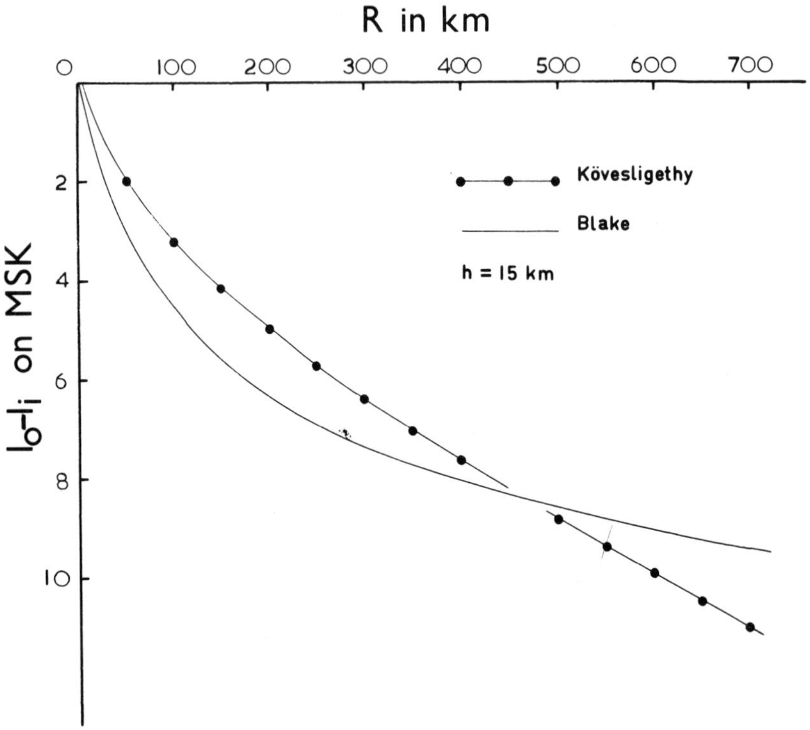

Fig. 9. Schematic intensity-epicentral distance plot for a hypothetical earthquake of 15 km focal depth using the constrained values of observed attenuation obtained in this study.

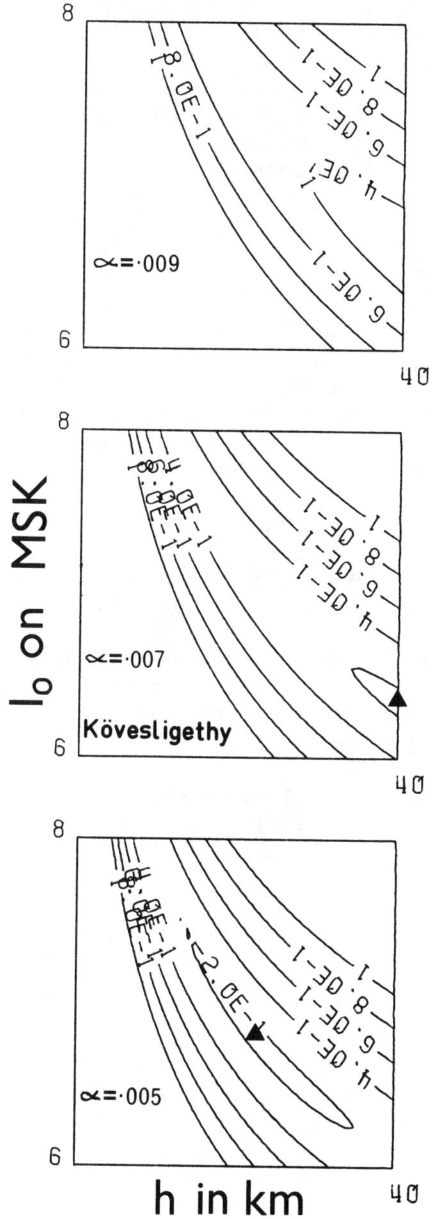

Fig. 10. Variable attenuation. Contoured h and I_o surfaces from Koveslighethy for the 1906 June 27 Swansea earthquake, with the attenuation coefficient α constrained at .005, .007 and .009 respectively. The solid black triangles indicate alternative solutions.

4 and 5 MSK. The isoseismal 3 MSK was sketched in to the east (not illustrated here, see ref. 9 for inspection of the macroseismic data and isoseismals). Fig. 11 shows the optimal constrained solution (black dot) using all four isoseismals. Four further sets of solutions obtained by omitting one isoseismal from each optimisation are illustrated; the best of these was found to be that omitting isoseismal 3 MSK with solution indicated by the black triangle. This latter solution obtained omitting isoseismal 3 MSK might be preferred to that using all the data, although it is not far removed, the main point emerging in this instance being the low confidence of this particular isoseismal rather than in the overall h and I_o solution.

CONCLUSIONS

22. Macroseismic intensity attenuation may be analysed using the methods of Kovesligethy or Blake. Insight into both these methods, and the quality and properties of the data analysed, may be obtained by modernising the graphical approach of Sponheuer to computer means and considering the link between h and I_o.

23. Focal depths from Blake exceed those from Kovesligethy; the latter are to be preferred. A macroseismic data set analysed using Blake formalism will usually tend to overestimate the focal depths and thereby underestimate the local seismic hazard (and risk).

24. Analysis of a selection of isoseismal maps from a larger data base indicates macroseismic focal depths throughout the crust in Britain, ranging in this sample from 5 km to 36 km (here quoting mean h values obtained using extrapolated I_o).

25. The constrained value of attenuation using the Kovesligethy formalism is .0068, that is equation (2) becomes

$$I_o - I_i = 3\log(r_i/h) + .0089(r_i - h) \ . \qquad (5)$$

ACKNOWLEDGEMENTS

This work was supported by the Natural Environment Research Council and is published with the approval of the Director of the British Geological Survey (NERC).

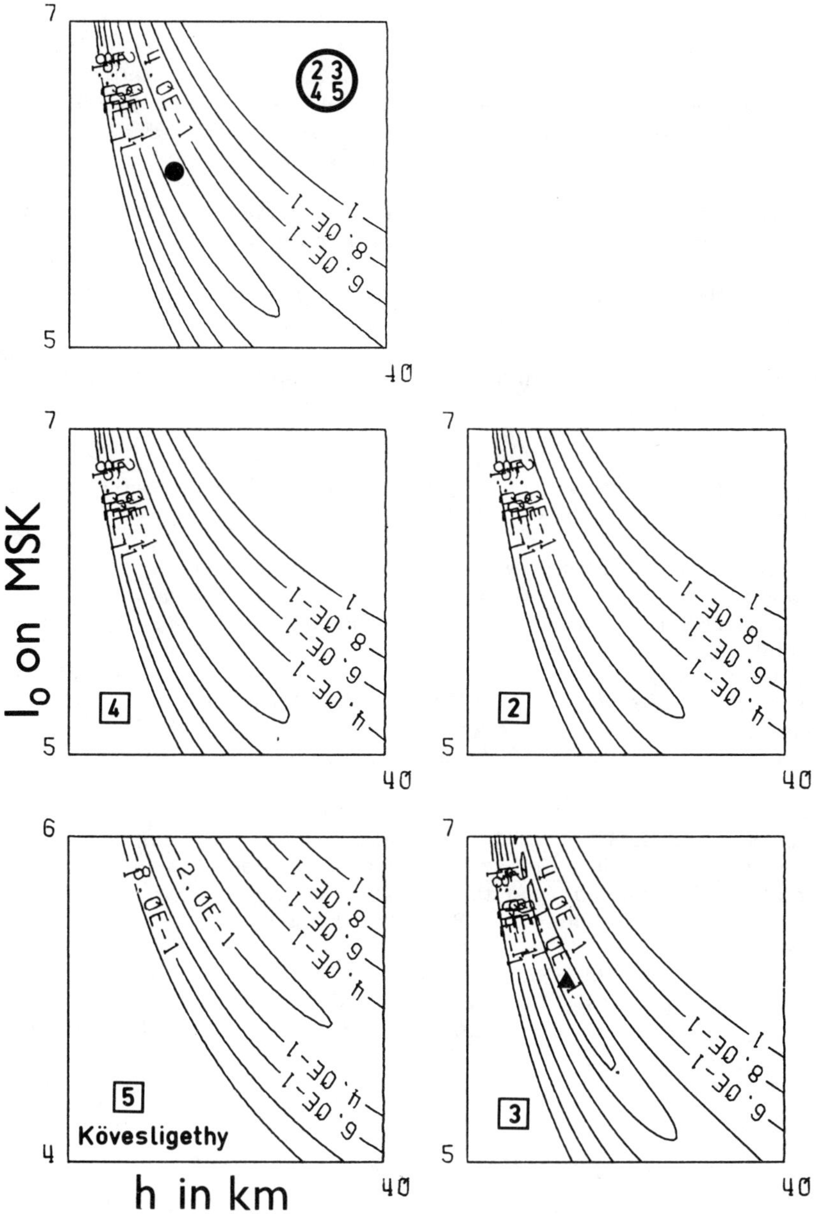

Fig. 11. Isoseismal subsets. Contoured h and Io surfaces from Kovesligethy for the 1957 February 12 Derby aftershock. The computed optimal solution using all four isoseismals, 2, 3, 4 and 5 MSK, is shown by the solid black dot in the upper left diagram. The bold number on the other diagrams indicates which isoseismal has been omitted from the computation. Omitting 3 MSK (bottom right plot) produces the alternative solution shown by the solid black triangle.

REFERENCES

1. BURTON, P.W., MUSSON, R.M.W. and NEILSON, G. Studies of historical British earthquakes. British Geological Survey Global Seismology Unit, Edinburgh, 1984, Report No 237.

2. KOVESLIGETHY, R. von. Seismicher Starkegrad und Intensitat der Beben. Gerlands Beitr. z. Geophysik, 1907,vol. 8, 363-366.

3. SPONHEUER, W. Methoden zur Herdtiefenbestimmung in der Makroseismik. Freib. Forsch.-H., 1960, No 88, 117pp.

4. BLAKE, A. On the estimation of focal depth from macroseismic data. Bull. Seismol. Soc. Am., 1941, vol. 31, 225-231.

5. BOLT, B.A., HORN, W.L., MACDONALD, G.A. and SCOTT, R.F. Geological hazards . Spring-Verlag, Berlin Heidelberg New York, 1975.

6. NEILSON, G., MUSSON, R.M.W. and BURTON, P.W. Macroseismic reports on historical British earthquakes V: Midlands. British Geological Survey Global Seismology Unit, Edinburgh, 1984, Report No 228(a).

7. MUSSON, R.M.W., NEILSON, G. and BURTON, P.W. Macroseismic reports on historical British earthquakes VII: North Scotland. British Geological Survel Global Seismology Unit, Edinburgh, 1984, Report No 232.

8. MUSSON, R.M.W., NEILSON, G. and BURTON, P.W. Macroseismic reports on historical British earthquakes I: Northwest Englandand Southwest Scotland (cont). British Geological Survey Global Seismology Unit, Edinburgh, 1984, Report No 207(b).

9. NEILSON, G., MUSSON, R.M.W. and BURTON, P.W. Macroseismic reports on historical British Earthquakes V: Midlands (cont). British Geological Survey Global Seismology Unit, Edinburgh, 1984, Report No 228(b).

Discussion

DR B.O. SKIPP, Soil Mechanics Ltd
Would Dr Burton comment on focal depths now deduced for the Hereford Earthquake of 1863 for which application of European formulae gave incredible depths?

DR BURTON
If one inverts an isoseismal map of the Hereford 1863 earthquake (e.g. that published by Principia Mechanica Ltd, which provides data for four isoseismals) to joint estimates of focal depth and I_0 then the optimum computed values obtained by the method described (using unconstrained attenuation) are approximatey $h = 32$ km and $I_0 = 6.2$ MSK.

9. Seismicity associated with Longwall coal mining

N. J. KUSZNIR, T. G. BLENKINSOP, University of Keele, and
R. E. LONG, M. J. SMITH, University of Durham

SYNOPSIS. Mining seismicity, and the related phenomenon of rock bursts, occurs throughout the world in association with the mining of coal, gold and other ores. Mining seismicity has occurred with magnitudes as great as 4.5 (South Africa), 4 (Poland) and 5 (Canada) and therefore constitutes a potential hazard, especially since focal depths are shallow. In Britain mining seismicity has been reported in several coalfields with magnitudes as great as 3.5. Various rock failure mechanisms have been identified. In this paper mining seismicity case studies from the North Staffordshire and South Wales coalfields are reported. Pillar failure, due to the super-position of stresses from active mining on residual stresses from previous workings, is responsible for the larger magnitude events, and presents the greatest seismic hazard.

INTRODUCTION

1. Seismicity associated with the extraction of coal, gold and other ores is a relatively common phenomenon occurring throughout the world, and in many of the reported case histories a direct casual relationship has been established between mining and seismicity. The seismicity is almost always associated with the occurrence of rock bursts during which rock in the vicinity of the mine excavation fails rapidly resulting in the sudden release of stored strain energy and the violent expulsion of rock from exposed faces. The occurrence of rock bursts during mining (ref. 1), which can cause significant damage to faces and roadways and fatalities has provided the main incentive for mining seismicity studies. However mining seismicity is also sometimes associated with the occurrence of earth-tremors of sufficient magnitude to be felt extensively at the surface. In South Africa and Canada, earth-tremors of maximum magnitude 4.5 and 5 respectively have been associated with mining (refs 2-5). Mining therefore can generate a significant seismic hazard. In Table A the occurrence of mining seismicity around the world is summarized together with estimates of the maximum magnitude and its nuisance potential. The potential for surface damage arising from mining seismicity is often enhanced since mining areas are frequently industrialized, corresponding to population concentrations,

COUNTRY	MATERIAL MINED	ROCK BURSTS	ENVIR. NUISANCE	MAX M_L
S. AFRICA	GOLD	✓	✓	4.5
U.S.A.	COAL	✓	?	4
U.S.S.R.	COAL	✓	?	?
POLAND	COAL	✓	✓	4
U.K.	COAL	✓	✓	3.6
CANADA	GOLD	✓	✓	4.5-5
INDIA	GOLD	✓	?	4
CZECHOSLOVAKIA	ZINC ORE	✓	?	2.8
GERMANY	COAL	✓	?	?
JAPAN	COAL	✓	?	?

TABLE A

and because the focal depths are shallow, lying at or above the level of the mine workings.

2. For rock bursts and their associated seismicity mining activity is usually responsible for both the initial generation of the stored elastic strain energy and the release of this stored energy by brittle failure to generate the seismicity. However not all mining seismicity may be related to mining in such a way. In Silesia, Poland, although rock bursts wholly generated by mining do occur, seismicity extensively felt at the surface has been attributed to the release of natural stored strain energy by the mining process (refs. 6,7). Similarly in Utah, USA, the relationship between mining and adjacent seismicity is in part tectonically controlled, although some seismicity can be directly attributed to mining (refs 8,9).

3. In Britain rock bursts and mining seismicity are associated mainly with coal extraction. The following rock failure mechanisms have been identified.

(a) Failure of rock and coal in the working area on faces or roadway sides e.g. South Staffordshire during 1930's (ref. 10).
(b) Pillar collapse between workings e.g. Barnborough (ref. 11) and Hickleton - main colliery (ref. 12) in the Yorkshire coalfield.
(c) Pillar and abutment collapse remote from the active face e.g. North Staffordshire coalfield (refs 13,14)
(d) Activation of old fault planes due to adjacent mining activity e.g. North Staffordshire coalfield (refs 14,15).
(e) Outbursts of gas and coal during the working of anthracite e.g. Cynheidre colliery, South Wales Coalfield (ref. 16).

4. Seismicity of sufficiently large magnitude to be felt at the surface has been experienced in the North Staffordshire, Yorkshire and Midlothian coalfields (refs 13,11,17) and has been attributed to mechanisms (b), (c) and (d). In this paper seismicity associated with longwall coal mining, the main coal extraction method used in British mines, is described and compared for mining seismicity case studies in the North Staffordshire and South Wales coalfields.

MINING SEISMICITY IN THE NORTH STAFFORDSHIRE COALFIELD

5. Small earth-tremors, associated with coal mining have occurred throughout the last century in the North Staffordshire coalfield. On 15th July 1975 a large earth-tremor was felt over an extensive area of Stoke-on-Trent. This had a magnitude, m_b, of 3.4 (as determined by the Institute of Geological Sciences Lownet seismometer network) and a maximum Mercalli Intensity of VI, and was the largest of a swarm of tremors which had been felt throughout early 1975 and continued to be felt until August 1977. During this period some 55 events were felt at the surface. This seismicity caused considerable concern to many people; the epicentral area being within the southern part of the Stoke-on-Trent conurbation. The seismicity was shown to be generated by the active mining of coal.

6. During the time that this seismicity occurred the epicentral area was being mined in the Ten Feet seam at a depth of 900 m below the surface. This area had previously been extensively extracted in the Moss, Bowling Alley, Rowhurst and Great Row seams and these earlier workings, especially those of the Moss seam, played an important part together with the active Ten Feet seam workings in generating the seismicity. A plan view of the Ten Feet seam workings and the earlier workings in other seams is shown in figure 1. The Ten Feet seam was extracted using the longwall mining method with total caving; the height of the extraction was approximately the seam thickness (2.4 m), and the face width was between 150 and 200 m. The geological sequence in the mined area consists of numerous coals occurring in a typical cyclical sequence with mudstone roofs and floors and occasional intervening sandstones (ref. 13). The dip is 8° to the South. The Ten Feet panels under extraction were situated between two major faults; the Newcastle and Apedale faults.

7. In response to the public concern, the Department of Geology, Keele University, in collaboration with the Global Seismology Unit of the Institute of Geological Sciences were commissioned by the National Coal Board to undertake a study of the seismicity in order to determine the cause. Seismic activity was monitored for 2 years using a 6 component seismometer network deployed in and around Stoke-on-Trent (ref. 13) and during this time panels 205 and 206 were extracted in the Ten Feet seam. The seismicity is shown in figure 2 together with the position of the Ten Feet panels, previous workings and faults. Plan and vertical sections are shown.

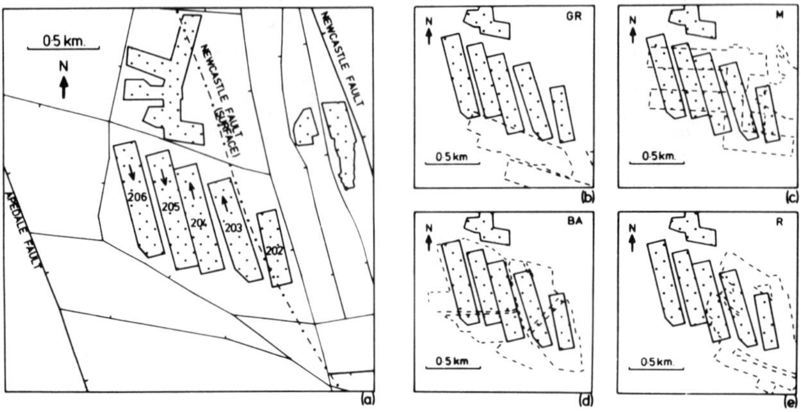

Figure 1
(a) Ten Feet seam panels 205 and 206 and faults at Ten Feet seam datum. Old workings in (b) the Great Row seam, (c) Moss seam, (d) Bowling Alley seam, (e) Rowhurst seam are alwo shown.

The tremors lie in close proximity to the active mine workings, mainly between the depths of 400 and 1000 m. The vertical E-W section shows that the seismic events do not lie along the Newcastle fault, although the workings are under the fault.

8. For one month during the mining of panel 206 a small aperture seismometer network was placed above the panel and used to augment the data of the larger network (ref. 14). The seismic events occurring during this month are shown in figure 3 on a N-S vertical section. These events have a higher accuracy of location (±50 m) compared to the events of figure 2 and show a smaller depth scatter. From first motion analysis on these events both implosional and shear source mechanisms were identified. The implosional sources were found to correspond to the smaller magnitude events while the shear sources had larger magnitudes and were those felt at the surface. The shear events were mainly located in the pillars left between the Old Moss workings.

9. Analysis of the relationship between the hypocentres and the position of the face (ref. 13) showed that the events occurred just in front of the face and moved forward with the face. This suggests that the front abutment pressure played an important part in generating the events. The larger shear events within the Moss seam pillars occurred when the Ten Feet seam face passed under the pillars and were generated by shear failure triggered by the interaction of the Ten Feet front abutment pressure with the residual Moss pillar stresses

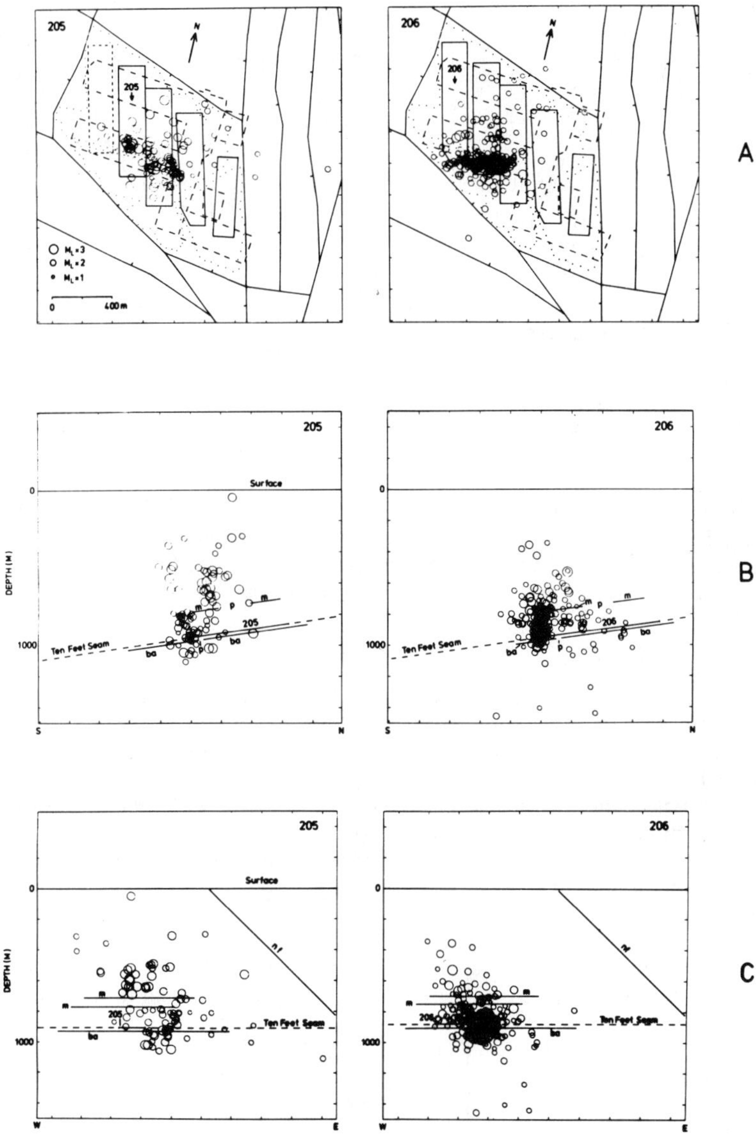

Figure 2
Tremor hypocentres occurring during the mining of panels 205 and 206 of the Ten Feet seam.
A) Plan view
B) Vertical section parallel to face advance direction
C) Vertical section parallel to face.
Old workings in the Moss and Bowling Alley seams together with faults at the Ten Feet seam datum are shown.

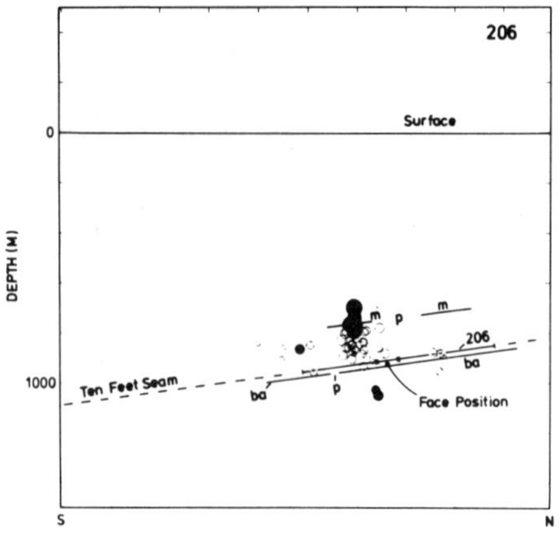

Figure 3
Vertical section parallel to the direction of face advance showing higher accuracy tremor locations for one months mining of panel 206. Open circles represent implosional source mechanisms - solid circles, shear mechanisms.

(ref. 14). In contrast the smaller implosional events occurring mainly at the level of the Ten Feet seam were most probably caused by waste collapse behind the advancing face (ref. 14).

10. Examination of the seismicity by day of week for tremors occurring during the mining of panels 205 and 206 (figure 4) showed that most seismic activity occurred through the working week falling between Monday and Friday. Seismicity at the weekend was much reduced and corresponded to no mining activity.

11. In figure 5 frequency-magnitude relationships are shown. The frequency-magnitude relationships follow the relationships

$\log(N) = 3.18 - 0.61\ M_L$ for panel 205

$\log(N) = 4.61 - 1.19\ M_L$ for panel 206

12. The magnitude distribution has been examined using Gumbel's third statistical distribution (ref. 18), and upper magnitude levels of 3.20 and 3.25 were predicted for panels 205 and 206 respectively. These values are only slightly greater than the maximum measured.

13. During the monitoring period 1975-1977 seismicity associated with mining was located elsewhere in the North Staffordshire coalfield. The seismicity together with the active mine workings is shown in figure 6. None of the seismicity was felt or exceeded magnitude, M_L, 2.5.

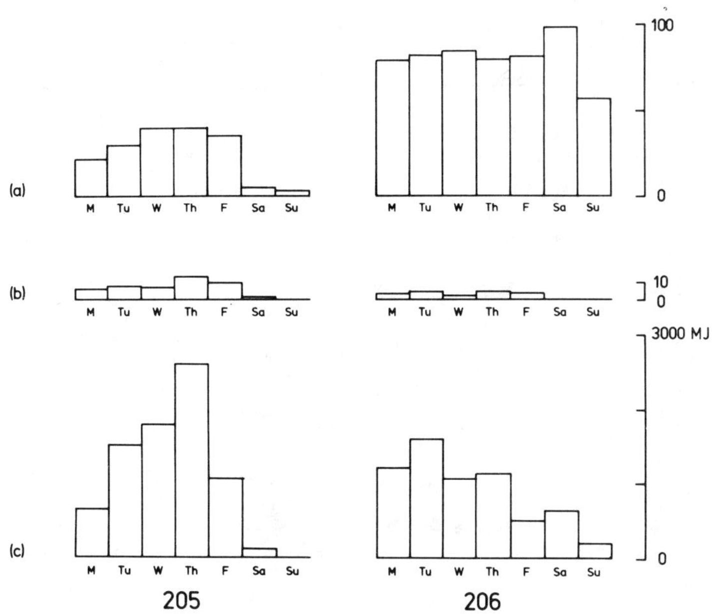

Figure 4
Histograms of cummulative number of tremors per day of week for panels 205 and 206
(a) All located events
(b) Felt events only
(c) Total seismic energy

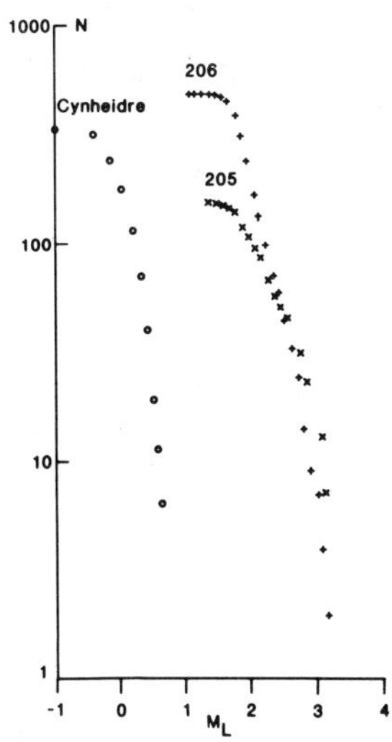

Figure 5
Frequency-magnitude relationships for tremors generated by mining of panels 205 and 206 of the Ten Feet seam, North Staffordshire and by mining at Cynheidre colliery South Wales Coalfield. Each relationship corresponds to mining over a similar time period.

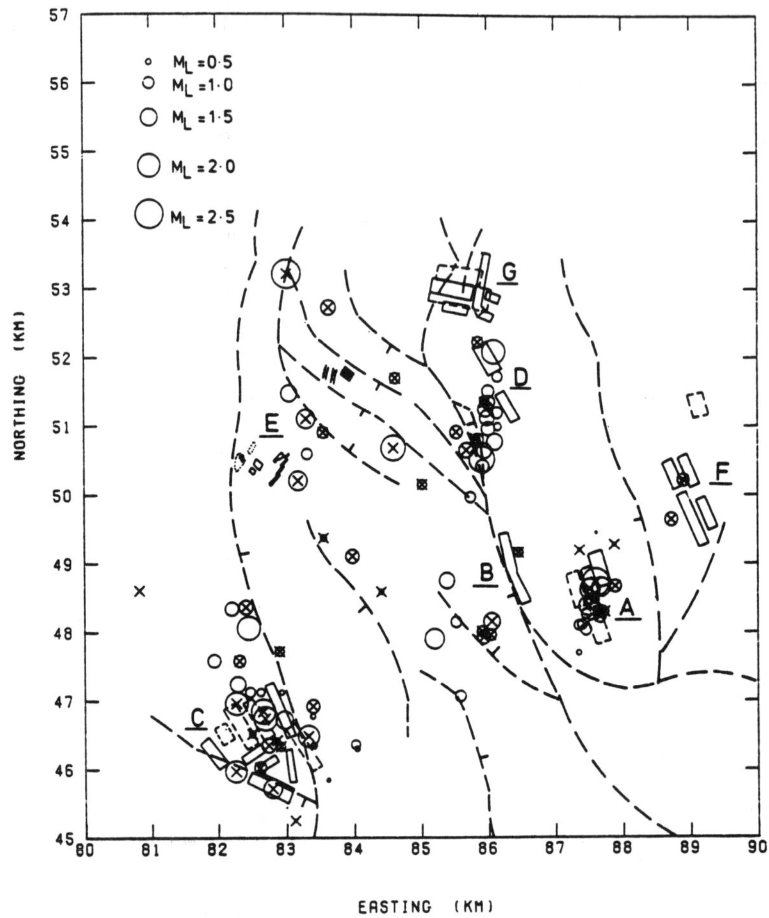

Figure 6
Tremors occurring in the North Staffordshire coalfield during 1975-1977 together with active mine workings. Open circles represent implosional source mechanisms - circles with crosses, shear mechanisms.

Seismicity in areas A, C and D was shown to have a similar mechanism to that of panels 205 and 206 of the Ten Feet seam; the seismicity again being generated by the interaction of active workings with pillars in old workings. In area B the seismic activity is thought to have been generated by subsidence induced slip as a pre-existing fault plane. During 1980-81 the seismicity of areas A and E increased in magnitude (maximum magnitude, 3.6) and was felt extensively (ref. 19). The mechanism responsible for the seismicity in area E is not known; it is not associated with active mining. The epicentral area for E is however underlain by old mine workings in a large number of seams.

MINING SEISMICITY AT CYNHEIDRE COLLIERY, SOUTH WALES COALFIELD

14. Seismic activity was investigated (ref. 20) at Cynheidre colliery*with the principle objective of studying activity associated with active longwall coal mining in a situation where there was no interaction with previous mine workings. This is in contrast to the Stoke-on-Trent mining seismicity case history where the active mining was in an area extensively mined in seams above and below the active panel. None of the seismicity at Cynheidre was felt. A subsidiary objective of the Cynheidre monitoring program was to seek seismic precursors to outbursts of coal and gas which have caused a number of fatalities and present a severe production problem at Cynheidre.

15. Cynheidre colliery, lying some 15 km north of Llanelli, is situated in the western most part of the South Wales coalfield and produces anthracite at depths up to 900 m below surface from seams within the Lower and Middle Coal Measures. The coal measures are here also typical cyclothemic delta deposits with sandstones comprising about 10% of the total. The geological structure (ref. 20) of the area is extremely complex owing to the close proximity of the Variscan front. As a consequence the coal seams are within variably dipping strata frequently interrupted by both thrust and normal faults.

16. Seismic activity was monitored in the Cynheidre area for 9 months between January and September 1983 during which time mining of the Big Vein seam occurred in 3 panels; BV24, BV11 and BV23. The majority of the seismicity occurred in the immediate vicinity of the active coal workings. Plan and vertical sections showing seismicity together with the active panels' layout, the Big Vein seam and faults are shown in figure 7. Seismicity clusters around the 3 active panels and is largely restricted to depths shallower than 1000 m below surface. The greatest number of events and released energy occurs at a depth of 0.5 km below OD, corresponding to the depth of the active coal extraction. The locations have an accuracy of ±50 m.

17. The existence of the 3 distinct clusters of events at the location of the three active panels indicates that this activity is related to the coal extraction. Levels of activity in each of these regions varied considerably throughout the monitoring period corresponding to variations in the mining

*commissioned by the Department of Environment

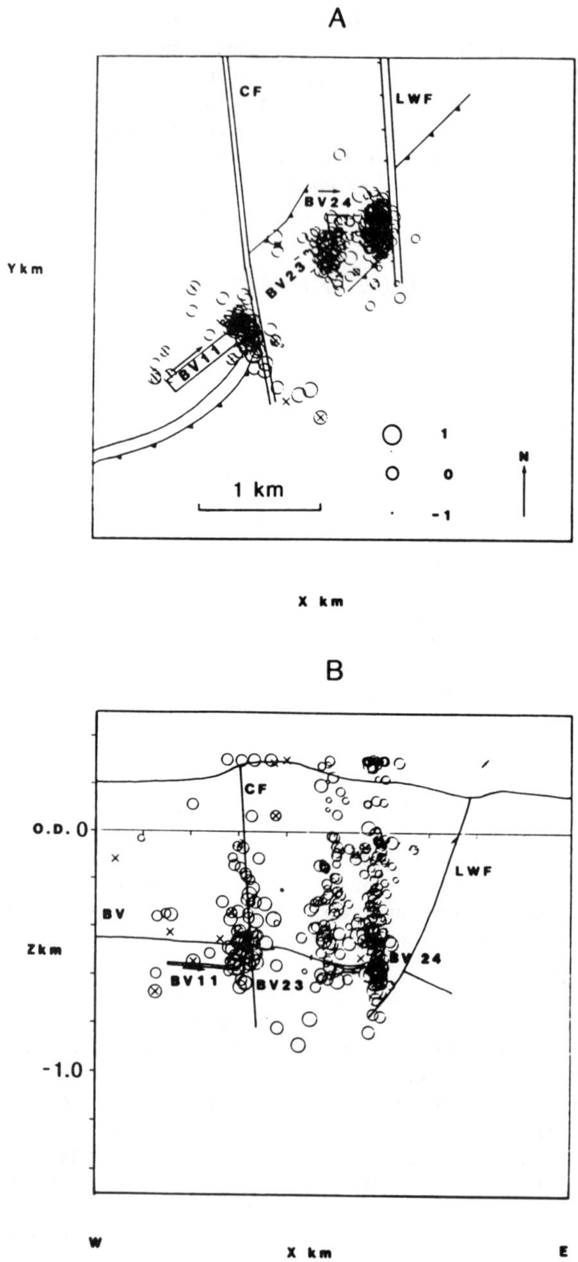
A
CF
LWF
BV24
BV23
BV11
Ykm
1 km
1
0
-1
N
X km
B
O.D. 0
-1.0
Zkm
CF
LWF
BV
BV11
BV23
BV 24
W
X km
E

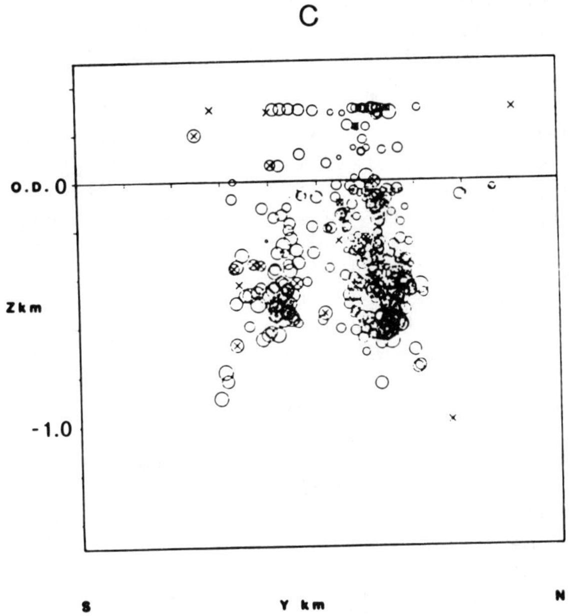

Figure 7
Seismic hypocentres located during the monitoring period February-September 1983 in the vicinity of Cynheidre colliery. Major faults and the location of the active panels are shown. Crosses are events of undetermined magnitude. C.F. Coalbrook Fault, L.W.F. Llety-Wilws Fault.

A) Plan

B) Vertical section (E-W)

C) Vertical Section (N-S)

Vertical Exaggeration 2:1

activity of each area and the geological structure. The activity of BV23 was associated with development work only.

18. Seismicity associated with panels BV24 and BV11 is shown in figure 8 where event hypocentres are plotted relative to the position of the moving face of each panel. Plan and vertical sections are shown; the vertical section being parallel to the direction of face advance. The position of the face is marked with a cross. Plan views show that for each panel the seismicity clusters within approximately 100 m of the face (the loose cluster of events in the bottom left quadrant of figure 8(a) is associated with BV23). The seismicity has also been shown to move forward in unison with the face (ref. 20). Vertical sections show that the majority of the events are distributed between the level of the face and the ground.

19. The seismicity associated with panels BV24 and BV11 is most probably caused by rock failure in the region above and in advance of the face, induced by the front abutment pressure. In the case of panel BV24 this panel worked towards, and came close to the Llety Wilws fault. This may have induced some failure on the fault or differential subsidence across it, although there is no direct evidence for this. Some of the observed seismicity behind the face may be caused by waste collapse. Focal mechanism studies show both shear and implosional source mechanisms.

20. The seismicity described in figures 7 and 8 lies in the magnitude range M_L = 1 to -1. The frequency-magnitude relationship for the whole of the mining related seismicity occurring during the 9 month monitoring period is shown in figure 5. Although this curve has been conventionally modelled as a straight line for both natural and induced seismicity no such relationship is evident in figure 5. This may be due to the mixing of three distinct sources, or the event selection procedure.

21. The mining of anthracite at Cynheidre colliery is occasionally interrupted by violent outbursts of gas and suspended coal dust from the face or headings. These present a serious hazard to life and considerable interruption to the production schedule. During the monitoring period 4 outbursts occurred on BV24; 2 spontaneously and 2 induced by shot firing on the face. The occurrence of these outbursts does not relate in any obvious way to the seismicity already described in the magnitude range M_L = 1 to -1. In figure 9(a) a seismogram of one of the outbursts is shown, recorded on the surface immediately above BV24. The outburst itself lasted about 20 seconds and was preceded by a number of very small events for which there is an increase in frequency of occurrence prior to the outburst until they become indistinguishable from each other. The outburst itself appears to consist of a large number of these smaller events which are estimated to have a magnitude less than -1. In figure 9(b) a histogram of the number of these very small events occurring per 20 seconds, plotted against time for the two spontaneous outburst is shown. A clear increase in the activity level at these low magnitudes (M_L < -1) can be seen

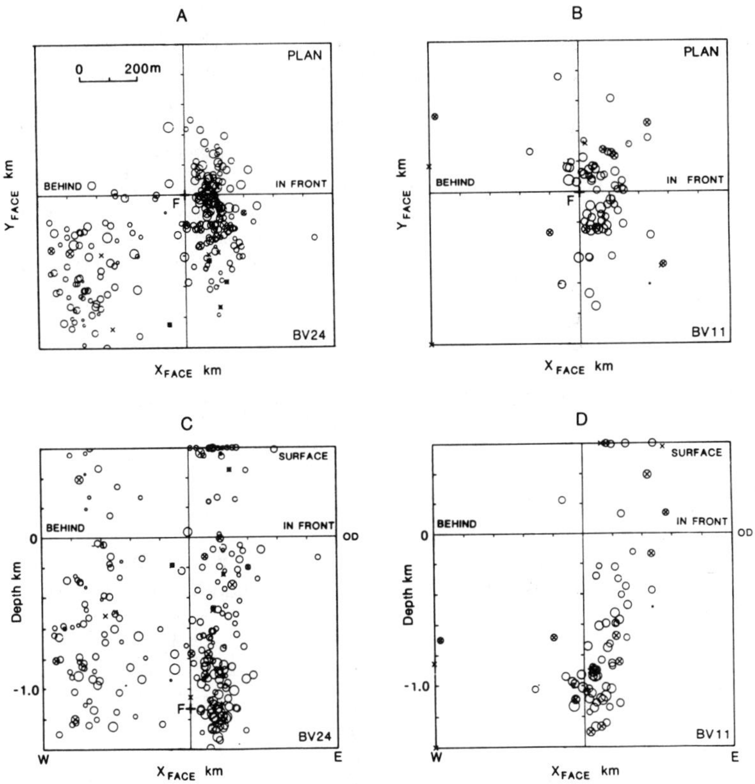

Figure 8

Seismic events associated with panels BV24 and BV11 at Cynheidre Colliery plotted in the coordinate frame of the advancing face. The cross at F marks the mid-face position. Other crosses are events of undetermined magnitude.

A) Plan, BV24. Events in the south east quadrant relate to BV23
B) Plan, BV11
C) Vertical Section, (E-W), BV24, parallel to face advance direction. Events behind face relate mostly to BV23.
D) Vertical Section, (E-W), BV11, parallel to face advance direction.

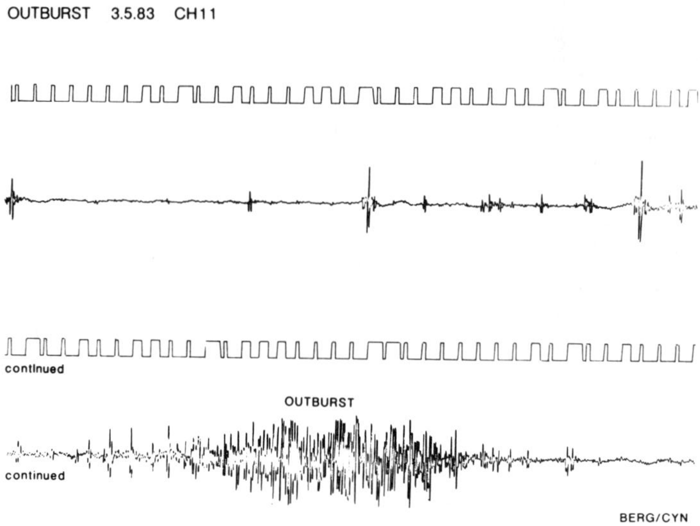

Figure 9(a)

Seismogram recorded above BV24 prior to and including the spontaneous outburst of 3rd May 1983 in BV24. Each tick on the time code represents one second.

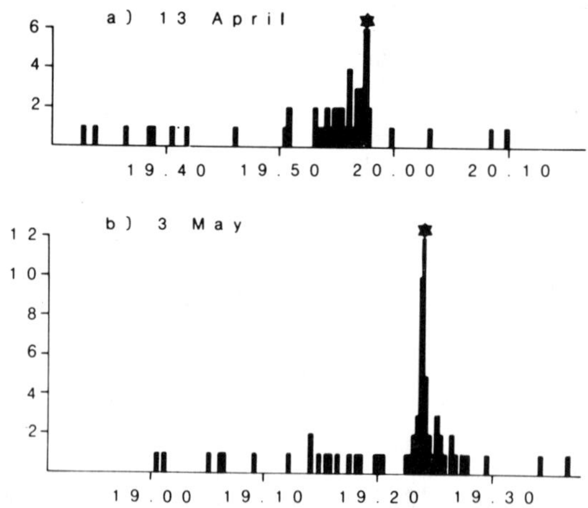

Figure 9(b)

Histograms of number of events per 20 seconds recorded above BV24 during two spontaneous outbursts in that panel (stars).

a) 13th April 1983

b) 3rd May 1983

during the minute or so prior to the outburst suggesting that this precursive activity may provide a useful predictive tool for outbursts.

DISCUSSION

22. Seismic activity has been shown to be generated by longwall coal mining in both the North Staffordshire coalfield and at Cynheidre colliery, South Wales. Various rock failure mechanisms have been identified, including waste collapse, failure within the front abutment, subsidence induced slip on faults, and pillar collapse in old workings as they are undermined. Mining seismicity does not present an environmental problem at Cynheidre, in contrast to North Staffordshire where the seismicity has been felt extensively at the surface. The difference in the maximum magnitude of the mining seismicity generated in the two areas (M_L = 1 for Cynheidre, M_L = 3.5 for N. Staffs) can be attributed to the occurrence of seismic events with pillar failure mechanisms in the case of North Staffordshire.

23. The pillar failure events of panels 205 and 206 of the Ten Feet seam occurred predominently within the Moss pillars when the Ten Feet face passed under the Moss seam pillar. Pillars in the Moss seam, as a result of serious extraction of this seam, would have been left with vertical stresses considerably greater then lithostatic and with lateral stress greatly reduced. Especially if these pillar cores had failed during the original mining of the Moss seam large vertical residual stresses would remain (ref. 21). While these pillars would have been stable prior to the mining of the Ten Feet seam, the superposition of the Ten Feet front abutment pressure caused renewed failure of these pillars. Failure of these pillars in compression generated the observed shear failure source mechanism. The compressional failure also enabled more strain energy to be stored compared to the tensile rock failure of smaller magnitude implosional events generated by waste collapse. The role of pillars in adjacent seams in generating mining seismicity has been reported elsewhere (ref. 22).

24. Mining seismicity, generated by pillar failure arising from the interaction of the active face with old workings, has been seen to have a magnitude, M_L as great at 3.5. By contrast, the Cynheidre data suggest that generally seismicity of magnitude <1 is associated with longwall mining in an area unaffected by previous workings. Pillar failure mechanisms therefore constitute a potential seismic hazard, particularly since the focal depth of this seismicity is shallow (800 m in the case of panels 205 and 206 North Staffordshire). Earth tremors with this type of mechanism may however be controlled or moderated by panel layouts which minimize the effects of interseam interaction.

ACKNOWLEDGEMENTS

25. Much of the work described in this paper was financed by the Department of the Environment and the National Coal Board.

However, the views expressed herein are the views of the authors and not necessarily those of the D. of E. or N.C.B.

REFERENCES

1. RICHARDS L.R. et al. Rockbursts : prediction and control. Institute of Mining and Metallurgy, 1983, 173 pp.
2. WOOD H.E. The Witwatersrand earth tremors. Journal of the Chemical, Metallurgical and Mining Society of South Africa, 14, 1914, 423-427.
3. CRANE, P.G. et al. A seismic investigation of the Witwatersrand earth tremors. Bull. Seismol. Soc. Am., 36(2), 1946, 49-80.
4. DEMPSTER E.L. Regional Aspects of mining induced seismicity : theoretical and management consideration, in Rockbursts - prediction and control. ed. RICHARDS L.R. et al. Institute of Mining and Metallurgy, 1983, p. 37-52.
5. BUCKLE F. The rock-burst hazard in Wright-Hargreaves mine at Kirkland Lake, Ontario. Canadian Mining Journal, 86, 1965, 81-87.
6. JANCZEWSKI, E.W. Earth Tremors in Upper Silesia, Archizoum Gornictura i Hutnictwa, 3(2), 1955, 205-225 (British Lending Library, Translation No. RTS 8899).
7. WIERZCHOWSKA Z. New views on the origin of earth tremors in Upper Silesia. Przeglad Gorniczy, 18(7-8), 1962, 417-422. (British Lending Library, Translation No. 8894).
8. OSTERWALD F.W. et al. Field instrumentation studies of earth tremors and their geological environments in central Utah coal mining areas. United States Geological Survey, Professional Paper No. 693, 1971, 20 pp.
9. SMITH R.B. et al. Source mechanisms of microearthquakes associated with underground mines in eastern Utah. Bull. seism. Soc. Am. 64, 1974, 1295-1317.
10. SOUTH STAFFORDSHIRE AND WARWICKSHIRE INSTITUTE OF MINING ENGINEERS. The occurrence of bumps in the thick coal seam of South Staffordshire, Transactions Institution of Mining Engineers, 85, 1933, pp 116-147.
11. PHILLIPS D.W. Rock bursts or bumps in coal mines. Transactions of Institution of Mining Engineers, 104, 1944, 55-84.
12. SHEPHERD R and KELLET W.H. Strata behaviour, a study of faces and roadways in workings under a sandstone roof liable to rock bumps. Colliery Guardian, 221(3), 1973, 93-102.
13. KUSZNIR N.J. et al. Mining Induced seismicity in the North Staffordshire Coalfield, England. Int. J. Rock Mech. Min. Sci. & Geomech. Abstr., 17, 1980a, 45-55.
14. KUSZNIR N.J. et al. Induced seismicity resulting from roof caving and pillar failure in long wall mining, in Strata Mechanics, Developments in Geotechnical Engineering, Vol. 32, edited by I.W. Farmer, Elsevier, 1982, 7-12.
15. KUSZNIR N.J. and AL-SAIGH N.H. Some observations on the influence of pillars in mining induced seismicity. J.S.A. Instn. Min. Metall. In Press.

16. KUSZNIR N.J. and FARMER I.W. Rockburst Phenomena in British Coal Mines, in Rockbursts - prediction and control, ed. RICHARDS L.R. et al. Institute of Mining and Metallurgy, 103-115, 1983.
17. MASHKOUR M. A seismological study of a mining area, Ph.D. Thesis, University of Strathclyde, 1976.
18. GUMBEL E.J. Statistics of Extremes. New York and London: Columbia University Press, 1958.
19. BROWITT C.W.A. et al. Earthquake swarms in the Potteries, 1980-1981. Geophysics. J.R. astr. Soc., 65, 1981, p. 250.
20. LONG R.E. et al. Final report on seismic investigation at Cynheidre colliery, South Wales Coalfield. British Earthquake Research Group report to the U.K. Department of Environment, 1984, 113pp.
21. WILSON A.H. and ASHWIN D.P. Research into the determination of pillar size. Mining Engineer, Vol. 131, 1972, 409-430.
22. SALUSTOWIC H.A. and PARYSIEWIC W. The problem of rock bursts in mines. Proc. Third Int. Mining Congress, Salzburg 1963.

Discussion

DR B.O. SKIPP, Soil Mechanics Ltd
Did the list of mining induced events given include surface mining?

MR T.G.BLENKINSOP, Keele University
All the events mentioned were from underground workings.

10. The seismotectonics of North-West Europe

R. MUIR WOOD, PhD, Principia Mechanica Ltd

SYNOPSIS. Observations on stress, strain and seismicity from throughout NW Europe can be combined to reveal the regional seismotectonics. While post-glacial isostatic crustal rebound obscures the long-term tectonic control of crustal deformation in Scotland and Fennoscandia, over much of the region the dominant tectonics appears to be NE-SW extensional. A band of relative aseismicity can be traced from northern Poland, through northern Germany and across the Central North Sea, around Scotland into Ireland. The dominant WNW Tertiary dextral transcurrent faults passing through the region continue to sustain an important seismotectonic control.

INTRODUCTION

1. Within the past few years considerable advances have been made in the collection of data relating to seismicity and neotectonics in North-West Europe (away from the Alpine chain). Remarkably however, given the accessibility for research and the large number of available scientists, in comparison with many other developed regions (including much of North America) there has existed no detailed and scientific programme to integrate data into a comprehensive picture of the seismotectonics over the whole region. While in Germany and to a lesser degree in France and Sweden there has existed an initiative for local and regional seismotectonic studies, national objectives have continued to obstruct a wider and more appropriate overview of the problems. Tectonics however respects neither national nor land-sea boundaries.

2. Seismotectonics requires the integration of a disparate collection of observations into a picture of crustal deformation and the resultant seismic energy release. The component observations of the seismotectonic model include: studies of historical seismicity; instrumental monitoring; in-situ stress determinations; focal mechanism determinations; neotectonics, and regional crustal strain determinations. From the synthesis of these

data, through the use of models of activity, and from analogy with other tectonic provinces it is possible to gain a seismotectonic understanding of the causes and location of continuing seismicity. Within this paper an attempt is made to comprehend the broad seismotectonic regionalisations of North-West Europe.

REGIONAL SEISMOTECTONIC PARAMETERS

Current tectonic regime

3. In situ stress determinations have been carried out in a number of countries. Results from Germany, northern France and SW England have been remarkably consistent in revealing a NW-SE to NNW-SSE orientation to the maximum horizontal compressive stress (refs. 1, 2 & 3). Results from Norway and Sweden have tended to confirm this orientation, but with an E-W orientation along coastal Western Norway (ref 4.). This has led a number of authors to propose that the current tectonics of NW Europe is compressional, powered by 'ridge push' from the North Atlantic spreading ridge.

3. Focal mechanism determinations from recent earthquakes have also been used to justify regional models for stress-regime, but there are very few well constrained solutions from the countries around the North Sea. By far the most reliable observations are geological because they reveal long-term tectonic strain.

4. Within the Rhenish Massif the pattern of recent volcanoes reveals a NW-SE orientation indicating major NW-SE dykes, implying NE-SW dilatation. In southern Holland and the neighbouring German section of the Lower Rhine Graben several important NW-SE Quaternary extensional faults have been described (ref. 5), and focal mechanism determinations from this region also indicate NW-SE striking normal faulting (ref. 6). The association of these normal faults with several hundred metres of Quaternary subsidence in northwest Holland implies that such extension may be continuing through a trough of enhanced subsidence found continuing to the north through the North Sea (ref. 7) and this has been confirmed by the discovery of up to 150m post Upper Pliocene normal faulting along the NNW-SSE Øygarden Fault at the boundary of the exposed crystalline basement in the Norwegian North Sea. Evidence for normal faulting along NW-SE faults is found off the northeast coast of Norfolk (ref. 8) and similar extension along NNW-SSE faults has been reported from Brittany (ref. 9).

Strain

5. The rapid rise in Flandrian sea-levels and the great changes in land-elevation brought about by the removal of ice-loads at the end of the last glaciation conspire to

obscure all observations of tectonic strain-rates from long-term measurements of sea-level. Thus while for many regions there are copious examples of changes in the relative land-sea elevation measured over long (radiocarbon dating of beaches, bogs and lagoons) and short (tide gauges) time periods, the tectonic component of any measurement has to be carefully extracted from the data.

6. Around Fennoscandia and Scotland the rebound is far too dominant a 'noise' to determine any tectonic signal, but around the margins of the southern North Sea there is a greater possibility for uncovering tectonic strain. Among those observations that are probably recording tectonic strain are those from Kent that reveal a discrepancy within a general model for regional subsidence of southern England (ref. 10), and from Essex where 4.3m of differential sinking since Neolithic (5000 BP) times (ref. 11) found corroboration in land-level changes measured from tide-gauges (ref. 12).

7. The most striking inheritance of Tertiary tectonics across the region has come from episodes of regional uplift, that in contrast to uplifts associated with the early stage of rifting have involved non-transient lithospheric density changes. Thus the uplifted topography remains elevated with no evidence of ensuing extension. These uplifts have been centred around Western Norway and across Western Britain and have provided the mountains that ascend to more than 1,300m in Britain and 2,400m in Norway. The modern outcrop of Upper Cretaceous chalk around these regions demonstrates the regional pattern of uplift. The contemporary coastline of Western Britain and Western Norway is chiefly the product of the relative erosibility of the uplifted Mesozoic basins and the neighbouring crystalline basement formations. Some central regions that have now become shallow seas, such as the Irish Sea off Cumbria, may have lost more than 1000m of overlying Mesozoic sediments.

8. Evidence of the timing and form of uplift is always hard to find. However the presence of warped river terraces, incised meanders, raised gravels and recent volcanic activity in the Rhenish Massif of northern Germany encouraged the Deutsche Forschungsgemeinschaft to initiate in 1976 a multidisciplinary 6 year research programme on epeirogenesis: 'Vertical movements and their causes using, as example, the Rhenish Massif'. Results have now been combined (ref. 13) to indicate the distribution of contemporary uplift, seismicity and vulcanism of the region. The most important discovery within this programme was of a body of low P-velocity material in the western part of the Rhenish assif between 50 and 150km in depth, with a volume of about $5 \times 10^5 km^3$.

9. Further into Belgium, England and France there is much evidence for differential land-movements, that have yet to

be studied in detail. While the rias of southern Cornwall and southwest Wales indicate long-term subsidence, incised river-valleys in the Derbyshire dales, in the Wye Valley and north Devon provide landscapes with similarities to those found within the rising Rhenish Massif. In East Anglia Pliocene and Early Pleistocene marine Crag deposits are found onland at altitudes of up to 40m while offshore to the east similar deposits descend to several hundred metres below sea-level (ref. 14). Isolated relicts of late-Neogene marine deposits are to be found at elevations up to 200m around SE England while contemporary rapid subsidence in the Thames Estuary has required the construction of the Woolwich Flood Barrier. The inter-relation of this evidence for both recent uplift and subsidence has yet to be unravelled.

10. While Tertiary uplift raised Britain and Norway the North Sea sedimentary basin continued to subside. This subsidence has not however been as simple or as passive as regional models have tended to emphasise (see for example ref. 15). Certain areas around the outer Moray Firth have shown almost two kilometres of subsidence in the early Tertiary and little since. The inner Horda Platform in the NE North Sea became uplifted and deeply eroded during the later Tertiary. Most critically for current seismotectonics, a study of the pattern of Quaternary deformation reveals a new istribution of tectonically enhanced subsidence following three en echelon NW-SE sections passing from Holland, through the Central Graben, into the Eastern Horda Platform (ref. 7). Within these zones subsidence rates are up to ten times faster than those in operation during the Tertiary.

11. Relatively few examples of Quaternary faulting are known from the region, and although a large number of claims for neotectonics exist in the literature these have often not been substantiated. The best studied onland Quaternary faults are found in the Lower Rhine Graben where Ahorner estimated there had been 200m of extension (ref. 6). The largest single displacement of any fault in the region is along the Øygarden Fault (ref. 7). The most spectacular examples of recent faulting onland are in northern Scandinavia, predominantly in northern Sweden and Finland but also passing into the Alta district of Norway. These NNW striking thrust faults (the longest of which runs for 150km and the highest displacement being 30m) moved following the rapid unloading of the 3000m thick Fennoscandian ice-sheet, less than 10,000 years BP. The faults reveal a tectonic component to the regional deformation distinct from that found further towards the southwest in and around the North Sea. (refs 16 & 17).

Seismicity

12. The analysis of seismicity data requires considerable discrimination as much of the relevant published information is misleading. All the various NW European nations have produced catalogues of historical seismicity, generally more than 50 years old, interspersed with errors of exaggeration, mislocation, duplication and ommission. Instrumental catalogues of seismicity have been published (and in some instances continue to be published) with a range of magnitude scales, with no sieving of artificial seismic events, and with highly variable locational accuracy. Before such information can be employed to demonstrate seismotectonic relationships both historical and instrumental data-sets require detailed re-researching in order to ensure that they are accurate, consistent and comprehensive.

13. The many centuries of local earthquake reporting as preserved in chronicles, diaries, newspapers and scientific reports, provide the foundation of a homogeneous picture of earthquake distribution. For regions around the southern North Sea the complete database of the larger earthquakes is known for a period of five to eight centuries and for northern lands one to two centuries. Up to now only for detailed local investigations or for offshore regions such as the Viking Graben in the northern North Sea can instrumental data offer a significantly better picture of seismic distribution than can be gained from historical studies. However as every years monitoring improves so this balance also alters.

14. Before instrumental information can be employed to demonstrate the regional pattern of seismicity it is necessary to investigate spatio-temporal clustering, whereby a short time-sample may not provide a random picture of the whole pattern. For two seismically active regions of NW Europe, the Lower Rhine Graben (ref. 18) and offshore Western Norway (ref. 8) instrumentally monitored seismicity reveals a very similar configuration to that indicated by the historical seismicity. However for seismicity in England, Belgium, much of France and Denmark instrumentally monitored seismicity fails to reveal the spread of seismicity indicated by the historical record. The long-term historical sample for these areas does, in contrast, appear to reveal significant regional patterns of seismic energy release.

15. The pattern of seismicity across the region is far from uniform. More than half of onland and offshore NW Europe appears to be effectively 'aseismic', having almost no felt or instrumentally detected seismicity. The most important onland 'aseismic' regions pass as a band from northern Poland through northern Germany and southern Denmark to the Zuyder Zee in Holland, through north-eastern

England and north-eastern Scotland and much of Ireland. When added to the less easily defined 'aseismicity' of the offshore regions, this band is found to be continuous, through the southern North Sea, around the north of Scotland and down to Ireland, with the only significant break in the zone possibly occurring between northern Ireland and the Outer Hebrides (see fig. 1).

16. The range of areal seismic energy release between these regions is at least two orders of magnitude. During the past two centuries no earthquakes have been felt in western Ireland even at intensity II, while parts of western Norway have been shaken at intensities between IV and VI more than twenty times. The lighthouse keeper at Ytteroen, an island located close to the NW corner of Western Norway, mentioned in 1887 that 'since the island was inhabited it has been like a rolling wagon'. It is to be noted within this simple regionalisation of seismicity that the land-sea boundary is of little consequence.

SEISMOTECTONIC MODELS

17. Throughout the Tertiary and Quaternary NW Europe has sustained the same location, as a peninsula of continental crust between two plate boundaries. To the west, and from the Eocene the north-west, there have been a series of spreading ridges, and to the southeast and south a complex and rapidly changing pattern of collision zones running through the Mediterranean and Alpine chains. For much of its length the Alpine collision zone has been oblique, with movement directed towards the NW. This orientation of plate movement has periodically coincided with the transform faults of the spreading North Atlantic Ocean. Communicating between the strike-slip faults associated with both plate boundaries there have been a series of intermittent 'sub-plate' boundaries passing from the Carpathians, Alps and Pyrenees, out into the North Atlantic.

18. Such strike-slip faults are often difficult to trace and the horizontal movement is generally apparent only where the orientation of the fault changes, forcing vertical compensatory deformation. Where these dextral transcurrent faults step to the right, transtensional pull-apart basins were formed, often filled with thick lake sediments; where they step to the left transpression structures are produced. Tertiary pull-apart basins within North-West Europe include a series of Eocene-Oligocene basins passing from Brittany up through the western coast of Britain as far north as the Isle of Lewis, the largest of which lies in Cardigan Bay off the west coast of Wales. Transpression structures are most notable around the Bay of Biscay, in the Western Approaches to the south of Ireland and in Southern England. Other transpression anticlines are also known from parts of the margins and centre of the North Sea.

19. In a traverse from Norway to France across the North Sea one crosses three of these major transcurrent fault-zones. The northernmost is the Tornquist Zone (ref. 19) a fracture zone of NW-SE and WNW-ESE faults marking the southern boundary of the Fennoscandian shield, which since the Caledonian has been the site of many hundreds of kilometres of horizontal displacement. Mesozoic and Tertiary movement can also be traced along this structure (ref. 20).

20. A similar zone of faulting can be traced from Holland into the Dowsing and South Hewett Faults, off the north-east coast of Norfolk. Further to the northwest a complex series of faults pass off from the main WNW line towards the NW and NNW. The deep transtensional Mesozoic Sole Pit Basin became the location of considerable transpressional late Cretaceous-Early Tertiary inversion in the Sole Pit region and North York Moors (ref. 21). Evidence from Holland indicates renewed mid-Tertiary movement through this zone (ref. 22). This fault-zone marks the northern boundary of the London-Brabant microcraton and was the site of significant transcurrent faulting during the Variscan orogeny. In the Permian it marked the southern boundary of the Zechstein Sea (the northern boundary was the Tornquist Zone). Onland through Lincolnshire a series of WNW faults define a continuation of the zone. These pass into the Stublick and Pennine Boundary Faults and appear to continue on into the Southern Uplands along the transtensional basins at Dumfries and Sanquhar. The Variscan phase of dextral transcurrent faulting along this zone caused the complex pattern of transpressional and transtensional deformation known, but previously unexplained, across northern England. The existence of a fundamental WNW lithospheric fracture at the northern margin of the zone is indicated by the line of the Early Tertiary Armathwaite-Cleveland Dyke.

21. The southernmost of these three major fault-zones is associated with offsets along the Variscan Front, and passes into Britain across the Dover Straits. The Artois-Kent axis has also been the site of Variscan and Mesozoic transcurrent faulting and associated inversion structures both in the Early and Middle Tertiary (refs. 23 & 24). Between the Artois-Kent axis and the Dowsing - Hewett Fault belt there is (at least) one further WNW axis passing from Zeeland to the south-east Suffolk coast, which formed the axis of subsidence and the margin of the North Sea in the Pliocene and Lower Pleistocene. Through the Upper Pleistocene the Artois-Kent axis marks a boundary of the subsiding North Sea (and probably the Thames Estuary region) and has also been associated with poorly known transtensional basinal subsidence in the Plaine de la Lys (ref. 25).

22. These three zones provided the major 'arterial' faults along which the Tertiary displacements passed into the

fig. 1

Seismic Provinces of N.W. Europe

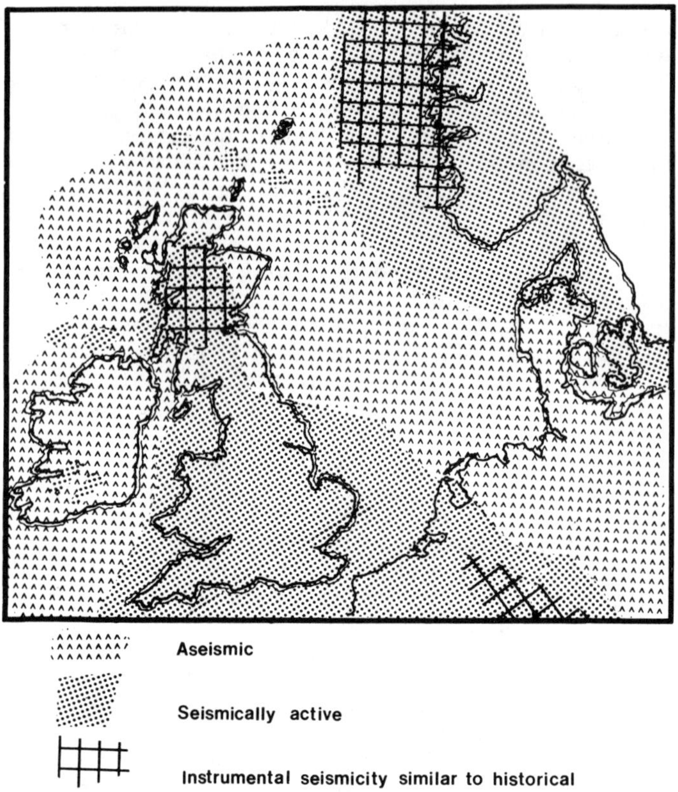

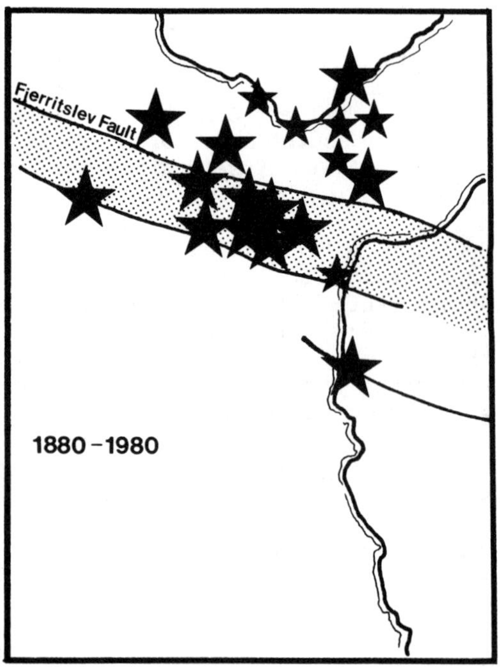

fig. 2

Historical seismicity of southern Norway, western Denmark and adjacent North Sea: 1880-1980 (events over 50km felt area radius). Tornquist Zone stippled.

region. (Other similar NW-SE faults can be traced through the Western Channel.) As continuous fracture-zones with a relatively unrehealed mineralogy, they remain as primary candidates for current reactivation.

23. The reanalysed historical seismicity to the south of Norway indicates that the Tornquist Zone marks the southern boundary of continuing seismicity in the north-eastern North Sea (ref. 8), and fig 2. To the north-west of Denmark instrumental monitoring has revealed that seismic activity is also concentrated along the fault-zone (ref. 26). North of Scotland a few events, instrumentally located by BGS (in a region generally without seismicity), may also lie on the continuation of the Tornquist Zone out towards the North Atlantic.

24. The south-eastern end of the Holland - Dowsing - Pennine fault-zone does not appear to be active but through northern England and the adjacent North Sea this wide faul belt (which includes the neighbouring Sole Pit basin) marks the north-eastern boundary of historical seismicity (see fig. 3 & ref. 27). It is to be noted that a WNW fault at the south of the fault-zone passes close to Lincoln where a highly damaging earthquake occurred in 1185. The Carlisle earthquake of 1979/12/26 may lie on the 'same' fundamental tectonic fracture zone as the 1931/6/7 Dogger Bank earthquake: the largest in the NW Europe so far this century.

25. The distribution of historical seismicity around the southern margin of the North Sea reveals an E-W band of important damaging earthquakes passing out from the Lower Rhine graben (ref. 28), that have all been characterised by a WNW elongation of both felt area and area of damage. At its western end this 'band' of seismicity follows the London-Artois axis. Other significant damaging earthquakes within the London-Brabant microcraton are also concentrated around important WNW axes, to the north of the major Variscan overthrust front (see fig. 4).

CONCLUSION

26. Away from the post-glacial rebound domes of Scotland and Fennoscandia NW European seismicity is a product of tectonic crustal deformation. However just as the tectonic history of the region is extremely complex, involving a series of overlapping orogenies, phases of rifting and involving patterns of deformation and block rotation related to major transcurrent fault-zones, so the continuing pattern of seismicity is itself complex. However it is possible to elucidate some important seismotectonic constraints.

27. The major WNW transcurrent fault-zones that have been active during the Tertiary continue to dictate important boundaries of seismic provinces and axes of continuing deformation. The regional tectonics appears to be chiefly

fig. 3

Historical seismicity (1800-1980) and major faults of the Holland - Dowsing - Pennine fault-zone.

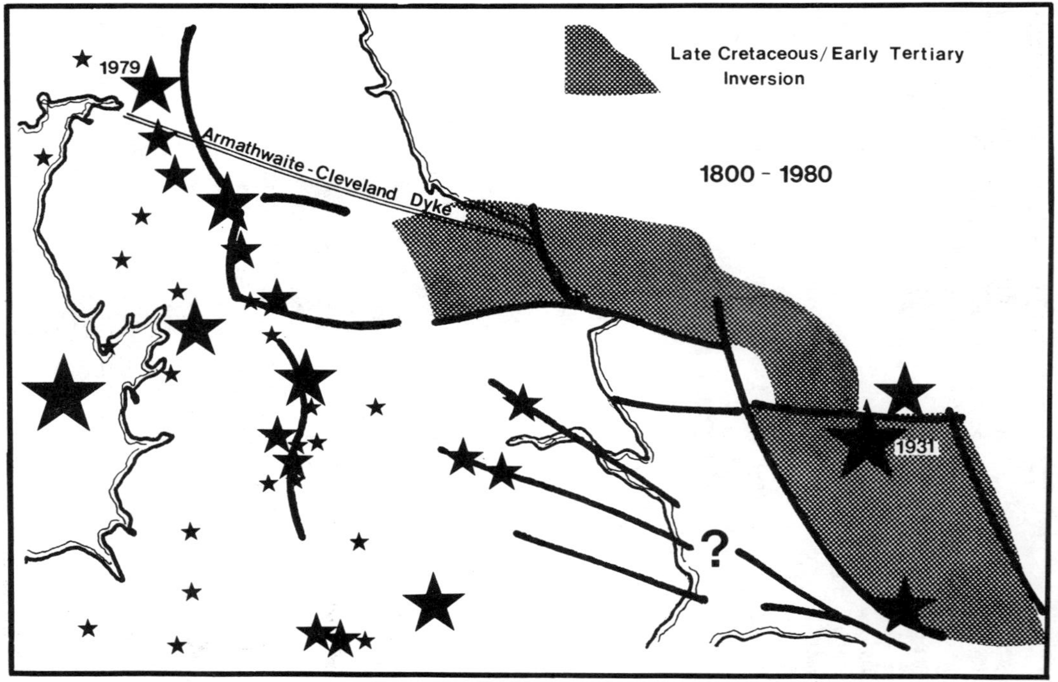

Key to historical earthquake size for figs. 2, 3 & 4.

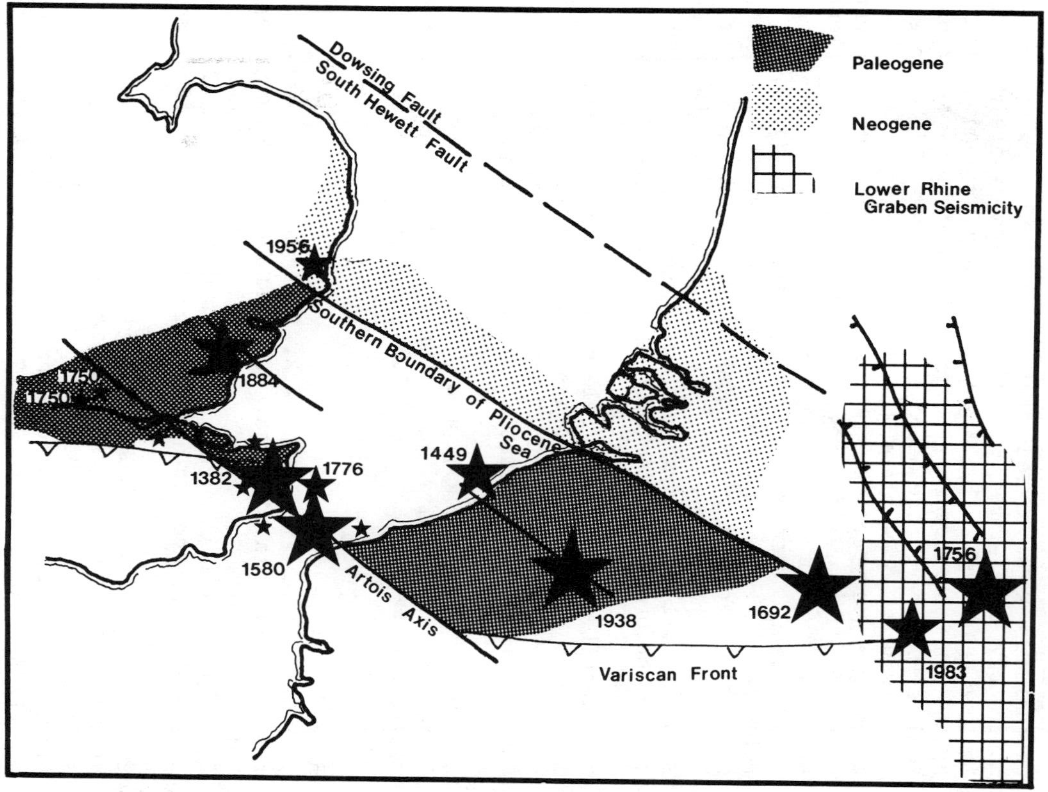

fig. 4

Historical seismicity and major WNW tectonic axes of the London - Brabant Massif.

extensional with allied strike-slip faulting. Local and apparently short-lived tectonic strain-rates are more than 1mm per year (refs. 11, 12 & 13) although long-term rates of uplift and subsidence throughout the Quaternary are closer to 1 x 10-4 m/yr. The macrotectonic context of NW Europe in the midst of two plate boundaries has provided a fast changing recent tectonic history offering few parallels with cratonic 'intraplate' regions such as Eastern North America.

29. Local seismotectonics research is not just of relevance for the analysis of seismic hazard at critical facilities. Following the catastrophic floods in Holland in 1953 a major programme of geodetic, neotectonic and seismological investigations was undertaken so better to define the problem. Remarkably, amidst the enormous quantity of money that it has cost to prevent London being flooded, no detailed geodetic relevelling surveys have been instituted, nor neotectonics investigated in order to attempt to learn more of the causes and locations of this manifestation of the crust's continued mobility.

Acknowledgements: Background research on the regional historical seismicity has been undertaken in fruitful collaboration with Dr. C.P. Melville and Prof. N. Ambraseys.

REFERENCES

1. GREINER G. and ILLIES J.H. Central Europe: active or residual tectonic stresses. Pure and Appl. Geophys., 1977, vol. 115, 11-26.

2. FROIDEVAUX C., PAQUIN C., and SOURIAU M. Tectonic stresses in France: in-situ measurements with a flat jack. J. Geophys. Res., 1980, vol. 85, 6342-6.

3. PINE R.J., TUNBRIDGE L.W. and KWAKWA K. In situ stress measurements in the Carnmenellis Granite: 1. overcoring tests at South Crofty Mine at a depth of 790m. Int. J. Rock Mech. Min. Sci. and Geomech. Abstr., 1983.

4. RANALLI G. and CHANDLER T.E. The stress-field in the upper crust as determined from in-situ measurements. Geol. Rundschau, 1975, vol. 64, 653-674.

5. AHORNER L. Untersuchungen zur Quartären Bruchtektonik der Niederrheinischen Bucht. Eiszeitalter Ggw., 1962, vol. 13, 24-105.

6. AHORNER L. Present-day stress field and seismotectonic block movements along major fault zones in Central Europe. Tectonophysics, 1975, vol. 29, 233-49.

7 MUIR WOOD R. The Quaternary tectonics of the North Sea, J. Geol. Soc. Lond., 1985 in press. (Abstract in Proc. Geol. Soc. Lond. November, 1984).

8. PRINCIPIA MECHANICA LTD. The seismicity of the North Sea. Report prepared for the UK Dept. of Energy, HMSO, 1985.
9. PAREYN C. Mise en évidence d'une activité néotectonique pliocène et quaternaire dans le Cotentin, le bassin de Carentan et la Bessin (Manche et Calvados). Bull. Soc. géol. France, 1980, vol. 22, 695-704.
10. FLEMMING N.C. Multiple regression analysis of earth movements and eustatic sea-level changes in the United Kingdom in the past 9,000 years. Proc. Geol. Assoc., 1982, vol. 93, 113-125.
11. GREENSMITH J.T. and TUCKER E.V. Evidence for differential subsidence on the Essex coast. Proc. Geol. Assoc., 1980, vol. 91, 169-174.
12. ROSSITER J.R. Sea-level observations and their secular variation. Phil. Trans. R. Soc. Lond., 1972, vol. A272, 131-9.
13. FUCHS K., GEHLEN K.von, MÄLZER H., MURAWSKI H. and SEMMEL A. (eds) Plateau Uplift, the Rhenish Shield - A case history. 1983, Springer Verlag, Berlin, 411p.
14. WEST R.G. Relative land-sea-level changes in South-Eastern England during the Pleistocene. Phil. Trans. R. Soc. ond., 1972, vol. A272, 87-98.
15. CHRISTIE P.A.F. and SCLATER J.G. An extensional origin for the Buchan and Witchground Graben in the North Sea. Nature, 1980, vol. 283, 729-32.
16. LAGERBÄCK R. Neotectonic tructures in northern Sweden. Geol. För. Forhandl., 1978, vol. 100, 263-9.
17. LAGERBÄCK R. and MUIR WOOD R. Post-glacial faulting and earthquakes in northern Fennoscandia, (in litt.)
18. AHORNER L. Historical seismicity and present-day microearthquake activity of the Rhenish Massif, Central Europe. in Plateau Uplift...(see ref. 13), 198-221.
19. PEGRUM R.M. The extension of the Tornquist Zone in the Norwegian North Sea. Norsk Geologisk Tidsskrift, 1984, vol. 64, 39-68.
20. PEGRUM R.M. and LJONES T.E. 15/9 Gamma gas field offshore Norway, new trap type for North Sea basin with regional structural implications. Am. Ass. Petr. Geol. Bull., 1984, vol. 68, 874-902.
21. GLENNIE K.W. and BOEGNER P.L.E. Sole-Pit inversion tectonics. in L.V. Illing and G.D. Hobson, eds., Petroleum geology of the continental shelf of north-west Europe: London, Institute of Petroleum, 1981, pp110-120.
22. KEIZER J. and LETSCH W.J. Geology of the Tertiary in The Netherlands. Verh. Kon. Ned. Geol. Mijnbouwk. Gen., Geol. Series, 1963, vol. 21-2, 147-72.
23. CURRY D. The palaeogene beds of South-east England. Proc. Geol. Assoc., 1965, vol. 76, 151-1965
24. MÉGNIEN C. Tectogenèse du Bassin de Paris: étapes e l'évolution du bassin. Bull. Soc. géol. France, 1980, vol. 22, 669-680.

25. COLBEAUX J-P., LEPLAT J., PAEPE R. and and SOMME J. Tectonique récente dans le Nord de la France et le Sud de la Belgique. Ann. Soc. Geol. Nord., 1979, vol. 98, 179-88.
26. GREGERSEN S. Earthquakes in the Skaggerak recorded at small distances. Bull. Geol. Soc. Denmark, 1979, vol. 28, 5-9.
27. PRINCIPIA MECHANICA LTD British Earthquakes, Report for CEGB, BNFL & SSEB, 1982.
28. MUIR WOOD R. What happens to the Rhine Graben 'sub-plate boundary' where it meets the south North Sea? in A.R. Ritsema and A. Gürpinar eds., Seismicity and seismic risk in the offshore North Sea area, Proceedings of the NATO Advanced Research Workshop, held at Utrecht, 1983, D. Reidel Publ., Dordrecht, 35-42.

This research has been supported through projects undertaken for the CEGB and the Department of Energy.

Discussion

MR C.R.SMITH, National Nuclear Corporation
To what extent have the theories presented in Dr Muir Wood's paper been validated by analytical studies such as would be carried out for a structural engineering problem?

DR MUIR WOOD
It is important that engineers should understand the nature of the earthquake problem and the difference between earthquake hazard and some other potential failure internal to the structure which they have designed. The earth has inconveniently provided no blueprints of crustal structure. It is known however that the crust has been multiply deformed through various cycles of deformation, providing faults which have often moved within different periods of activity. Secondly the cause of crustal deformation is to be found within the earth's mobile mantle, hidden many tens of kilometres underground and moving at imperceptible speeds. As dwellers on the planet's surface all we can observe is the effect of movement: the earthquakes generated by the deformation of the brittle crust. Given our ignorance of some of the most important parameters affecting seismicity, the most useful beginning to comprehending the problem comes from collecting reliable scientific data, and attempting to show the overall seismotectonic patterns which emerge. In engineering terms we are in the position of a structural engineer attempting to comprehend potential failure of a large ancient building, for which no plans exist, observed from a distance at night.

11. Fault activity and palaeoseismicity during quaternary time in Scotland—preliminary studies

C. A. DAVENPORT, MSc, DIC, MIGeol, FGS, and P. S. RINGROSE, BSc, University of Strathclyde

SYNOPSIS. Reconnnaissance investigations of surface faulting have identified a) historical and geological displacements in the Loch Linnhe area of the Great Glen, and b) structural lineaments associated with seismic activity in the Kinloch Hourn region of the Western Highlands. Quaternary sand,silt,clay and peat sequences at four localities exhibit intraformational deformation interpreted as palaeoseismic structures by reference to seismites documented elsewhere. Fault activity potential is discussed in terms of the release of tectonic stress during post-glacial isostatic rebound, and in terms of a displacement chronology giving a basis for the estimation of slip-rates.

INTRODUCTION

1. In 1983 engineering geologists at the University of Strathclyde initiated a research project aimed at characterising neotectonic movements in Scotland. Traditionally, engineering geologists have utilised such data in the siting of, and derivation of design earthquakes for, major engineering facilities in continental areas of high seismicity. More recently, offshore areas and regions of low seismicity have been assessed for earthquake-hazard potential (ref. 1). In the U.K., earthquake-hazard assessment has been primarily based upon seismological data: however, there are regions, such as Scotland, which should be expected to yield geological data for an improved basis for the assessment of earthquake potential.

2. In modern engineering studies, appropriate geological information can be expected to:

a) establish the location of surface faults
b) establish capability to produce earthquakes
c) establish fault-movement histories
d) estimate pre-historic earthquake recurrence (palaeoseismicity)
e) provide an alternative basis for near-source attenuation,

all of which are used in the estimation of design levels of vibratory ground motions, and potential for surface displacements and liquefaction of foundation formations.

A geological alternative to seismological data is particularly important for nuclear facility siting and design where the safety earthquake is usually taken to have a return period of approximately once-in-10,000 years (refs 2,3).

3. This paper constitutes the first progress report for the current project which involves the investigation of seismotectonic and palaeoseismic phenomena in Western Scotland and adjacent areas. Two forms of evidence are sought:

a) surface faulting involving recent tectonic movements
b) palaeoseismic soft-sediment deformation structures in Quaternary lake sediments and other susceptible deposits.

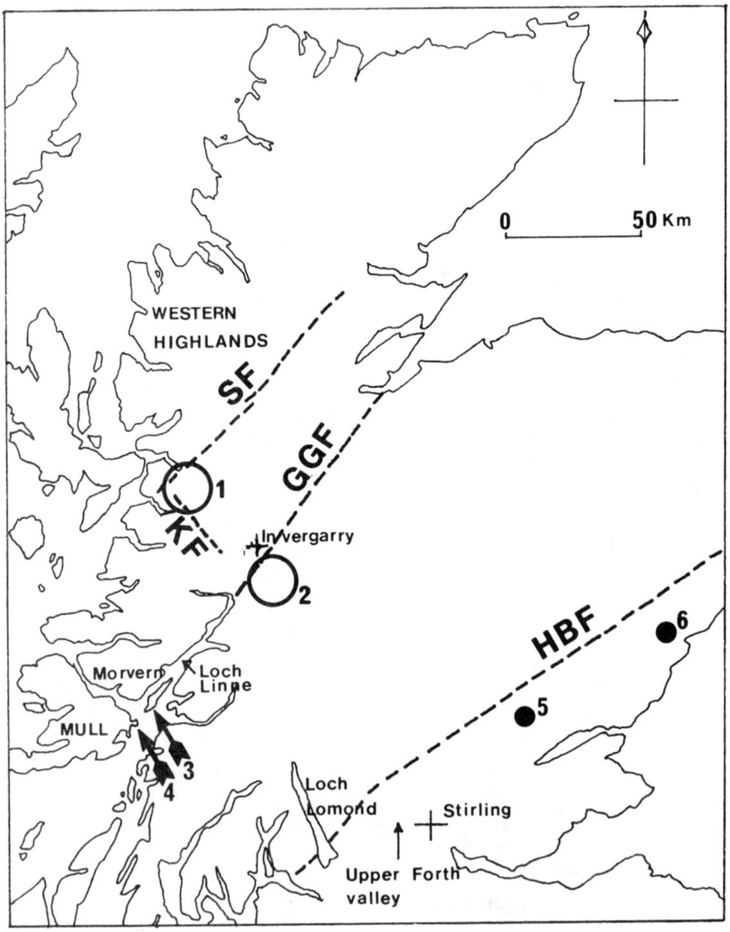

Fig. 1. Location Map.
Localities: 1-Kinloch Hourn, 2-Glens Roy,Gloy,Spean, 3-Lismore, 4-Port Donain, 5-Blairgowrie, 6-Brechin.
Faults: SF=Strathconon KF=Kinloch Hourn
GGF=Great Glen HBF=Highland Boundary.

Methods used include aerial photography(at 1:25,000 and 1:10,000 scales), satellite imagery, and geomorphological,structural and geotechnical field mapping techniques (including mechanical excavators). Microearthquake locations acquired by the networks of the British Geological Survey(BGS) have also been used to identify target areas. By mid-1984 a number of lineaments had been identified as field targets in Kinloch Hourn, Morvern, Mull and islands of Loch Linnhe. Sediment deposits were identified in the Western Highlands and northeast Lowlands. Accordingly, the field evidence for active surface faulting and late Quaternary palaeoseismicity is presented. The localities described below are shown in Fig. 1.

FIELD OBSERVATIONS

Fault Displacements

4. Selective reconnaissance mapping and levelling surveys have been made in the Western Highlands in an attempt to locate and measure surface-fault displacements and establish the magnitude of offsets suspected from earlier levelling surveys (refs 4,5). Three regimes of fault displacement are demonstrable.

5. Vertical displacements. A vertical offset of 2.7±0.3 m has been confirmed by levelling. The offset occurs in the Main Rock Platform of lateglacial age (around 11,000 years b.p. [before present]) at Port Donain,East Mull. A straight,sharp 3km long fault linear intersects the platform at the point of offset; however, unbroken peat cover up to a metre thick means that the offset is not directly observed.

6. Lateral displacements. Numerous lateral offsets of near-vertical Tertiary dykes were mapped on the south end of the Isle of Lismore, Loch Linnhe (within the Great Glen Fault zone). In each case tectonic (as opposed to emplacement) offset could be established. Offsets of uncertain origin were disregarded. In one location a closely-spaced series of offsets, within a zone less than 200 m wide, involve a total displacement of 61.2 m ; 55.3 m of a sinistral sense and 5.9 m of a dextral sense. All offsets occur on faults orientated approximately northeast. The dykes have an age of 55 to 60 Ma (million years) (ref. 6). The temporal distribution of the displacements since the intrusion of the dykes is unknown.

7. Historical displacements. Two offsets in historical features and three in geomorphological features were observed on one 2.3 km long fault trace - the Loch Fiart Fault, Lismore. These indicate 0.20 m of sinistral offset since about 500 years b.p., and 0.80 m since glaciation. The two historical features occur 300 m apart, and all five features are encountered within a kilometre-length section of the fault. The historical features comprise:-

a) 0.20±0.05 m of sinistral offset of a pre-1700 A.D. turf-dyke.
b) 0.20±0.10 m of sinistral offset of a mediaeval stone wall.

The other features are sinistral offsets of between 0.2 and 0.8m observed in the rock outcrop morphology, presumed to be moulded and plucked by the last ice. Together the features suggest a recurrently active strike-slip fault whose most recent movement occurred in the last 500 years. The rupture zone itself is narrow. At a coastal exposure of the fault, clay gouge and microbrecciated material were observed in a fracture not more than 30 mm wide.

Seismicity and Fault Activity Associations.

8. East Mull vertical offset. This region of the Great Glen zone has relatively high historical seismicity (compared to the rest of the U.K.), though source-location lineations in instrumentally recorded seismicity have not been observed. A magnitude M_L =1.3 event is recorded on LOWNET (Nov. 1969) as occurring within 1 km of the fault trace (ref. 7).

9. Isle of Lismore lateral offsets. The island is located ten km N.E. of Mull in the same region of the Great Glen, so that the general observations above apply. The presence of an historical surface rupture raises the question of possible correlation with historical seismicity. One possibility would be a large event felt over much of the U.K., and possibly in N.W. Europe. Surprisingly Davison reports an event in November 1880, with a "felt area of 50,000 sq.miles", and a centre "about one mile W. of Lismore lighthouse"(ref. 8). This would place it very close to the Loch Fiart fault. However Davison's locations are frequently poor, and a relocation based upon the primary evidence made by Muir-Wood (pers. comm.) places the epicentre about 100 km to the west in the Atlantic. There are several other poorly-reported possibilities for a large event before 1800, but no obvious association has been established to date. A re-assessment of the 1880 event, and other earlier events, is in progress.

10. Seismicity and Field Studies in the Kinloch Hourn Region. Microearthquake locations recorded since 1967 by the B.G.S. are observed to form a broad NW-SE zone of seismicity across Scotland (refs 7,9). Where this zone intersects major NE-SW Caledonian lineaments there are concentrations in the activity. The most prominent of these occurs in the Kinloch Hourn area of the Western Highlands, at intersection with the Strathconon Fault (Fig. 1). Several micro-swarm sequences appear to be associated with a surface-fault feature at Kinloch Hourn. A spate of historical activity between 1888 and 1899 ('the Invergarry earthquakes'), very apparent along the same fault linear, has been reported by Davison (ref. 8). Photogeological and field studies of this fault-area have revealed the following:

a) an extremely sharp surface linear for a length of 14km.
b) a straight surface trace despite strong topography.
c) drainage and geomorphological features all deflected and truncated against the fault.
d) fault gouge consisting of two phases of clayey material up to 30 mm thick observed over a 50 m exposure of the fault, and passing through a wedge of peaty soil thought to have infilled a deep fissure-opening of the fault surface.
e) deposits containing soft-sediment deformation structures close to the fault, discussed in paragraph:14.

Although no definite evidence for displacement has been found, the evidence for recent fault activity along the lineament is believed to be significant.

Palaeoseismites

11. Excavations have been made at four sites having soft-sediment deformation structures. Various of these structures are compared to similar structures known to have been produced insitu by earthquake activity or experimentally by induced shock. Ball-and-pillow, fluidization, dish, load,flame, slump fold, and 'fault grading' structures are all observed. These structures show remarkable similarities to:-

a) structures observed in silty sediments of the lower Van Norman reservoir, correlated with known earthquakes in the San Fernando area,California (ref. 10)
b) load structures experimentally produced by shock inducement (ref. 11)
c) a 'fault grading' stratigraphy observed in the Miocene Monterey shales, Santa Barbara, California, interpreted as a seismite (refs 12,13)
d) dish and fluidization structures related to liquefaction processes (refs 14,15).

12. Brechin,Perthshire and Blairgowrie,Angus. At these two sites, large excavations have allowed detailed evaluation of fabric, stratigraphy, and deformation structure relationships. From these studies it is possible to discount alternative means of formation, and demonstrate a high degree of consistency with mechanisms associated with cyclic loading. Both deposits consist of fine sands and silts with occasional clay layers. The deposits occur in outwash sands of late Devensian age (about 13,000 years b.p.)(ref. 16).In the Brechin deposit deformation is observed in a lens-shaped channel or lake fill with exposed dimensions of 100 by 200 m. The maximum thickness of deformation is 4 m. The deposit near Blairgowrie is only exposed for 30 m laterally, but the thickness of the deformed material is at least 9 m thick. Similar structures were observed 1.5 km away where a gas pipeline excavation cut through the same deposit (Paterson, pers. comm.).

13. Glen Roy, Central Highlands. Small excavations in glacial lake sediments were made throughout an area of 150 sq.km in the glens Roy, Gloy and Spean. Numerous forms and degrees of deformation and slumping were observed. Preliminary studies indicate the following:-

a) several structures similar to documented seismites, as noted in paragraph:11.
b) a 'fault grading' stratigraphy.
c) an areal distribution of exposures showing decreasing degrees of slump size and deformation type, outwards from a central zone into peripherally undisturbed sediments.

Sissons and Cornish (ref. 17) reported a large landslip in the central zone, a vertical displacement of up to 2 metres in the 'parallel roads' of Glen Roy (former lake beach levels), and a sharp surface-fault trace. They attributed these features to an earthquake event at the end of the Loch Lomond Re-advance (10,000 years b.p.(ref. 16)). Our study of deformed sediment and the discovery of a 37.7 m sinistral offset of a Tertiary dyke along a sharp fault linear in the area supports and develops their interpretation. To date we have not established whether the 2 m displacement in the 'parallel roads' results from generative fault movement or landslip processes.

14. Kinloch Hourn, Western Highlands. Deposits containing soft-sediment deformation structures occur in glacial sands and postglacial peats at two locations 7.5 km apart and within 2 km of the Kinloch Hourn fault trace. The style and extent of deformation are suggestive of formation by ground shaking. The deformation event must have occurred after the Loch Lomond Re-advance (i.e. after 10,000 years b.p.), as a truncated top to the deformation occurs within post-glacial peat at both locations. The deformed horizon is between 0.5 and 1 m thick and is observed throughout exposures extending for 50 and 100 m.

DISCUSSION

15. Much of the evidence presented above has yet to be fully exploited and evaluated in terms of fault activity and earthquake recurrence. However at this stage the following are presented with some confidence regarding implications for assessing earthquake potential.

Faulting

16. Lateral Movements. It seems clear that significant surface faulting took place during Quaternary times. Fault displacements have only been found in the Loch Linnhe area of the Great Glen, although beach offsets have been reported for the Forth valley (ref. 4). The discovery of these features in the Great Glen area must be partly influenced by the high degree of exposure afforded in the Western Highlands. This is in contrast most of the U.K., and therefore the evidence must be considered to be of general importance. The fact the fault movements involve both large lateral displacements as well as

smaller vertical displacements suggests that, when considering U.K. seismotectonics, the role of isostatic recovery can only be considered alongside longer term active tectonic processes. Fault plane solutions presently available for the U.K. suggest a general north-south compression (refs 18-20). Consistent with this is the fact that most displacements measured are sinistral on faults orientated between north and N090. The principal exception is the Glen Roy dyke offset which is faulted sinistrally along N160. The presence of an historical lateral displacement among large post-Tertiary lateral offsets strongly suggests strike-slip movements are continuing. The following points are presented regarding the significance of the 0.20 m historical fault displacement:-

a) because surface rupture normally represents only a small component of total slip along earthquake faults, the documented displacement is taken to indicate larger displacement at depth
b) this displacement is of a similar order to that inferred at depth for the Carlisle event of Dec. 1979, i.e. 0.30 m (ref. 18)
c) if a slip rate of 0.06 mm/yr is assumed for typical U.K. faulting (refs 18,21) then surface displacements of this order represent return periods of around 3000 years.

17. Vertical Movements. The confirmation of at least one of the vertical offsets of raised beaches inferred from earlier levelling work suggests that differential block movement occurred in post-glacial times. Exactly what place displacements of this kind have in the overall tectonic framework will be difficult to establish, however the following are put forward for discussion.

a) Maximum crustal stress due to ice loading and recovery occurs at the edge of the ice mass.
b) These stresses will decrease with time as recovery continues.
c) Thus there will be a critical interval, both spatial and temporal, during which stresses could be sufficient to cause movements along existing faults.
d) Existing tectonic stress will form some contribution to movements generated by this mechanism.
e) Where zones of tectonic stress coincide with the critical interval, the potential for large fault movements is increased.

It is possible that the Loch Linhhe area has been subject to this kind of tectonic activity. Two pieces of evidence are presented in this context:-

a) the two large offsets reported in beach levelling studies (refs 4,17) both occur near the margin of standing ice (Loch Lomond Readvance) at the time of offset. Sissons noted that one of the displacements occurred close to a well known fault; the Abbey Craig fault near Stirling (ref. 4)

b) there is a marked anomaly in the Main Rock Platform levelling data of Gray (ref. 5) in the region of the Great Glen Fault. Data for Mull are relatively clear from anomalous levels beyond 12 km east of the fault. The position of the Devensian ice at the time of offset is uncertain.

18. Relative Importance of Isostatic Recovery Strain. The discussion above raises questions concerning the relative contribution of isostatic strain and longer-term tectonic strain in faulting of this kind. Two end-member situations are suggested; either,

a) these vertical movements are entirely a result of additional loading and recovery stresses, and their existence on major faults is purely a matter of re-activation, or,

b) tectonic stress is a large and significant contributor to these movements, and loading and recovery strain is performing largely as a local trigger.

We may consider what effect these two end-member situations are likely to have on frequency-magnitude curves for glacial times. It is supposed that the dominance of loading/recovery strain would have the effect of enhancing the low-magnitude end of the spectrum (Fig.2A). The effect of additional triggering of tectonic strain could manifest itself either by an overall change in slope (an increase in b-value) (Fig.2B), or by an increase in seismicity throughout the spectrum (Fig.2C). The implications, here, for the estimation of large-event recurrence are important, and clearly more field data are required, along with the development of rigorous theoretical models.

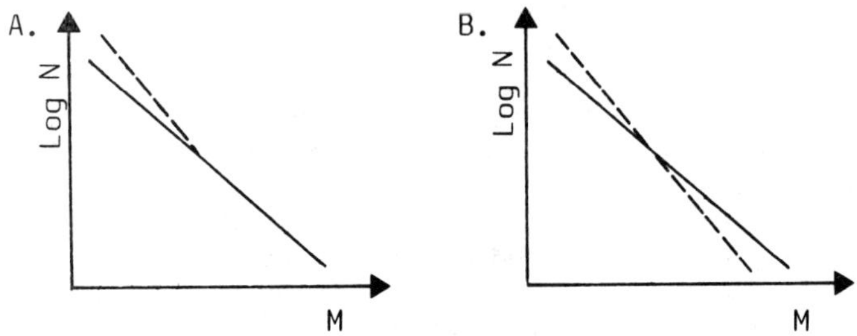

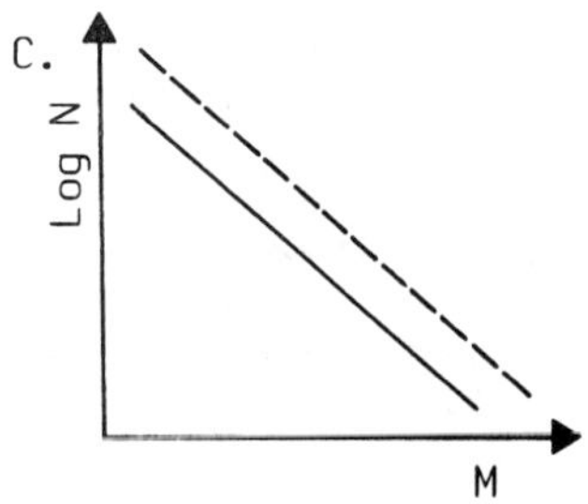

Fig. 2. Frequency-Magnitude Graphs.
solid curve - 'background' relationship.
dashed curve - 'critical interval' effects.

Palaeoseismites

19. Detailed studies on these deposits are in progress. Some general points can be made at this stage as to their probable significance for evaluating earthquake recurrence. The importance of deformation structures in geologically young sediments for evaluating large-earthquake recurrence has been emphasized by Ambraseys(ref. 22),Sims(ref. 10) and Davenport(ref. 23). Their testimony is perhaps most important in low to moderate seismicity zones where large events are rare and less evident in historical records. Sims(ref. 10) noted seven criteria for the interpretation of deformation structures for earthquake recurrence. All his sedimentological criteria are met in our deposits: however his first criterion that 'the study area is in a presently seismically active region, with earthquake intensities of MM VI or greater' is of particular note when considering the U.K. area.

20. Sediment Type. Only the Glen Roy sediments are unequivically lacustrine and hence similar to the sediments reported by Sims(ref. 10). The deformation structures further enhance the similarity. The Kinloch Hourn deposits are probably also lacustrine, but have a much more varied composition reflecting a variable short-lived glacio-lacustrine environment. The other two deposits near Brechin and Blairgowrie have sediments of a higher energy fluvio-glacial environment with perhaps periodically temporary lacustrine conditions. They are however sufficiently fine-grained for liquefaction by cyclic loading to be expected. The structures observed in these deposits tend to differ in scale and type from those reported by Sims, reflecting this difference in environment.

21. Lateral extent. All four geological formations contain soft-sediment deformation with significant lateral extent, a factor contributing to their interpretation as seismites. The most extensive deposit is in the Glen Roy area where deformed glacio-lacustrine sediments are found over an area of approximately 100 sq.km.. It is interesting to note that this is of a similar order to the area of deformed sediment in the Lake Cahuilla deposits, Imperial Valley, California (ref. 10). The Kinloch Hourn deposits are an order less extensive (7.5 km between the two sites). The Brechin and Blairgowrie deposits are large single exposures, but could have been influenced by a common seismic source; the sites being 56 km apart.

22. Depth of deformation. The maximum observed thicknesses of deformed sedimentary sequences are:

a) one metre in Kinloch Hourn,
b) three metres in the Glen Roy area,
c) four metres near Brechin, and
d) nine metres near Blairgowrie.

Whereas in Glen Roy the largest deformation thicknesses involve slump material, all the other exposures reveal insignificant lateral movement. The presence of undisturbed stratigraphy above and between deformation zones, the preservation of stratigraphic order within deformation structures, the small amount of rotation and slide apparent in structures, and the lack of slope failure stuctures (except in the Glen Roy area) all suggest in-situ liquefaction by cyclic load mechanisms. It is therefore supposed that even considering any inherent susceptibility to liquefaction, extensive deformation by cyclic loading in deposits of this size must require strong ground motions. A number of mechanisms of soft-sediment deformation have been recognised, including liquefaction by static, dynamic and cyclic means, fluidization and pore fluid migration, and slope or structure collapse by gravity. Work on the deposits is directed towards quantification of the relationship between the structures and causative mechanisms, in anticipation of evaluating the size of causative ground motions.

23. Fault Sources. Fault sources capable of generating events which could be responsible for the formation of these seismites are in evidence for two of the deposits. The two Kinloch Hourn exposures lie within 2 km of the Kinloch Hourn fault, and the Glen Roy area has a fault showing large offset. However, within the vicinity of the other two sites, nearby faults have not been located. Both these sites lie within 20 km of the Highland Boundary Fault, which could therefore be a more distant source. However another possibility is that, considering their easterly location and substantial volumes of deformed material, large distal events, possibly in the North Sea, are responsible. If the latter were the case, similarly deformed deposits could be expected in the coastal areas of E.Scotland and N.England and, perhaps, Scandinavia and the Netherlands.

WORK IN PROGRESS

24. All the data presented here are the product of detailed reconnaissance surveys during one field season, as part of a three year research project. A more quantative displacement chronology for faults shown to have surface displacements will be possible by further excavation and trenching. Chemical and X-ray analyses, and dating by carbon and other methods, of fault gouges and displaced features comprise the next stage of this work. More comprehensive mapping and studies of Landsat imagery should allow extrapolation of these initial data to a regional significance and extent. Levelling survey work will be completed when permanent stations are referenced with Electronic Distance Measuring (EDM) equipment. Several more sites could be surveyed at a later stage. It is hoped that a study of records of historical seismicity will provide location data which can be related to historical displacements. Dates of last movement on these features will be vital to such studies.

25. The interpretation of the soft-sediment deformation deposits is of great significance. Initial field studies strongly suggest an earthquake origin. A clear demonstration that these structures can be formed uniquely by appropriate levels of vibratory ground motion is required. Our approach involves mathematical modelling of the deformation stratigraphy expected in these deposits under conditions of cyclic loading, and natural-scale simulation of the structures on triaxial

shaker tables. Carbon dating of the deposits is in progress and is anticipated to establish the dates of deformation and permit some correlation between sites. The extensive Glen Roy sites may permit the compilation of a palaeo-intensity map.

CONCLUDING REMARKS

26. The results of our studies indicate that in Scotland there are a number of features suggesting:-

a) active surface faulting
b) historical displacements
c) seismogenic sources, and
d) pre-historic earthquakes (palaeoseismicity).

27. To date only the most promising features have been reconnoitred. Many other seismogenic zones and active surface faults exhibiting historical and pre-historical movements could exist.

28. The role of tectonism in Scottish activity, and the extent of reactivation of ancient fault lines has yet to be established. The evidence for lateral displacements contests the common view that fault movements during late- and post-glacial times must be interpreted as differential isostatic rebound across old (reactivated) faults.

29. Future fault studies will incorporate the results of radiocarbon age-dating and extensive trenching; in an attempt to establish a chronology of movements along the fault zones most amenable to investigation. Palaeoseismicity studies will focus upon elaboration of the deformation mechanisms required to produce the complex geometries and textures seen within the deformed sediment layers. Both field logging and sampling, accompanied by natural-scale simulation using shaker tables, will be employed. Carbon-14 isotope dating is also expected to provide a chronological framework for estimating earthquake-recurrence over a span of at least 10,000 years.

30. In drawing attention to these preliminary geological observations we endeavour to encourage engineering research and design organisations to support such investigations,to reappraise the role of fault-oriented studies in seismic risk assessment in the U.K., and to consult with experienced engineering geologists whilst formulating guidelines, codes, and regulations for the siting and design of potentially hazardous engineering structures in Britain.

ACKNOWLEDGEMENTS

We gratefully acknowledge N.E.R.C. for funding this project (research studentship - PSR). Special thanks are offered to our colleagues in various departments of the University of Strathclyde. The co-operation of the Global Seismology Unit of the British Geological Survey, Edinburgh, along with Lord Lansdowne and North Angus County Council, has been most valuable.

REFERENCES

1. RITSEMA A.R. and GURPINAR A. (eds.) Seismicity and Seismic Risk in the Offshore North Sea Area, D.Reidel 1983.
2. United States Nuclear Regulatory Commission (USNRC). Nuclear Power Plants - Seismic and Geological Siting - 10 CFR 100, Appendix A, Federal Register of the United States of America, 1973.
3. International Atomic Energy Agency (IAEA). Earthquakes and Associated Topics in Relation to Nuclear Power Plant Siting - A Safety Guide, 1979, Safety Series No. 50-SG-SI, Vienna.
4. SISSONS J.B. Dislocation and non-uniform uplift of raised shorelines in the western part of the Forth valley. Transactions of the Institute of British Geographers, 1972, Vol.55, 145-159.
5. GRAY J.M. The main rock platform of the firth of Lorn, western Scotland. Transactions of the Institute of British Geographers, 1974, Vol.61, 81-99.
6. HOLGATE N. and MACINTYRE R.M. K-Ar age determinations on dykes of the Fishnish Bay area, Isle of Mull. Scottish Journal of Geology, 1980, Vol.16(4), 267-273.
7. BURTON P.W. and NEILSON G. Annual catalogues of British earthquakes recorded on LOWNET (1967-1978). Seismological Bulletin of the Institute of Geological Sciences, 1980, no.7, HMSO.
8. DAVISON C. A History of British Earthquakes, 1924, Cambridge University Press.
9. TURBITT T. Catalogue of British earthquakes recorded by the BGS Seismograph Network (1979,1980,1981), 1984, British Geological Survey, Global Seismology Unit, Report No. 210.
10. SIMS J.D. Determining earthquake recurrence intervals from deformational structures in young lacustrine sediments. Tectonophysics, 1975, Vol.29, 141-152.
11. KUENEN P.H. Experiments in geology. Geology Magazine, 1958, Vol.23, 1-28.
12. SEILACHER A. Fault-graded beds interpreted as seismites. Sedimentology, 1969, Vol.13, 155-159.
13. SEILACHER A. Sedimentary structures tentatively attributed to seismic events. Marine Geology, 1984, Vol.55, 1-12.
14. LOWE D.R. Water-escape structures in coarse-grained sediments. Sedimentology, 1975, Vol.22, 157-204.
15. ALLEN J.R.L. Sedimentary Structures. Volume II, ch.8,9. Elsevier 1982.
16. SISSONS J.B. The Quaternary in Scotland: a review. Scottish Journal of Geology, 1974, Vol.10(4), 311-337.
17. SISSONS J.B. and CORNISH R. Rapid localized glacio-isostatic uplift at Glen Roy, Scotland. Nature, 1982, Vol.297, 213-214.
18. KING G. A fault plane solution for the Carlisle earthquake, 26 December 1979. Nature, 1980, Vol. 286, 142-143.
19. ASSUMPCAO M. The NW Scotland earthquake swarm of 1974. Geophysical Journal of the Royal Astronomical Society, 1981, Vol.67, 577-586.

20. MARROW P.C. and ROBERTS G. Source mechanisms of aftershocks of the 1979 "Carlisle" earthquake. British Geological Survey. Global Seismology Unit, 1984, Report No.217.
21. MAIN I.G. and BURTON P.W. Physical links between crustal deformation, seismic moment and seismic hazard for regions of varying seismicity. British Geological Survey. Global Seismology Unit, 1983, Report No.199.
22. AMBRASEYS N.N. Value of historical records of earthquakes. Nature, 1971, Vol. 232, 375-379.
23. DAVENPORT C.A. North Sea seismotectonics and the role of geology in offshore risk assessment. In: A.R.Ritsema and A.Gurpinar (eds), Seismicity and Seismic Risk in the Offshore North Sea Area, 15-30. 1983, D.Reidel.

Discussion

DR B. O. SKIPP, Soil Mechanics Ltd
How did the requirement that sediments to be used as indicators of palaeoseismicity should have had no overburden come to be made?

MR D. T. SHILSTON, Soil Mechanics Ltd
What relationship exists between the offset field boundaries and the local topography?

MR DAVENPORT
In reply to Dr Skipp, 'no overburden' refers to a lack of overlying sediments. In the deposits discussed, there is envisaged a series of relatively thin sediments recently deposited in a local body of water such that would be expected to have a water/sediment interface, or 'mud-line' at the time of liquefaction.

The demonstrable absence of pre-deformational overlying sediments favours the elimination of liquefaction induced by sediment loading.

The described displacements are essentially identical but occur in different topographic situations. The post-seventh century stone walls are frequently aligned parallel to contours and are seen to be disturbed by slope movements. In the case described the wall is on more-or-less level ground where slope movement has not taken place. The turf-dyke trend down the hill slope is normal to the contours and exhibits displacements contrary to those which would be expected by slope movements.

12. Satellite imagery applied to investigations of geological structure as part of recent seismic hazard studies in the UK

J. C. GUTMANIS, MSc, MIGeol, FGS, Soil Mechanics Ltd, R. W. HOLT, PhD, Phillips Petroleum, and R. A. WHITTLE, BSc, FIMM, Rock Mechanics Ltd

SYNOPSIS. Applications of remote sensing from space include the interpretation of structural geology expressed at the earths surface. An understanding of structural geology is fundamental to the assessment of seismic hazard.

Over the last few years digital image processing of satellite scenes has become a powerful new technique available to the earth scientist. This, together with the advent of the Thematic Mapper sensor on board Landsats 4 and 5, adds a new skill to the geologists repertoire. The paper describes how interactive image processing and structural interpretation have been applied to structural geology fieldwork in a complementary role.

INTRODUCTION

1. Over the last few years seismic hazard studies commissioned from within the UK nuclear industry have included the application of satellite remote sensing techniques to structural geology. A state-of-the-art understanding of the geological structure, both locally and in terms of major regional tectonics, is considered to be a fundamental part of the assessment of the seismic hazard at a particular site. Remote sensing is one of a number of techniques available to the structural geologist who must study not only the site vicinity but a large region around it, up to 11,000 km^2 in one recent case.

2. Satellite remote sensing techniques have been available to the earth scientist for over 10 years, but their application to seismic hazard studies world wide has met with limited success. This is particularly so in regions where the solid geology is masked by drift deposits, vegetation and the effects of cultural activity, such as much of the UK. Mistakes have been made, and remote sensing from space has been presented as an accomplished technique prior to its technological coming of age and prior to the development of soundly based interpretative methods.

3. The high resolution imagery provided by the Thematic Mapper (TM) on Landsats 4 and 5 represents a major advance in the quality and quantity of information that is now becoming available to users of remotely sensed data. For some time there has been doubt as to whether imagery from Landsat can really help the field geologist in places such as the British Isles, where what can be seen from space is so obviously largely influenced by man. With the Thematic Mapper there are no such doubts. In appearance the imagery is not unlike that of high altitude aerial photographs, but having the added advantage that images are available from narrow wavebands within the non-visible infra-red. The full value of this TM imagery can only be realised when it is viewed on an interactive processor.

4. This paper attempts to show how interactive image processing and structural interpretation can be successfully applied to seismic hazard investigations, even in regions such as lowland UK, where the solid geology can be masked. It also discusses limitations, a number of specific results, and problems yet to be tackled.

DEFINITIONS

5. A review of literature dealing with the application of satellite remote sensing techniques to structural geology, carried out prior to these studies revealed serious ambiguities and misconceptions in use of terminology. To avoid further confusion the terms used in this paper are defined:

Linear	: a line-like feature, of any origin, visible on imagery.
Structural lineament	: a line-like feature, visible on imagery, that is known to be directly related to geological structure.
Topographical lineament	: a line-like feature, visible on imagery, that can be correlated with topography (it might also have a structural origin).
Interactive digital processing	: the computerised manipulation of digital data in a satellite image, controlled throughout by an experienced operator.

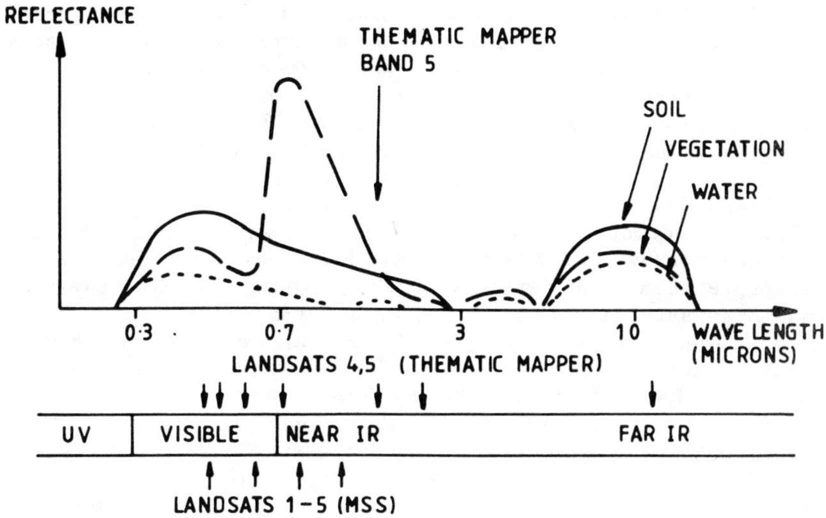

Fig. 1. Landsat MSS and TM spectral bands in relation to the spectral reflectance curves of soil vegetation and water

OBJECTIVES OF THE STUDY

6. The remote sensing study had five main objectives:-

i) to provide geologists with a means of inter-relating visible structures, topographic features and other geologically related phenomena at regional scale; to provide a regional overview of the surface features.

ii) to use imagery to help answer specific questions raised during fieldwork, such as the lateral continuity of certain faults and fault zones

iii) to examine imagery for unmapped or previously unrecognised features that might have structural significance and to input this information to the fieldwork.

iv) to identify the major structural lineaments in the study regions, particularly those of regional extent.

v) to examine imagery for evidence of geologically young surface faulting, where possible, taking account of the limitations of resolution.

7. At the outset it was decided that the study would not be merely an exercise in finding lines and plotting orientations, either manually or aided by computer. On the

contrary, the overriding concept was held to be a structural interpretation carried out in concert with fieldwork and taking into account other sources of information (maps, publications etc). This decision was arrived at on the basis of a large amount of previous experience and on the literature review referred to in Paragraph 5. This indicated that previous studies based on line finding and plotting alone have not been particularly successful.

IMAGE PROCESSING AND ENHANCEMENT

8. Interactive digital processing and enhancement was carried out by two structural geologists experienced in image interpretation and with an understanding of the physical basis to satellite remote sensing and image processing.

9. The GEMS Image Display System at the National Remote Sensing Centre, Farnborough, was used to process and enhance numerous Landsat 3, Landsat 4 Thematic Mapper scenes and experimentally processed Seasat scenes selected to give multi-seasonal coverage. GEMS consists of an image processing mini-computer and high resolution screen, linked to a host computer. It enables the user to sample and manipulate the digital data in a scene according to his requirements and to assess the information content as he proceeds.

10. Most of the study was undertaken using Landsat 3 multi-spectral scenes, supplemented by Thematic Mapper and experimentally processed Seasat scenes as these became available during the course of the study. Initially only Landsat 3 scenes of the required areas were available. A period of experimentation on GEMS produced a general sequence of algorithms which gave consistently good images for structural interpretation. This sequence, which is outlined in Figure 2, was adhered to in general, although some flexibility was built into it and the applicability of other algorithms or altered sequences was under constant review. In many cases deviation from the sequence was essential because of subtle differences in image characteristics from area to area.

11. Some comments are appropriate for each step in the sequence. Steps 1 and 2 are standard preparatory operations whereby the required sub-scene is defined together with a sampling ratio, because each satellite scene contains too many pixels (picture elements) to fit on the monitor screen and some trade-off between the size of sub-scene required, image distortion and frequency of pixel sampling must be achieved. The crux of the image processing sequence is held in steps 3 and 4. The third step, Principal Components Analysis, is a statistical technique, which correlates, to varying degrees, the information present in the different wavebands of a scene in order to bring out differences not normally apparent to the eye. Four levels of correlation will produce four different images, but only the first two were used.

12. The fourth step, Edge Enhancement, is a spatial filtering technique designed to maximise contrasts in pixel brightness across boundaries, the result being an image in which the boundaries are sharpened and linears enhanced. A filtering algorithm, whose size can be varied, was passed through the digital data. For the particular regions investigated and the sampling ratios used a filter size of 16 pixels by 16 pixels was found to give the most useful enhancements. During the period of experimentation, linear filters were employed and systematically rotated in an attempt to enhance line-like features of whatever strike. This technique was abandoned when it was found that it generated line-like artifacts.

13. The last step in the processing sequence was the application of a manual contrast stretch, giving instantaneous operator defined redistribution of the pixel brightness values. In this way exceptional enhancement of images by varying contrast could sometimes be achieved. As a general rule, and particularly when producing images for mosaics, similar stretches were employed.

14. At an advanced stage in the studies a number of Landsat 4 Thematic Mapper scenes became available, and these were interactively processed in a similar manner to that described above. The Thematic Mapper data was often not significantly improved by principal components analysis or edge enhancement largely due to the high quality of the basic data.

15. Each image generated by interactive processing was assessed for its value in structural interpretation and, if thought suitable for detailed interpretation was photographed onto colour slide film using a Honeywell 3000 Colour Graphic Recorder.

16. Most of the images generated were monochrome but a number of standard and special false colour composite images were also produced and used in the interpretation.

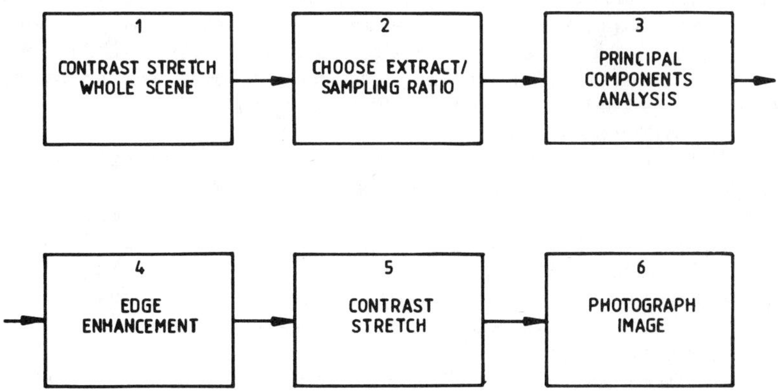

Fig. 2. The general sequence of interactive image processing developed for the study. See text for further elaboration.

IMAGE INTERPRETATION

17. The geologists who carried out the image processing were also responsible for the structural interpretation of the data and its presentation to the field geologist. Much of this took place in the field study area, to ensure that the geologists doing the sensing gained an appreciation of the terrain and the nature of the geological problems likely to need answering. The field geologist also took part in image processing sessions at the NRSC. In this way a two-way flow of information took place as the fieldwork proceeded. Information obtained from the Thematic Mapper scenes was passed to a geomorphologist to aid in the search for neo-tectonic features.

18. The generation and interpretation of imagery was carried out in phases over a period of about 6 months, generally in response to questions raised by the field geologist. Each processing session at the NRSC was extended so that after having concentrated on trying to answer specific questions, additional work could be carried out to identify areas for further fieldwork, in accordance with the objectives of the studies.

19. To carry out the structural interpretation, undistorted images were optically projected onto base maps to allow accurate positioning of features, all of the scenes having been geometrically corrected.

20. The images generally contained a large number of linears of varying length and contrast with the background, some of which were immediately discounted as being due to processing artifacts or man-made structures of no geological significance. The remainder were visually assessed in terms of their length, contrast and relationships with neighbouring linears. Emphasis was placed on determining the nature of persistent linears, especially those apparent at a regional scale. Isolated linears less than about 5 km long were, as a general rule, discounted except in a few special cases. Images of different scale, or from different scenes, or those produced using a different processing method, were used to help assess the credibility of linears. This initial stage in the identification of linears produced a series of overlay sheets of features, and necessarily involved a degree of structural interpretation.

21. There are a number of things that can bias or adversely affect processing and interpretation of imagery, some connected with the imaging system, some with the methods and algorithms used to process the data, and some which are related more to personal preferences or habits of the interpreter. Throughout the studies the interpreters remained aware of these factors and wherever possible removed or alleviated them.

22. A more detailed interpretation of the linears was then carried out, aimed at identifying structural and topographic lineaments. To this end, the BGS 1:50,000 and 1" to 1 mile geological maps and the Ordnance Survey 1:50,000 and 1:250,000 topographic maps were used. Reference was also

made to many publications dealing with the structural geology of the study area. These maps and publications were not used as "ground truth" or more correctly "ground reference data" but to acquaint the interpreters with what was already known about the region being studied.

23. Using these literature sources, each linear identified on imagery was correlated with known features, as far as possible, for their entire length. This back-analysis data was tabulated on the final maps produced for the study, and an extract is included here in Figure 4.

TYPICAL RESULTS

24. Dorset. There are several interesting points to emerge from the structural interpretation of Dorset (Figure 3).

25. Firstly, in the Dorchester area, and particularly to the south, there is a family of NE-SW trending linears. Most of them are topographic lineaments, although at Bincombe, 5 km NNE of Weymouth, part of a linear can be correlated with a mapped fault which appears to be younger than the E-W trending Ridgeway/Poxwell fault (BGS Sheet 342). Imagery suggests that the Ridgeway/Poxwell fault, which is known to have undergone reverse displacement during the Tertiary is post-dated by NE-SW linear structures whose length and frequency has not been recognised previously. This structural trend has been seen on images of other parts of south central and south east England, again with relationships suggesting that it could be the youngest present.

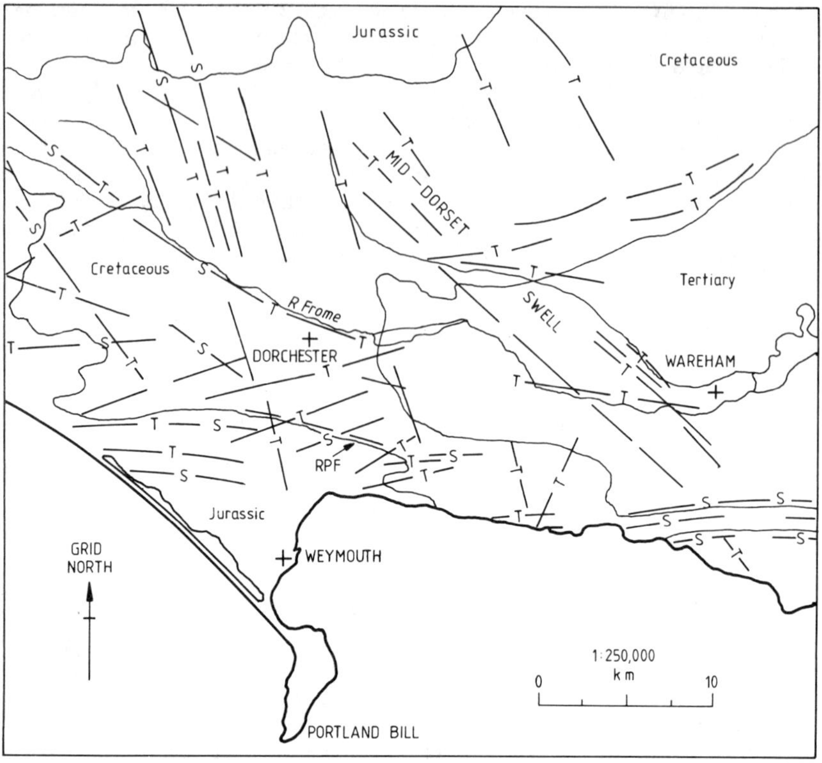

Fig. 3. Structural interpretation based on Landsat 3 MSS imagery of Dorset. The Mid-Dorset Swell is an axis of sedimentary thinning within the Cretaceous sequence (Drummond 1970). S=structural lineament T=topographic lineament. RPF = approximate line of the Ridgeway/Poxwell fault.

26. Secondly, there is a linear passing from the top left corner of Figure 3 down the valley of the River Frome to Wareham. It is part of a linear of regional scale, visible on imagery of SW England, that can be traced from the Somerset coast, along the west side of the Quantock Hills past Taunton and thence into Dorset, admittedly with some gaps. For much of its length it is parallel to the sun azimuth at time of imaging, a circumstance that tends to subdue linears. On the Tertiary outcrop in Dorset it swings from a NW-SE trend towards W-E, becoming almost parallel with the structures of southern Dorset such as the Weymouth anticline and Purbeck monocline. In Somerset, this linear can be correlated with the Watchet, Cothelstone and Hatch fault zones, all of which show dextral displacement and together form a line proposed by some authors as a major structure (Whittaker 1972, Melville and Freshney 1982). This structure is parallel to dextral wrench faults which cut Devon and Cornwall, such as the Sticklepath, Coombe Martin

and Cambeak Faults.

27. A south eastward continuation of the Watchet/Cothelstone/Hatch fault line into Dorset, or some expression of it in the cover rocks, would be of considerable interest. Despite gaps around the Somerset/Dorset border and Crewkerne, the linear can be correlated with faults over several sections including parts of the Cretaceous outcrop. The existence of this structural correlation, plus the generally strong topographic correlation and the swing in trend towards W-E are suggestive of some form of influence in the Cretaceous and Tertiary rocks of Dorset, possibly in terms of basement-cover reactivation.

28. Thirdly, there is a group of NW-SE linears passing from the Cretaceous outcrop onto the Tertiary outcrop in the area between Dorchester and Wareham. These cannot yet be related to structure, although some correlate with topography. Their coincidence with the Mid-Dorset Swell, an axis of sedimentary thinning within the Cretaceous sequence, and their persistence into the Tertiary outcrop, is significant. The Mid-Dorset Swell is thought to indicate Mid - and Upper Cretaceous tectonic activity (Drummond 1970).

29. The Mere Fault Area. At the northern side of the Vale of Wardour in Wiltshire, Jurassic rocks are faulted against Chalk on a southward dipping listric reverse fault. The Mere Fault is thought to have been initiated as a pre-Albian growth fault during a Mesozoic extensional phase. Movement was subsequently reversed during Tertiary compression. The structure overlies and is probably rooted in a Variscan basement thrust (Chadwick et al 1983).

30. On Landsat 3 imagery, the surface trace of the Mere fault is barely visible, although its position can be inferred by the topographic and agricultural contrasts between the low lying Vale of Wardour to the south and the Chalk hills to its north. Some linears can be identified in the area, particularly in the hills, but in general the Landsat 3 imagery is of limited use.

31. With the Thematic Mapper the situation is more clear. In Band 5, the spectral reflectances of vegetation, soil and water, are low (see Figure 1). Topography is exceptionally well displayed, whereas field boundaries, woodlands and bodies of water are supressed. This is ideal for structural interpretation and represents a significant increase in image quality over snow covered Landsat 3 scenes.

32. On TM imagery the trace of the Mere Fault is clearly seen together with some of its known associated minor faults. The arcuate trace of the main fault plane is due to its listric profile. Many other linears are also indentified, mainly topographic lineaments on the hilly Cretaceous outcrop. Of particular note is a NW-SE trending topographic lineament (number 266 on Figure 4) which appears to truncate the Mere Fault at its eastern end. This lineament correlates with a strong ridge, with no suggestion, from published maps, of structural control.

33. Linears of similar NW-SE trend are common over the entire Chalk outcrop of Salisbury Plain (not covered by Figure 4). In places they can be designated structural lineaments, for example at Barford St Martin. Most are designated topographic lineaments, and it is likely that some of these may represent faults, bearing in mind the difficulty of mapping faults in a Chalk upland area. A family of NW-SE faults, geologically the youngest present, would be analogous to structural relationships further west in Devon and Cornwall, and some authors (eg Melville and Freshney 1982) have proposed major dextral wrench faults crossing south central England.

34. Linears on the Jurassic outcrop immediately west of Salisbury Plain are scarce. A 20-25 km wide N-S trending belt of country running from Bath southward towards the West Dorset coast has markedly different linear trends to its neighbouring country east and west. It is co-axial with the southern extension of the Malvern line, of which the Bath Axis (Wilson et al 1958) is a cover expression, and with the Hawkins line to the south (Wills 1978). It also aligns with a gravity anomaly (Howell in discussion of Stoneley 1982). On imagery, the Mere Fault itself cannot be traced westwards across this belt of country, and there is no visible evidence to suggest a connection with structures such as the Cannington Park Thrust, as proposed by some authors (Drummond 1970, Gardiner and Sheridan 1979).

LINEAR No	STRONG TOPOGRAPHY	WEAK TOPOGRAPHY	SMALL RIVER	LARGE RIVER	ROAD	RAILWAY	FAULT	LITHOLOGY	AGRIC / VEG
261		Δ	Δ	Δ	Δ		Δ	▲	
262		▲					▲	▲	
263	Δ	Δ					▲	▲	
264	▲		Δ					Δ	Δ
265		▲					▲	Δ	
266	Δ						Δ	Δ	Δ
267	▲								
268	Δ		Δ		Δ			Δ	
269	▲								
270	▲		Δ		Δ				Δ
271		▲						Δ	Δ
272		▲							▲
273	Δ							Δ	Δ
273a	Δ			Δ	Δ			Δ	
274	Δ	Δ						Δ	
275	Δ	Δ	Δ		Δ				Δ
276		Δ	Δ					Δ	
277	▲							Δ	
278	▲							Δ	
279	▲								
280		Δ							
281		▲							
289	Δ	Δ	Δ		Δ			Δ	

BACK ANALYSIS DATA
▲ = Full length Δ = Part length

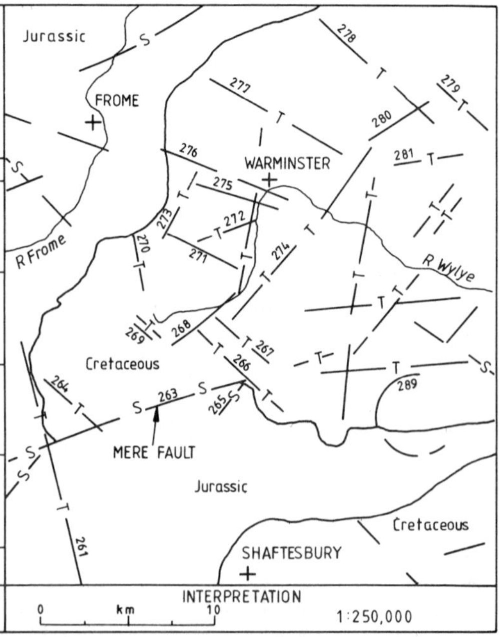

Fig. 4. Structural interpretation based on Landsat 3 MSS and Landsat 4 TM imagery of the Mere Fault area, with back analysis data. S=structural lineament T=topographic lineament.

CONCLUSIONS

35. Simple statistical treatment of the data showed that 20-25% of all linears could be directly correlated with faults shown on existing maps over part or all their length, and it was fairly safe to designate them structural lineaments. For linears greater than 5 km in length, the proportion of structural lineaments rose to 30-50%. Some 50-75% of all linears correlated with strong topography, and it is likely that many of these represent structural control. A similar percentage could be correlated with known lithological boundaries. Although there was some variation from area to area according to geology and terrain differences, these figures were found to be remarkably consistent.

36. As may be seen from these figures, a considerable proportion of the linears identified could not be directly correlated with known structure, particularly linears less than 5 km long. It was not unreasonable to suggest that a proportion of these might actually represent unmapped structure, hence a number of these linears of particular interest became the subject of further study on imagery, in the field and in the literature.

37. As the work proceeded it became clear that many scenes from Landsat 3 were of limited use for structural interpretation, owing mainly to limitations of resolution (59 m x 79 m) and the limited spectral range of the Landsat 3 wavebands. The exceptions were snow scenes with a low sun angle, in which the effects of vegetation were masked and topography was enhanced. Although snow cover obviously masks other information, little is lost in practice because of the coarse resolution involved anyway. Care was taken to consider the bias that could result from effects of shadows, particularly with winter scenes.

38. With the advent of the Landsat 4 and 5 Thematic Mapper the increase in resolution (30 m x 30 m in 6 of the 7 wavebands), provided a dramatic increase in image quality over the Landsat multi-spectral scenes (see Figure 1). Band 5 is of sufficiently high quality to be used to select areas which might benefit from close inspection by field geologists and geomorphologists searching for evidence of neotectonic features. Selection of such areas would normally involve extensive use of aerial photographs. The Thematic Mapper can effectively reduce the time spent searching through photographs.

39. The digitised Seasat scene was noisy and of limited use, generally providing far less information than the Landsat images.

SOME REMAINING PROBLEMS

40. Even as this study was proceeding, new technology and new developments in image processing software were rendering parts of it obsolete. The quantity of remotely sensed data becoming available is threatening to inundate date handling centres. For each Thematic Mapper scene there is one

computer tape for each of the 7 wavebands recorded. The accessibility of this data, in terms of the time it takes to obtain the scenes needed and the cost of tapes and processing facilities, will become a significant factor influencing its usage.

41. Once the data is mounted in an image processing system it can be an extremely useful and cost effective tool in structural geology. However, there does remain a large gap between what is interpreted on an image and what is proven in the field. Further work is needed to understand how crustal structure manifests itself at the earth surface, particulary basement structures re-activated during or after deposition of overlying sediments. There is currently great interest in deep geology in the UK and the rest of Europe, and it may be that in due course the fruits of such work will converge with those of the work underway into the physics of remote sensing itself, currently also a major research area.

ACKNOWLEDGEMENTS

42. The work upon which this paper is based was carried out during 1983 and 1984 as part of a preliminary investigation commissioned from the Seismic Hazard Working Party (SHWP) by the Central Electricity Generating Board. The authors wish to thank the CEGB for permission to publish this paper.

REFERENCES

1. CHADWICK R A, N KENOLTY and A WHITTAKER (1983) : Crustal structure beneath southern England from deep seismic reflection profiles. J Geol Soc Vol 140 pp 893-912.

2. DRUMMOND P V O (1970) : The Mid-Dorset Swell. Evidence of Albian-Cenomanian movements in Wessex.
Proc Geol Assoc Vol 81 pp 679-714

3. GARDINER P R R and D J R SHERIDAN (1981) Tectonic framework of the Celtic Sea and adjacent areas with special reference to the location of the Variscan Front.
J Structural Geol Vol 3 pp 317-332.

4. MELVILLE R V and E C FRESHNEY (1982) : British Regional Geology, The Hampshire Basin and Adjoining areas HMSO.

5. STONELEY R (1982) : The structural development of the Wessex Basin. J Geol Soc Vol 139 pp 545-554.

6. WILLS L J (1978) : A palaeogeological map of the Lower Palaeozoic floor below the cover of Upper Devonian, Carboniferous and later formations.
Geological Society London, Memoir No 8.

7. WILSON V, F B A WELCH, J A ROBBIE and G W GREEN (1958) : Geology of the country around Bridport and Yeovil.
Mem Geol Surv Gt Britain.

8. WHITTAKER A (1972) : The Watchet fault - a post-Liassic transcurrent reverse fault.
Bull Geol Survey G B No 4 pp 75-80.

9. WHITTAKER A (1975) : A postulated post-Hercynian rift valley system in Southern Britain.
Geol Mag Vol 112 pp 137-147.

13. A ground motion probability analysis for Britain based on macroseismic earthquake data

R. E. LONG, PhD, University of Durham

SYNOPSIS. Historical earthquake data is at present far more extensive than instrumental data in Britain. The traditional approach of converting intensities to magnitudes to fuse with instrumentally determined data for subsequent ground motion estimates is therefore premature. Rather than convert from intensity (which is a measure of ground motion) via magnitude and attentuation relationships to get back to expected ground motion for hazard estimates, this paper generates ground motion-probability curves directly from intensities. The advantage is that intensity-magnitude and attenuation relationships, which are poorly known for U.K., are not required by the direct method. The procedure is to generate an areal probability of intensity and then use data relating intensity to ground acceleration, (both for horizontal and vertical motions), to generate a set of mean ground motion-probability curves for Britain.

INTRODUCTION

1. Estimates of U.K. seismicity have been based predominantly on historical data (Lilwall, 1976; Irving, 1982). The proportion of instrumental to historical data is small. Of the catalogued U.K. events only about 5% have instrumentally determined locations or magnitudes. Because of the poor station coverage of U.K. the bulk of these instrumentally located events have poor epicentral locations with no determined focal depth. Hence in U.K. there is no body of instrumental data to form the backbone of a data set for hazard analysis. The major component of this data set is therefore historical seismicity and consists of macroseismic data.

2. This fact is unusual in earthquake prone areas. Normally the use of instrumentation makes a significant improvement in the number of events detected. Worldwide the rate of data collection has risen significantly since the

1930's and more particularly since the 1960's. As a consequence it is the norm in earthquake prone areas that instrumental data predominates. The traditional method of handling historical data is therefore to derive from it parameters corresponding to those instrumentally determined. Thus historical data is used to generate a location (epicentre and depth) and a magnitude. The magnitude scale is that applicable to the area and normally is derived from instrumental data for that area.

3. Current estimates of U.K. seismicity have adopted this 'standard' approach despite the relatively small amount of instrumental data, and the inherent problems that result. There is for example, no established magnitude scale for U.K. Data is too sparse to establish magnitude - intensity - depth relationships for the U.K. This approach is therefore problematic unless we accept relationships, based on larger areas, e.g. Europe, when their applicability to U.K. remain unknown.

4. The normal output required from a hazard study is some ground motion - probability relationship which can be used for engineering designs. The traditional approach is to generate a magnitude - probability function and event location function and use these with an attenuation law to generate the ground motion distribution function. From an engineering viewpoint the magnitudes, location and attenuation relationship are not in themselves of interest. They are generated as a stage in the calculation.

5. Using historical data as a starting point, this traditional approach converts from intensity via magnitudes, location and attenuation to ground motion which is intensity. It is therefore more logical to determine the intensity (ground motion) statistics appropriate to a hazard assessment directly from observed intensities. Such a direct approach is independent of intensity-magnitudes/focal depth relationships and attenuation laws which are unknown for U.K. The appropriate laws for U.K. are implicit in the direct calculation.

6. In this paper therefore, a ground-motion probability distribution is derived directly from macroseismic data without involving location estimates or assuming attenuation laws. An intensity-probability relationship is formulated for any point in the U.K. Subsequent sections translate this relation to acceleration-probability curves using empiracal intensity-acceleration data. Detail regionalisation is not attempted since the data is considered too sparse.

THEORY

7. Macroseismic data consists of a number of point estimates of intensity, I. Traditionally for a particular event these estimates are separated into areas and

isoseismals drawn at one unit of intensity difference. The drawing of isoseismals effectively smooths over the data which can show a wide variation (about 1 unit of intensity) within each area. The positioning of the isoseismal is usually done by eye although a statistical approach taking proper account of the scatter in the data is to be preferred. In U.K. data is often sparse and therefore estimates by eye is all that can be achieved.

8. Thus for a given event this approach yields a series of isoseismals each containing an area, A(I), over which the intensity equals or exceeds the assigned intensity I of the isoseismal.

9. From such a set of data for all events over a period of time T, a mean cumulative frequency, $F_c(I)$, per unit area of U.K., per annum may be calculated for intensity from the equation

$$F_c(I) = \sum_T A(I)/A_oT \qquad (1)$$

where A_o = total area under consideration and $\sum_T$ indicates a summation over all events over the period T.

10. This cumulative frequency is equivalent in intensity terms to the frequency normally assigned to magnitude. The frequency of obtaining an intensity greater than or equal to I over an area A is then $A.F_c(I)$.

THE DATA

11. The macroseismic data used in this study is that collected by Principia Mechanica Ltd. (1982) and by Soil Mechanics Ltd. (1982). The former (PM) is more extensive, being a fairly complete catalogue at least since 1850 and probably since 1800. SM treated selected events only but their selection was designed to be a representative sample. SM data is therefore used only to check PM's work. In the following PM is used exclusively.

12. Both compilations include some location of isoseismals for the majority of the larger events. In a relatively few cases the areas within the isoseismals were quoted. Where quoted these were checked and when not quoted they were calculated. In cases where isoseismals were not quoted or sketchy (which since 1800 is usually only for small events) these were estimated. In many cases data was strictly speaking inadequate to draw precise isoseismals, nevertheless estimates were made to ensure a complete data set and therefore avoid underestimating frequency. Any doubtful situations were interpreted on the generous side to ensure that frequencies may be overestimates (conservative) but never under-estimates.

13. The areas calculated and used in equation (1) were for U.K. mainland only. Thus A(I) was specific to mainland and

ignored areas within isoseismal covered by sea. A_o was selected accordingly. This procedure avoided the incomplete data set which must occur if sea is included due to the fact that only large events are historicals reported in these areas.

14. PM noted the empiracally determined highest intensity, SM calculated the macrocentral intensities. Generally these were equal to the appropriate intensity range of the smallest isoseismal area. Where this was not the case an appropriate sub-area was estimated to reflect the fact that a small higher intensity area must exist centrally but may not be easily identified.

15. These and other corrections to the data ensured that as true a frequency as possible can be calculated. Frequencies may be overestimated but it is unlikely by virtue of these corrections that they are underestimates.

16. Intensities used here are Modified Mercalli. Although SM quote MSK this is considered as equivalent to MM for the purpose of this study. SM data was only used to check the PM data.

PROBABILITY INTENSITY DISTRIBUTIONS

17. Frequencies generated via equation 1 for various observed intensities are plotted as non-cumulative frequencies, F, in Fig. 1. Up to I = 6 there are several events contributing to the calculations. For I = 7 and 8 only one event makes a contribution. Thus as normal with hazard calculation, the higher intensity events are rare so that data is sparse at the high more damaging levels of ground motion. It should be noted that sampling up to I = 6 is relatively good and therefore frequencies may be expected to be well defined between 3 (minimum considered) and 6. As discussed later this has an important bearing on the accuracy of this hazard estimate.

18. Fig. 1 shows clearly a strong decrease in frequency to higher intensities which needs some discussion.

19. For any particular event there is a maximum intensity, I_m. Thus the isoseismal area - intensity curve for any event must be truncated at higher intensities. Generally plotting these curves for the larger events which are the only events for which we have data, shows that these curves are concave downwards tending towards zero area at higher intensity. This is consistent with these curves tending to a macrocentral intensity for any given event.

20. When we sum these curves over all events, it is not surprising that the probability-intensity curves showns a similar concave form. (Fig. 1 plots against intensity squared to linearise the relation). We cannot however, immediately assume from this that they tend to a limiting intensity value. Rather if we had an unlimited I_m range up

to the max of the scale with higher values of I_m having the smaller probability the curve would be asymptotic to zero probability as I_m becomes large. The question which therefore arises is can we distinguish between such a curve and a curve asymptotic to a limiting I.

21. For this purpose a Gumbel extreme analysis is made of the data set (Gumbel, 1956), using five year periods. The minimum Imax intensity in any one five year period appears to be about 5.

22. Fig. 2 plots intensity against the appropriate Gumbel statistic. If the distribution tended to a maximum intensity then fig. 2 should flatten off to be asymptotic to that value. Fig. 2 clearly does not; the graph fits a linear relation rather well suggesting that no maximum intensity exists or more precisely that intensity is unbounded to higher intensities.

GEOLOGICAL APPLICABILITY

23. No account is taken of the variable near surface geologies over which intensities have been estimated. The analysis of the previous section averages over several geologies and therefore might be expected to give intensities which are relevant to a 'mean' geology. This is not however a true mean.

INTERPRETATION OF INTENSITY

24. We have now reduced the problem of assessing the hazard at a site in U.K. to the interpretation of Intensity as an appropriate parameter for design input. In this section we review the relation between intensity and peak ground acceleration.

25. Trifunac and Brady (1975) and Murphy and O'Brien (1977) review attempts to relate peak acceleration, a, to intensity, I. Considering all authors there is a broad band of equivalence rather than any generally acceptable relation. Nevertheless the trend of this band follows a relation of the form

$$\log a = \alpha + \beta I \qquad (2)$$

where α and β are constants.

26. Trifunac and Brady use the Californian set of strong motion records to develop their own acceleration-intensity relations of a similar form. They make two major points concerning this data. The first is that the data is inadequate below intensity 4 and above intensity 10. The second which is probably more important is that there is a wide range of values for any particular intensity, throwing into doubt the existence of a one-to-one relation between intensity and ground motion parameters. Their point is that

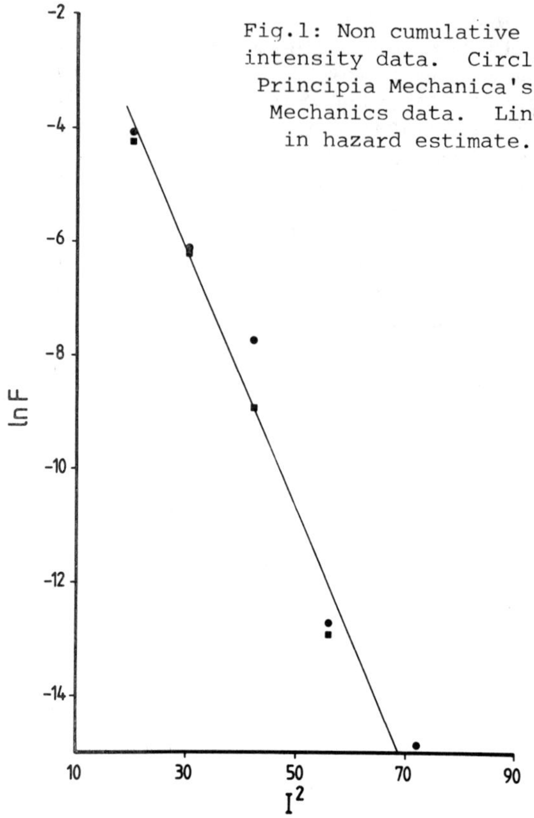

Fig.1: Non cumulative frequency - intensity data. Circles based on Principia Mechanica's data; Squares Soil Mechanics data. Line is relation used in hazard estimate.

Fig.2: Intensity plotted against Gumbel's probability function.

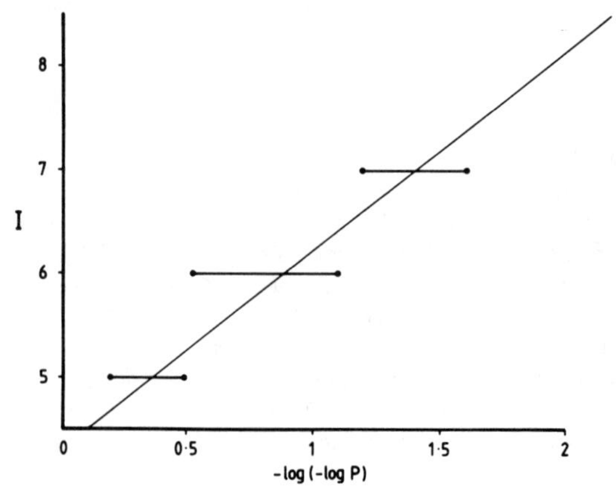

the larger the data set the wider the range. This point is taken up by Caputo (1980) who demonstrates that peak acceleration can assume a wide range of values for a given intensity. However, Caputo's data does have an apparent upper limit or envelope which has the same general form as equation 2, but lacks a lower limit.

27. Various workers have correlated intensity with other earthquake and site parameters.

28. Trifunac and Brady show little dependence on site geology. Dependence of the relationship on epicentral distance may be expected on the basis of a change in dominant frequency with distance. In this study we restrict the distance range to the near field where we might expect the larger intensities to occur. In the far field these equivalences will give an over-estimation of ground motion.

29. Having established that there is not a one to one relation between intensity and peak accelation it is necessary to treat the relation as a probability distribution. Trifunac and Brady present plots of these probability distributions for I = 5, 6 and 7. These distributions are clearly positively skewed with a tail to higher accelerations. Murphy and O'Brien suggest that they are log-normal distribution. The distributions for vertical and horizontal acceleration are plotted for I = 6 and 7 on log-probability scale in fig. 3. They form approximate linear relationships when plotted on log-probability scale which confirms their approximation to a log-normal distribution.

ACCELERATION - PROBABILITY ESTIMATES

30. This acceleration distribution P(I) is combined with the intensity-frequency distribution (equation 2) to give the cumulative frequency distribution of $f_c(a')$ from

$$f_c(a') = \int_{a'}^{\infty} \int_{I} F(I)\ P(a',I)\ dI\ da \qquad (3)$$

31. Values defining the log-normal distributions P(I) are taken from Trifunac and Brady. Fig. 4 plots cumulative frequency, $f_c(a)$, for various acceleration levels.

32. The integration windows are set to provide a calculation accuracy of about 1%. At this accuracy the frequencies for acceleration to up to 300 cm s^{-1} (about 0.3 g) can be determined from data within the intensity range 3 to 6, which is well defined both from U.K. field data and Trifunac and Brady's intensity-acceleration.

33. The intensity-frequency curve is well defined in this intensity range, only at levels I = 7 and 8 is the data dependent on a very small sample.

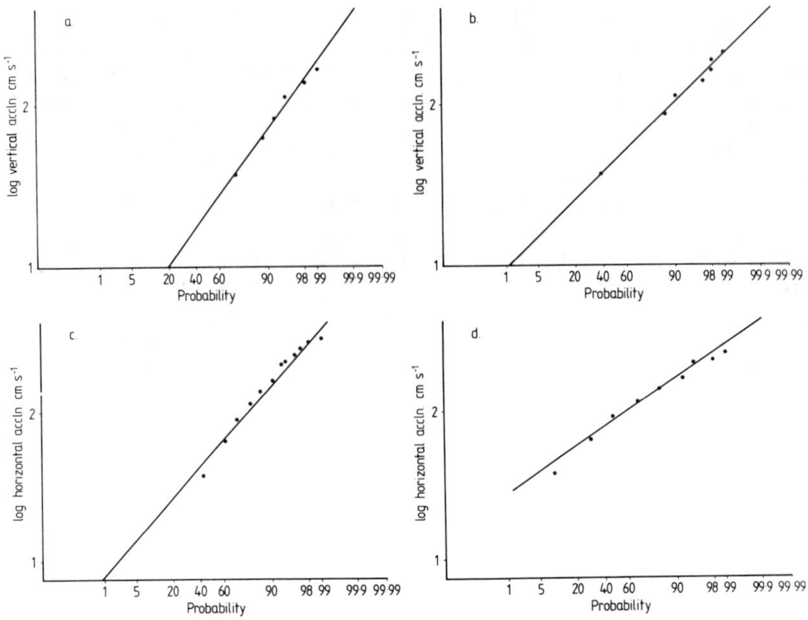

Fig.3: Peak Acceleration - probability distributions for various intensities (a and c, I=6, b and d, I=7)

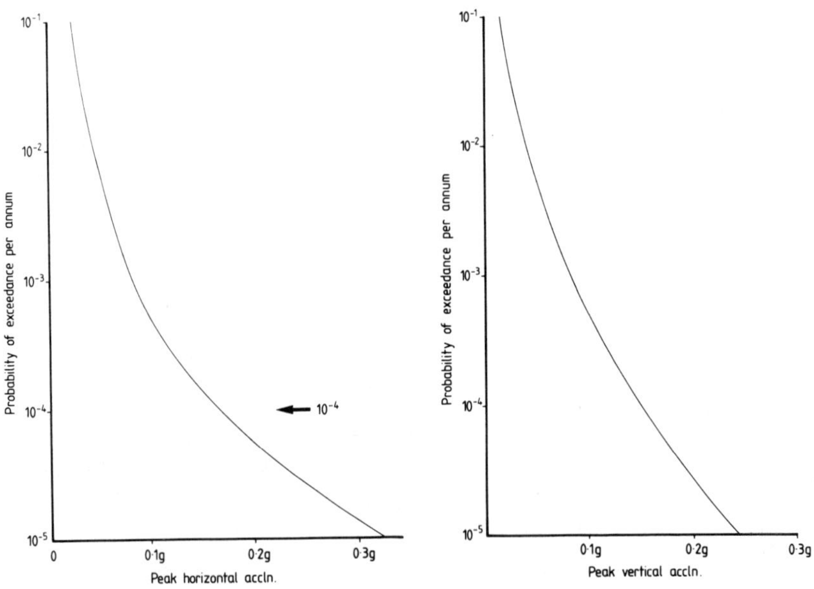

Fig.4: Mean Acceleration - probability curves for U.K.

34. Uncertainties in our results are therefore dominated by the uncertainties in the acceleration-intensity relations over the intensity range 3 to 6, with higher acceleration probabilities being controlled predominantly by the higher (better defined) end of this intensity range. The uncertainties are included in the range of acceleration assigned to each intensity and are therefore implicitly included in the calculation. Thus we can regard the probability levels given in fig. 4 as inclusive of uncertainty and therefore representing an envelope of the data at a particular set of probability levels.

35. In the whole of this paper it has been assumed that Trifunac and Brady's data is applicable to U.K. Certainly it is the most comprehensively presented set of data and is the only published analysis which can be included in a hazard assessment of the type reported here. Their analysis was extended by Murphy and O'Brien (1977) who also investigated regional variation in the relationship between intensity and acceleration. Their conclusion was that for the limited data set then available, there was a significant higher acceleration for a given intensity for Southern Europe as compared with Japan or the Western United States. Hence we may conclude that our acceleration-probability curve could be in error by a factor of two in acceleration. This would give acceleration levels for a given intensity which are probably unrealistic. The Trifunac and Brady data is therefore preferred.

36. No attempt has been made to include intensity-area uncertainities.

COMPARISON WITH PREVIOUS WORKERS

37. Fig. 5 compares the horizontal and vertical acceleration probabilities calculated here by the direct method (curve 4) with those calculated by Irving (1982) using the the traditional method of assigning magnitudes to the historic events and using various attenuation relationships. His estimates are also based on PM data. Curves 4 of these figures agree with the various determinations at the 10^{-2} to 10^{-4} probability levels but there is no exact correspondence. The shapes of the several curves are in agreement. However, Irving's curves assume a fixed ratio of vertical to horizontal acceleration whereas the present study uses independent data.

38. The magnitudes and attenuation relationships used by Irving are uncertain. The attenuation relationship for U.K. is unknown, Irving uses data from the U.S.A. and Europe. Focal depth estimates are also poor. The direct determination uses only the intensity-acceleration relationships which, although determined for U.S.A. would be expected to be useful worldwide since they are effectively point determination by different parameters of the same

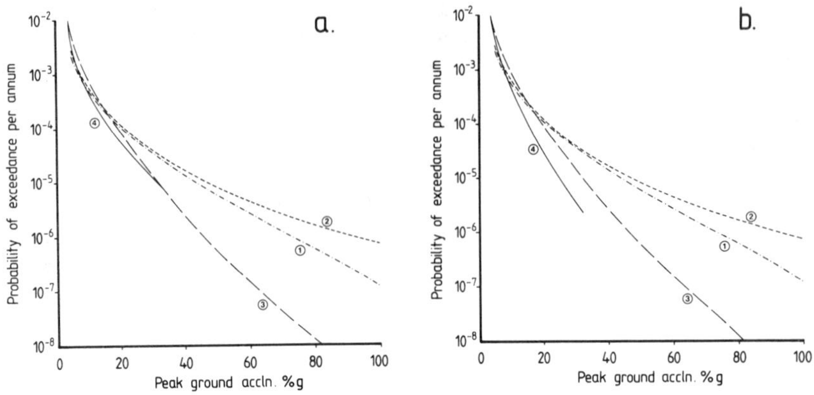

Fig.5: Comparison of acceleration - probability curves from this study (Curve 4 a) horizontal acceleration and b) vertical acceleration) with those of Irving 1982 using attenuation relations from Principia Mechanica, 1982 (Curve 1) Chiaruttini and Siro, 1981 (Curve 2) and Donovan, 1974 (Curve 3).

ground motion. The uncertainties in the calculation of Curve 4 are less than those used in deriving curves 1 to 3 of fig. 5. However, as noted in section 4 some doubt exists relating to the work of Murphy and O'Brien (1977) which suggests a variation worldwide in the Intensity, acceleration relationship. Consequently, curve 4 is justification for agreeing with Irving's conclusions, at least in the 10^{-2} to 10^{-4} range. A similar argument applies at lower probability levels.

ACKNOWLEDGEMENTS
This work was funded by the Nuclear Installation Inspectorate of the Health and Safety Executive. The opinions expressed in this paper are not necessarily those of the N.I.I.

REFERENCES

1. CAPUTO, M. Are there one to one relationships between magnitude, moment, intensity and ground acceleration. Geophys. J. Roy Astr. Soc., 1983, vol. 72, 83-92.
2. CHIARRUTTINI, C. AND SIRO, L. The correlation of peak ground horizontal acceleration with magnitude, distance and seismic intensity for Friuli and Ancona, Italy and the Alpide Belt. Bull. Seism. Soc. Am. 1981, vol. 71, 1993-2009.
3. DONOVAN, N.C. A statistical evaluation of strong motion data. Proc. 5th World Conf. on Earthquake Engrg., Rome, 1974.

4. IRVING, J. Earthquake Hazards. Sizewell Inquiry Report CEGB/S/400 (Saf/Env), 1982.
5. LILWALL. R. Seismicity and Seismic Hazard in Britian. I.G.S. Seismological Bulletin, No. 4, pp.9. 1976.
6. MURPHY, J.R. and O'BRIEN, L.J. The Correlation of peak ground acceleration amplitude with seismic intensity and other physical parameters. Bull. seism. Soc. Am., 1977, vol. 67, 877-915.
7. PRINCIPIA MECHANICA LTD. British Earthquakes. Report No. 115/82. 1982
8. SOIL MECHANICS LTD. Re-assessment of U.K. seismicity data. Report no. 7984. 1982.
9. TRIFUNAC, M.D., AND BRADY, A.G. On the correlation of seismic intensity scales with peaks of recorded strong ground motion. Bull seism. Soc. Am., 1975, vol. 65, 139-162.

Discussion

MR K. G. SMITH, Concrete Stuctures, Imperial College, London
I would like to offer some comments and questions on Dr Long's paper. The keynote address by Mr Key emphasized the need for effective liaison between the engineering and seismological professions. This is a point which has been treated in the literature by Housner and Jennings[10] and is not unfamiliar to engineers working outside the seismic field who deal with other building professionals[11]. As a civil engineer I am grateful that Dr Long is tackling the sparse data problem rationally and deriving mean (or to me the current structural engineering adjective characteristic) acceleration curves which will always be of use for preliminary designs even if one day there will be more instrumental data and reliable attenuation laws. In tackling this problem Dr Long is performing a useful service to engineers.

However, engineers should be expected to be interested in 'magnitudes, location and attenuation relationship' (paragraph 4). The reason is that engineers are concerned with the consistency of the accuracy of the various mathematical sub-models which service the design process. The seismological model is vital as the starting point for loading calculations and we need to avoid a 'garbage in garbage out' (GIGA) situation guarded against in computing science by not feeding low quality seismological signals into a relatively sophisticated structural analysis package without all parties being aware of what is happening. In the end the structure's economy and safety is the engineer's sole responsibility (unless his client has relieved him from this obligation).

Could Dr Long extend his analysis to include the derivation of characteristic acceleration amplitudes and rise times for those dominant peaks, typically found in accelerograms, which carry most of the potentially damaging energy. If this could be done then by borrowing the entropy maximization technique used in other branches of engineering where there is a problem of sparse infomation[12] and using a range of these characteristic peaks it would be possible for the engineer to synthesize an accelerogram which by trying different combinations and phases of peaks could be the worst conceivable one for a time history analysis of that particular structure. (The technique is not unlike finding the work pattern of wheel loads on a bridge [13] where here the wheel load is replaced by the characteristic spike on the accelerogram which has a characteristic range of amplitudes and rise times provided by an extension of his analysis).

I believe that this post processing of sparse seismological data is the most rational approach and could be attempted economically for certain classes of structure. How does Dr Long see the seismological sub-model developing in the next five years?

REFERENCES

10. HOUSNER G. W. and JENNINGS P. G. **Earthquake design criteria**, Monograph, 4, Earthquake Engineering Research Institute, Berkeley, Calif., 1982.

11. **Divided we fall**, Junior Liaison Organization, Conference at York Sept. 1973, ICE, London.

12. WILSON A. G. Use of entropy maximising models in the theory of trip distribution, mode split and route split. **Jnl Transp Econ. Pol. (London School of Economics)**, 1969, 3, No. 1, Jan 108-126.

13. CUSENS A. R. and PAMA R. P. **Bridge deck analysis**, Wiley, Chichester, 1975.

DR LONG

In reply to Mr K. G. Smith's comments and his suggestion for extending the method to generate synthetic accelerograms, the method as it stands can be extended by relating intensity to parameters other than acceleration. Using Trifunac and Brady's data[9] for example velocity and displacement probabilities can be provided in addition to acceleration probabilities described in the paper. Data does not exist exactly for what Mr Smith requires, but data does exist for relating envelope spectra to intensity. This data would begin to approach the problem of generating synthetic accelograms for the UK. However, the usefulness of such data is open to question. The data available to relate intensities to any of these parameters is derived mainly for the western USA. Its applicability to the UK remains uncertain. In the case of the acceleration study presented in the paper some overall justification is to be gained by the general agreement of acceleration probability relationship obtained by the direct or indirect method. As more complex analysis is approached the confidence that may be put on the data is reduced. Certainly I would support Mr Smith in worrying that engineers must fully appreciate the limitations on the data seismologists may give him. There is a danger here in squeezing the data too far.

As to Mr Smith's last point - the request for a five-year projection - it is unlikely that the data set for larger events (those significant for engineers) will have improved signifcantly. The occurrence rate of such events is only one or two per year. Thus it is important that as much data as possible be collected for each event. This requires a seismic network over the UK with adequate funding to allow a fair concentration of stations covering the whole UK. The data build will be slow but collection of such data is essential to the development of realistic seismic models.

More rapid returns may come from research into the neotectonics of the UK and its environment. Identification of active faulting and an understanding of the tectonic mechanisms causing the deformations responsible for UK earthquakes can provide the basis for more complete models of the intraplate seismicity which characterizes the UK. A programme on these lines was initiated two years ago and has produced its first papers.

14. Seismic ground motions in Great Britain

G. WOO, MA. PhD, Principia Mechanica Ltd

SYNOPSIS. Principia has carried out a comprehensive study of the full historical record of British earthquakes. The importance of this for the determination of seismic ground motions in Britain is demonstrated.

INTRODUCTION

1. In plate-margin regions of the world such as California where the recurrence interval between significant earthquakes is sufficiently low as to justify an extensive programme of strong-motion instrumentation, the task of specifying seismic characteristics for local engineering design can be exclusively based on data locally recorded.

2. In intraplate regions such as Great Britain, where the recurrence interval between major earthquakes is long and the strong-motion instrumentation programme just under way, seismic ground-motion specification is a difficult task, being dependent on resources other than local strong-motion data.

3. In the absence of adequate British strong-motion records, the onus rests on whatever other information sources on British seismicity are available to illuminate the principal characteristics of British earthquake motions. Such alternative sources include the full documented history of British earthquakes. In this paper, the application and value of such information to the issues of engineering design criteria in Britain are discussed.

THE LESSONS OF HISTORY

4. Prior to the turn of the century, information about British earthquakes consisted of a collection of assorted historical documents dating back to the first millenium, mentioning in varying degrees of detail and style the occurrence of locally felt earthquakes. Because of the comparatively low regional seismicity level, this information will for years to come be a basic resource of knowledge of the felt effects of significant British earthquakes, and hence of their ground motion characteristics.

5. The intrinsic value of the British earthquake archive for the characterization of ground motions depends on the extent to which the surface imprint of an earthquake on Man and his natural and artificial environment is sufficient to define it for

engineering purposes. For an academic seismologist or geophys--icist, high instrumental precision in the determination of seismological parameters is a prerequisite for mathematical seismic wave modelling; but for a practising earthquake engin--eer, the urgency of current seismic design requirements dic--tates the use of all available sources of information, even if not instrumentally based.

6. In the past, any attempt at employing macroseismic data to infer characteristics of ground motion has been scuttled by the absence of a reliable catalogue of British earthquakes. The problems of non-uniformity of data, reliance on secondary sources of information, factual errors and omissions have diminished the status of historical archive material to a level of a minority interest.

7. This situation has now changed with the construction of a completely new catalogue by Principia (ref. 1). This compre--hensive catalogue is based on a solid foundation of 8000 sheets of primary documented information on British earthquakes, obtained from diaries, personal letters, newspapers, books and journals written in the past thousand years. For each event in this catalogue, intensity and isoseismal maps have been drawn according to a standard systematic procedure, so as to provide a framework for quantitative analysis. The complete set of these isoseismal maps stands as a scientifc reference on felt effects of British earthquakes, which will be augmented by maps of future earthquakes occurring in this region.

8. Given this new macroseismic database, a fresh attempt has been made to determine important earthquake parameters from isoseismal felt-areas. The most critical parameter for which a functional dependence on felt-areas has been sought is that of earthquake magnitude. The stark mathematical definition of magnitude in terms of a seismometer displacement or velocity amplitude at a given period scaled according to epicentral dis--tance, seems far removed from the definition of intensity III involving qualitative descriptions of rocking chairs and birds falling from perches.

9. However, both magnitude and felt-area are measures of seism--ic wave energy in some form, so a correlation between the two can be anticipated. The actual usefulness of such a correla--tion of course depends on its quality. In regions such as California where the attenuation of seismic waves is severe, the surface area affected by an earthquake increases only mar--ginally with the magnitude of the event. Thus the felt radius of the 1906 San Francisco earthquake was only about 300 miles. For regions like this, knowledge of earthquake felt-areas does not serve to constrain their magnitudes very well.

10. In contrast, in regions of low attenuation such as Great Britain, the felt-area of an earthquake increases rapidly with magnitude, enabling a stable and quite precise assignment of magnitude to be made in the absence of instrumental data and even information on peak or epicentral intensity. This latter property of this method of magnitude estimation is particularly useful in cases where the epicentral region is offshore.

11. Thus far no qualification of earthquake magnitude has yet been made. In principle, correlations between felt-area and any definition of magnitude are possible, but empirically the best correlation appears to be between felt-area and surface wave magnitude, because of the dominant contribution of long period seismic waves to far-field observations of felt effects.
12. From twentieth century earthquakes in the region for which both felt-areas and surface wave magnitude values are known, the correlation derived is expressed succintly by the following two logarithmic relations:

$$M_S = -0.42 + 1.00 \log A_{III} \qquad (1)$$

$$M_S = 0.90 + 0.81 \log A_{IV} \qquad (2)$$

13. By means of this correlation, M_S assignments can be made for an earthquake of any age for which either A_{III} or A_{IV} is calculable. If both are known, then a consistency check on the magnitude values is then possible. Through the logarithmic stability of the correlation, this consistency check is satis- -fied for all British historical events.
14. With events being assigned M_S values according to this correlation, the statistics of magnitude-frequency occurrence are naturally expressed in these units, and for consistency of seismic hazard calculations, so too are the relations describ- -ing the attenuation of ground motion. The use of surface wave magnitude as a size scaling parameter for the parametrization of earthquake attenuation in Britain is thus enforced as a constraint imposed by the process of size determination itself. In early studies of seismic hazard in Great Britain, care in the definition of parameters such as magnitude was forsaken for expediency in producing engineering design values. Either no effort was made to discover the source of the magnitude data used, or else different types of magnitude were chosen for the different parts of hazard analysis, leaving the difficulties thus arising to be bridged by some form of linking equation between the magnitude definitions.
15. Such a linking equation between M_S and M_L or m_B might be applicable if it were stable; however there is in reality a large degree of variability associated with it, rendering its practical usage in seismic hazard analysis highly dubious and uncertain.
16. With problems like this being circumvented by the choice of surface wave magnitude throughout the computation of seismic hazard, the detailed parametrization of ground motion attenua- -tion becomes an issue of central importance. As with the catalogue of British earthquakes, information on ground motion attenuation in this region is provided in a descriptive non- -instrumental form by the documentation of felt effects.

ATTENUATION

17. The British historical record of earthquakes has been processed by Principia into a set of isoseismal maps whose contours constitute the basis for intensity attenuation analysis. For each map, the area A within each isoseismal contour has been measured, and an average epicentral distance calculated through geometrical smoothing. A hypocentral distance has then been calculated from the epicentral distance by allowing for the focal depth of the event. The hypocentral distance R is used as a regression parameter, together with M_S, the surface wave magnitude of the event. This latter parameter is preferred to I_0, the peak or epicentral intensity, because this is usually poorly estimated for earthquakes located offshore or in areas of low population density.

18. A multiple linear regression analysis of the intensity data from events in Britain and surrounding waters, yields the relation:

$$I = 1.36 + 1.80\,M_S - 1.18\ln R - 0.0019\,R \qquad (3)$$

The small absorption coefficient 0.0019 is indicative of the low attenuation characteristics of the region; a feature common to many intraplate zones, where the crust is less fractured than in the more seismic interplate ones.

19. Besides providing a general understanding of intensity-distance attenuation, the archive of macroseismic information provides the opportunity for studies of anisotropy and regional difference. These are particularly relevant to the assessment of ground motion generated by active faults close to a site of engineering interest.

20. Of course, the parameter of greatest significance for the seismic design of engineering facilities is not macroseismic intensity but ground acceleration. At present there is very little suitable strong-motion data upon which a direct evaluation of acceleration attenuation might be based, and recourse has to be made to surrogate sources of supply, such as Southern Europe. The alternative is to derive an acceleration attenuation relation through a correlation between intensity and acceleration, perhaps allowing for a magnitude and distance dependence (ref. 2).

21. Uncertainty though there may be over acceleration attenuation in Great Britain, it is encouraging that it is reduced with the occurrence of every instrumented local earthquake. So it was with the North Wales events of the Summer of 1984, which triggered strong-motion instruments and array seismometers alike in the near and far-field regions of the source. The level of ground tremors observed during the main shock was fairly modest, as would be expected of an event of around 20kms in depth, and $4.5M_S$ in size, and acceleration values seem to be in accord with attenuation relations scaled according to surface wave magnitude.

SUMMARY

22. The full historical record of earthquakes in Great Britain is of immense value in establishing reference points for inferences on local seismic ground motions. Information from this record on the magnitudes, depths, durations and intensities of past earthquakes serves to constrain the type of ground shaking experienced in this region, and therefore to delimit the bounds of typical vibration levels, and to refine engineering design criteria for British earthquakes.

REFERENCES

1. PRINCIPIA MECHANICA. British earthquakes. Report for CEGB, BNFL and SSEB. 1982.

2. McGUIRE R.K. Ground motion estimation in regions with few data. Proc. 8th World Conference on Earthquake Engineering, San Francisco. 1984.

Discussion

DR R. KUNAR, SDRC

In Dr Woo's presentation he showed a comparison of the UK design spectrum and a recent one from a recorded earthquake in the USA in a tectonic environment similar to the UK. The recorded earthquake shows a very large single acceleration spike and I feel that this is not very representative and if this spike is removed from the history, it is possible that the spectrum from the recorded history will be higher than the UK spectrum in some parts.

DR SKIPP

Would Dr Woo agree that having regard to the increasing amount of broad band near-field strong motion data from arrays, the ground motions and spectra formulated 5 years ago for the UK need revision?

DR WOO

In reply to R. Kunar, seismic hazard to engineering facilities in Britain is dominated by the contribution from near-field events of magnitude in the range 4.5 to 6.0 Ms. It is appropriate therefore that ground motion specification for this region should be based on strong motion records obtained near the epicentre of moderate earthquakes. Such records typically display high frequency spikes of acceleration, generated by compact seismic source zones, or asperities. As an example of this, the recently recorded New Hampshire earthquake of 1982 underlines the irregularity of near-field motion, with the level of disturbance being exceptional only for the occasional high acceleration peaks.

Answering Dr Skipp's point, in any science, progress in

understanding and modelling natural phenomena advances most rapidly with the accumulation of new data. In the field of earthquake strong motion, reliance on actual records is particularly great because of the complexity of the physics of earthquake faulting and wave propagation.

With ever-increasing deployment of accelerographs world-wide, and improvements in the quality of earthquake records from modern instruments and arrays, there is continuous need to re-evaluate seismic ground motion criteria for engineering plant. Through a programme of data analysis combined with theoretical stochastic ground motion modelling, Principia is committed to updating design criteria to the current state-of-the-art.

15. Colchester revisited

B. O. SKIPP, PhD, FICE, FGS, D. T. SHILSTON, MSc, MIGeol, FGS,
J. C. GUTMANIS, MSc, MIGeol, FGS, Soil Mechanics Ltd, and
R. D. ADAMS, PhD, International Seismological Centre

SYNOPSIS. This paper reviews the intensity assigned to the 1884 Colchester earthquake in the light of current procedures. It compares different sources of historical information and the performance of the different kinds of buildings in the epicentral area and considers the significance of local topography and soil conditions.

INTRODUCTION

1. On 22 April 1884 at 9.20 am the historic market town of Colchester, Essex together with all of eastern and central England was shaken by an earthquake with an assessed maximum intensity of VIII (MSK scale), (Fig. 1). It caused much alarm and sufficient damage to warrant setting up a relief fund although there was no loss of life. The earthquake stimulated a number of contemporary technical reports, notably the work of Meldola and White (ref. 1), and recently a popular account (Haining, ref. 2) and a technical appraisal (Scott, ref. 3).

2. The work upon which this paper is based involved the examination of reports in 70 contemporary newspapers and 51 technical references from which felt effects have been assessed for 249 and 357 localities respectively. Field visits were made to the epicentral area to examine building styles and individual buildings for which descriptions of the damage are available.

THE SOCIAL AND ECONOMIC SETTING

3. Macroseismic intensity scales in current use, for example the Medvedev-Sponheuer-Karnik (MSK) and Modified Mercalli (MM) scales, take into account differing building types but to employ these scales successfully it is also necessary to assess the character of the housing stock and its likely condition, that is its vulnerability. Some knowledge of the contemporary social and economic conditions is therefore required and this is especially so for the 1884 Colchester Earthquake.

4. Colchester in 1884 was at a turning point in its development. It had in the 18th Century and early 19th Century been a notable centre of the cloth trade but

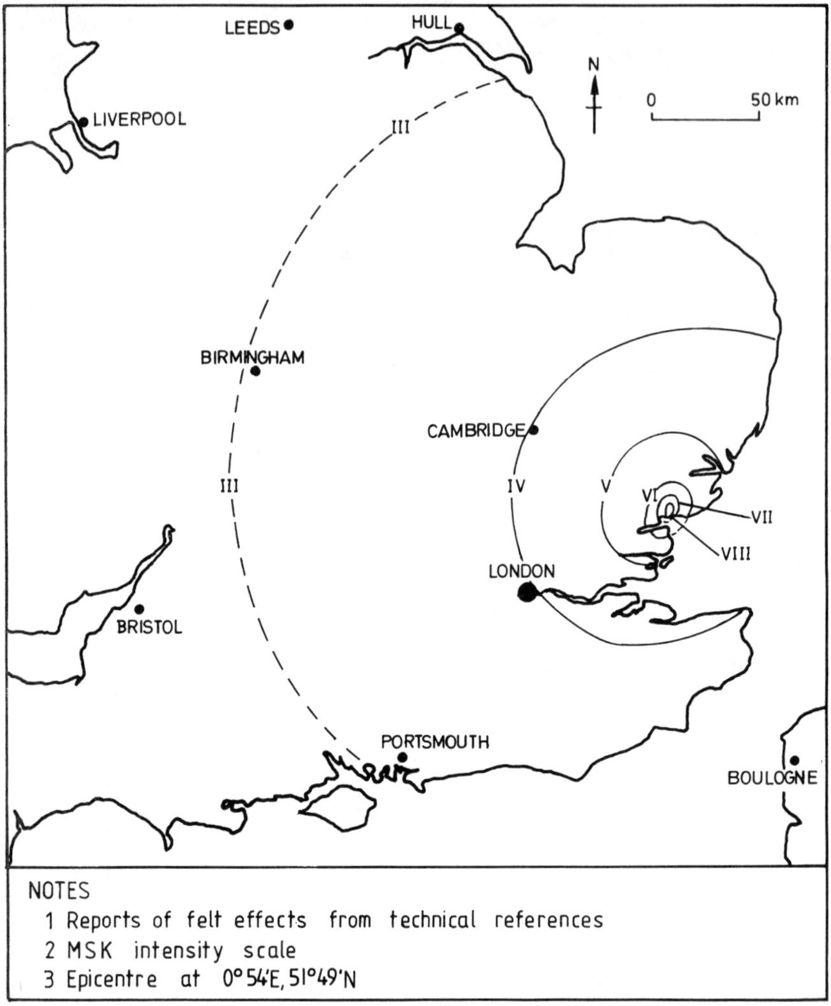

Fig. 1. Colchester earthquake 1884: Isoseismal map

this was in decline by the middle of the century. In the first quarter of the 19th Century the garrison provided an important economic stimulus but this was also much reduced during the long peace after the Napoleonic wars. Thus, in the latter half of the 19th Century Colchester became a rather sleepy country town. So much so that despite the advent of the railway in the 1860's the great preacher C H Spurgeon on hearing news of the earthquake remarked, "Well I am glad to hear that something has moved the old town at last".

5. The surrounding countryside was also in a state of economic decline. The agricultural depression arrived in

1875 and the rural population of north Essex was falling and had less money to spend (ref. 4). This slump also affected the fortunes of Colchester, the market town. The earthquake arrived after a long period of economic decline for the entire district and its impact was hard felt, particularly on the self employed artisans whose homes and means of production were damaged.

SECULAR BUILDING TYPES

6. The types of secular building present within the epicentral area and their state of repair in 1884 reflected the general economic malaise. In Colchester itself a substantial proportion of the housing stock originated before the 18th Century, often being timber framed with a brick skin. This was particularly so in the older parts of the town and in The Hythe, the wharves by the River Colne to the east of the town. Partly as a consequence of the arrival of the railway there had been some more recent development with the construction of brick cottages and villas on the outskirts of the 18th Century town together with the construction of barracks during and after the Crimean War (1854-6). There were in addition a few older brick workshops, some left over from the cloth trade of earlier years, a Norman Castle, the Town Hall (built in 1844) and 'Jumbo', the monumental water tower dominating the town which was completed in 1883.

7. By 1884 the effects of the agricultural depression in the surrounding countryside were marked and there were only a few modern load-bearing brick dwellings. Brick houses were often proudly announced as such (for example 'Brick Cottages') although, as always, the better off professional was equipping himself with new property. Wivenhoe, a port on the River Coln, was an exception. The railway had arrived in 1863 and brick terrace houses were built flanking the new railway cutting.

8. In the present study it was found that secular buildings could be classified as follows; a discussion of the relationship of these building types to the MSK scale categories (Table 1) is given later:

Type X Timber frame, with or without an external covering of plaster, rendering or weather boards. (MSK Type B)

Type Y Timber frame with a later veneer of brick designed to give the impression of a brick built house. The window openings are generally larger than in Type X and parapets are often present. (Approaching MSK Type A).

Type Z Brick buildings using essentially modern construction techniques. These were domestic 'villa' style houses, terraces and public buildings, such as schools. (MSK Type B).

Table 1. MSK scale interpretations used in this study.

Classification of secular building types

Type A	Buildings of field-stone, rural structures.
Type B	Brick buildings, structures of hewn stone, half timbered structures.
Type C	Well-built wooden structures.

Buildings were assumed to be of half timber (timber frame) or brick construction (ie MSK Type B). However, it was recognised that some were abnormally vulnerable owing to their state of repair or the presence of vulnerable architectural features.

Classification of building damage

Grade 1	Fine cracks in and fall of plaster. Slight cracks in chimneys. Loose bricks, tiles and architectural ornaments fall. Windows break.
Grade 2	Chimneys fall or severely cracked. Slight cracks in brick walls.
Grade 3	Large cracks in brick walls.
Grade 4	Fall of sections of brick walls.

Intensity Grades and effects most commonly reported

II	Felt only by those at rest, mainly on upper floors.
III	Duration estimated. May not be recognised as an earthquake. No rattling of loose household articles.
IV	Loose household articles rattle, furniture shakes (excluding beds). Clocks may stop. Vibration like heavy vehicle or train. Some people awake, not everyone in same household may be woken. Some people become concerned and may investigate cause.
V	Doors open, slam shut. Felt by all indoors. Loose household articles may fall. Many people awoken, many alarmed, some may run outdoors. Animals disturbed, particularly at night. Field walls may fall. Very occasional Grade 1 damage (if highly vulnerable).
VI	Fall of loose household objects is common. Everyone awoken. Many people fear collapse of house and run outdoors. Grade 1 damage to a few buildings.
VII	Most buildings sustain Grade 1 damage, many sustain Grade 2 damage and a few sustain Grade 3 damage.
VIII	Most buildings sustain Grade 2 damage, many sustain Grade 3 damage and a few sustain Grade 4 damage.

9. The sequence X, Y, Z is broadly chronological. Timber was not displaced as a construction material until long after the end of the Middle Ages and, although bricks were first used in the 13th Century, load-bearing brickwork did not become the dominant construction method for secular buildings until the 19th Century (ref. 5). Brick veneers were used to modernise old buildings during the period when brick construction was gaining in popularity and being used for buildings of increasingly lower status.

ASSESSMENT OF INTENSITY

10. The technical reports and contemporary newspaper reports were assessed separately using a slightly modified MSK scale with indicators as listed in Table 1. This enabled comparison of the styles of reporting and their influence on the maximum assessed intensity and pattern of isoseismals (Fig. 2 and 4).

11. Within the area of maximum intensity, assessed as VIII on the MSK scale, many chimneys fell or were severely cracked, often resulting in damage to roofs although some roofs were also untiled without the fall of a chimney. Some brick walls were cracked and parts of a few fell or had to be shored up. Lath and plaster walls were also cracked.

Technical references

12. The descriptions of felt effects and building damage given in the technical literature are considered to be more reliable than those in the newspapers. For this reason greater emphasis has been placed on the technical literature, especially the work of Meldola and White (ref. 1) who carried out a detailed study of the effects immediately after the earthquake. Other technical references add little to their reports of felt effects and building damage. Page numbers given below are those in Meldola and White (ref. 1).

13. The felt area of the earthquake in Great Britain is shown on Fig 1. The maximum intensity, assessed as VIII, occurred in an area south of Colchester at the villages of Peldon, Abberton, Wivenhoe, Rowhedge and East Donyland, both the latter being between Abberton and Wivenhoe (Fig. 2).

14. Damage to brick buildings within this area may be illustrated by an example from Abberton where at "a new and well built cottage .... about 20 feet of brickwork .... (was) .... thrown from the gable" (p 61); "this house .... is substantial .... some of the cracks in the walls were more than a quarter of an inch wide" (pp 61-62) and "hardly a chimney left standing" (p 59). Again, at Peldon, "a large modern strongly built house .... was shattered and rent in all directions" (p 68).

15. Strood Villa, close to the causeway linking Mersea Island to the mainland near Peldon was "one of the worst cases of destruction that came under our notice. The building, although modern (1860) and substantially built of brick was, without exaggeration wrecked .... every wall cracked and twisted, the door frames displaced .... 80 feet of the garden wall thrown

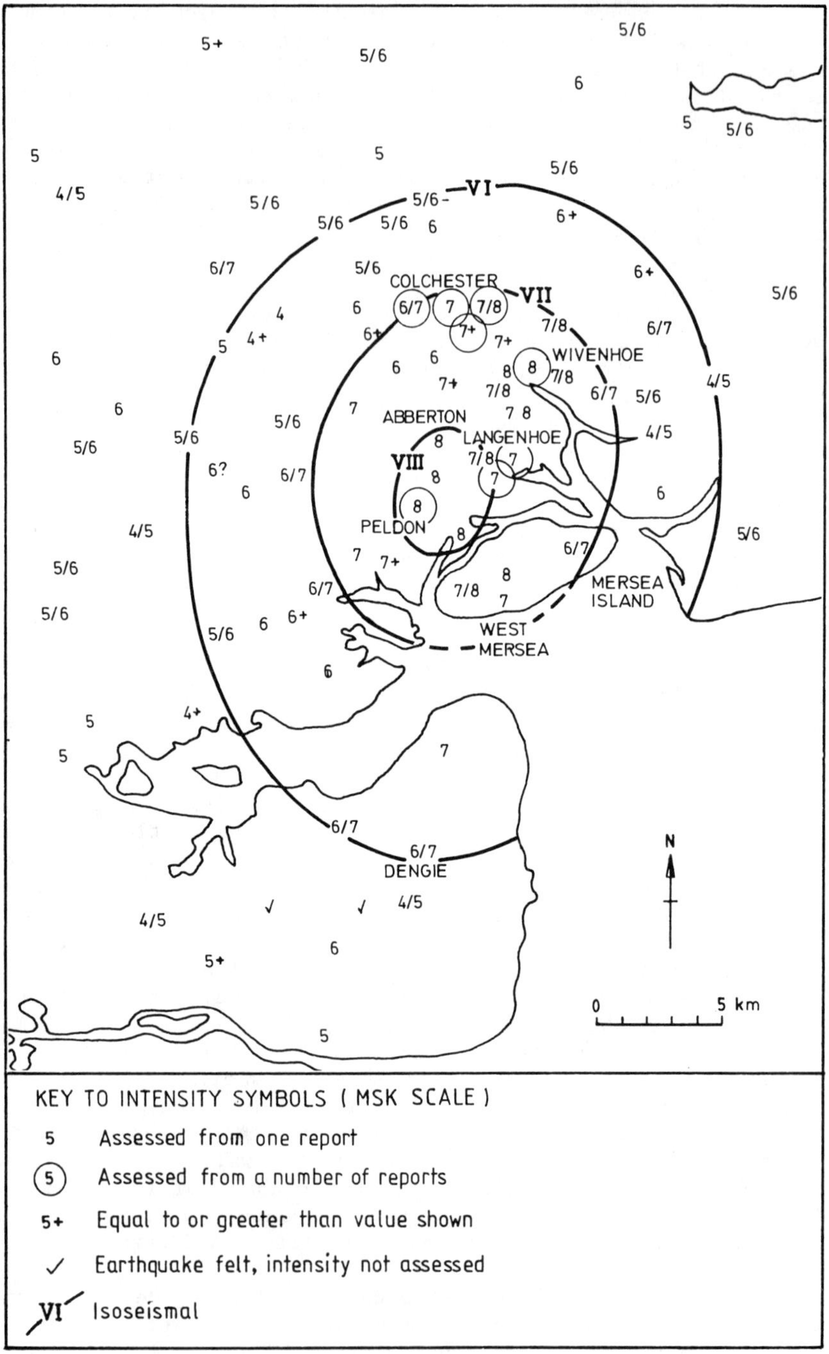

Fig. 2. Colchester earthquake, 1884: Intensity distribution assessed from technical references.

down .... the appearance of having been violently twisted round on its foundation .... another crack, nearly an inch wide in parts .... passing right through brick and mortar ...." (p 71).

16. Meldola and White (ref. 1) also provide summary details of damage for some of the most severely affected villages. At Peldon "every house and cottage sustained more or less injury, some of the buildings having been rendered temporarily uninhabitable" (p 67) .... "daylight could be seen through the walls" (p 68). At Wivenhoe "many houses along the Quay, some .... most substantially built, were practically wrecked" (p 81) and "along the river front .... 70 per cent of the chimneys were thrown down, and 50 or 60 houses in the village were unroofed" (p 81). At the neighbouring village of Rowhedge "perhaps 75 per cent (of chimneys) and one or two gables had been thrown down .... the new and good brickwork had suffered nearly as much as the old" (p 80).

17. These reports are consistent with Grade 3 damage to many and Grade 4 damage to a few buildings (Table 1). An intensity VIII is then obtained on the assumption that buildings are MSK Type B structures. This assumption, which may be questioned in the light of comments regarding the general state of the housing stock, is discussed later. Sorby (ref. 6) mapped out three damage areas in the epicentral area, the worst being where "nearly all the chimneys had been knocked down .... a large proportion of the house walls cracked and some boundary walls thrown down". His work supports that of Meldola and White (Ref. 1) and independently suggests an Intensity VIII, notwithstanding the above comments regarding the quality of the housing stock.

18. Within the epicentral area, the assessed intensity may be compared with the amount of damage sustained by means of data from the Mansion House Relief Fund (Meldola and White, ref. 1). The ratio of the number of buildings repaired to population provides an approximate measure of the proportion of buildings damaged by the earthquake at individual localities. Thus, it is an index of intensity which is independent of assessments made from observers' descriptions. This index is compared with our assessed intensities on Fig. 3 from which it is apparent that the two measures of intensity have a regular relationship. However, the town of Colchester, for which our assessed intensities range from VI/VII to VII/VIII, is anomalous and the intensity for the town as a whole may have been towards the lower end of this range.

19. Damage to churches receives considerable treatment in both technical references and newspaper reports and is discussed separately below.

Newspaper references

20. Following the same reasoning as that applied to the technical references the maximum intensity has been assessed as VIII for reports given in contemporary newspapers (Fig. 4). This occurred at the villages of Wivenhoe, Langenhoe, Abberton and Peldon in an area to the south of Colchester. The

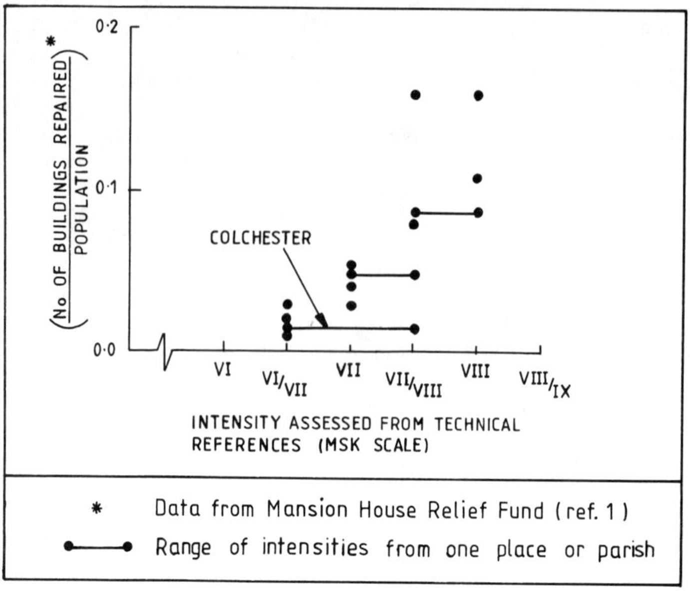

Fig. 3. Colchester earthquake, 1884: Comparison of proportion of buildings damaged and assessed intensity.

assessment is based on reports of building damage which was of Grade 3 in many cases and probably of Grade 4 in a few cases (Table 1).

COMPARISON OF INTENSITY ASSESSMENTS

21. A maximum intensity of VIII was assessed from both the technical references and contemporary newspaper reports. However, the newspaper reports tend to be exaggerated, and draw hasty, sometimes incorrect conclusions, particularly in the first editions following the earthquake. This is apparent both from a study of the newspapers in chronological sequence and from comparison with the technical references.

22. Phrases such as "at Langenhoe .... the fine old parish church was almost levelled to the ground" (Chelmsford Chronicle, 25 April) and "the Rose and Crown Inn (at Peldon) was literally wrecked" (Times, 23 April) give an exaggerated impression to the reader. Only when damage descriptions and surveys by individual observers appear in later editions is a more realistic impression gained and even then the reporting is not necessarily as critical as that in the technical literature.

23. The trend towards more balanced newspaper reporting of the earthquake becomes particularly apparent when the "Official

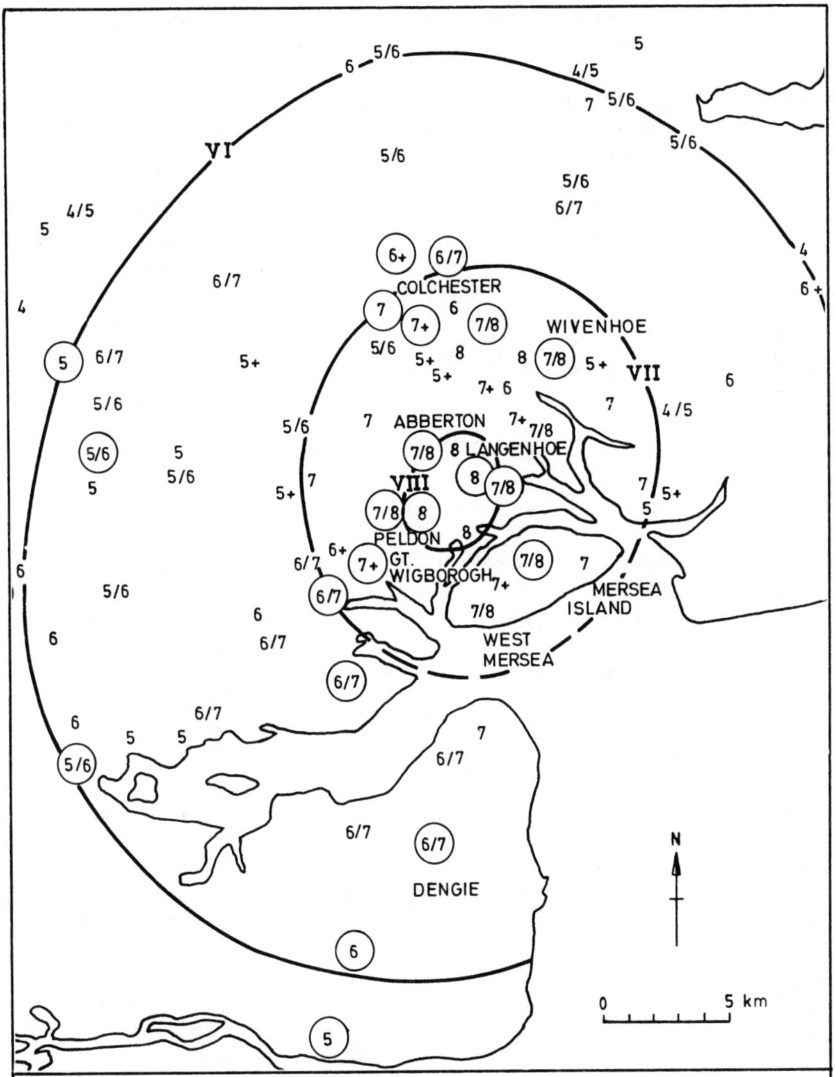

KEY TO INTENSITY SYMBOLS (MSK SCALE)

5 Assessed from one report

(5) Assessed from a number of reports

5+ Equal to or greater than value shown

✓ Earthquake felt, intensity not assessed

VI Isoseismal

Fig. 4. Colchester earthquake, 1884: Intensity distribution assessed from contemporary newspaper reports.

Report of the Damage Done" appears in many newspapers after 10 May. In this report, Langenhoe church ".... suffered very severely, all the battlements from the E side of the tower fell through the roof of the church, and from the W side were thrown to the ground .... there were several large cracks in the walls of the church, and part of the end of the Chancel was thrown outwards" (Essex Standard, 10 May).

24. A week later the Essex Standard prints a report titled "From the Engineer" which states that "the various reports which have been published in metropolitan and local journals give nothing like an accurate account of the earthquake .... the tendency all round is to exaggerate .... Langenhoe church was a small building crumbling to ruin, and we should say unfit for use ... the houses (at Wivenhoe) are ... of very poor type ... only one well built house in the place has really suffered much damage ..." (Essex Standard, 17 May). The tenor of the "Engineers" report suggests that the writer may have gone out of his way to play down the effects of the earthquake and contrast his professional viewpoint with that of the popular press.

25. The newspapers are also guilty of drawing hasty conclusions, perhaps the best example being a statement in the Times (24 April) which attributes to the earthquake "long fissures .... in the ground" at Langenhoe rectory. It is not until 10 May, in the "Official Report of the Damage Done" that "the Rector states the cracks .... mentioned by the newspapers are due to the dry weather, and had occurred before the occurrence of the earthquake" (Essex Standard, 10 May). Ground fissures reported at Great Wigborough church may also be explained in this way. Similarly, it is only some time after the event that the true circumstances of a fissure in a coastal road near St Peter's Well, West Mersea are reported in some papers. The road is reported as being "on a sort of embankment raised a few feet above the level of the shore, and hence any disturbance would more readily affect it" (East Anglian Daily Times, 29 April).

26. An extreme example of probable newspaper exaggeration is provided by the East Anglian Daily Times (23 April) which states that at Mersea Island "the sea is said to have rushed with resistless force over the marshes". This phenomena is not referred to in the technical literature.

27. Newspaper exaggeration is also confirmed by comparison with the technical reports, in particular Meldola and White (ref. 1). Although it should be noted that technical writers often use newspaper reports (or other secondary material) in their studies and are often not entirely free from the risk of exaggeration.

28. The isoseismal patterns arrived at on the basis of the two types of information are similar (Fig. 2 and 4). The isoseismals from technical references are thought to be more reliable and accurate than those derived from the newspapers and are in general closer to the epicentral area. This is particularly apparent with Isoseismals V and VI although in the

epicentral area the isoseismals from both sources are very similar. This suggests that the newspapers do indeed contain some exaggeration which is difficult to eliminate outside the area of building damage where people's descriptions of the felt effects cannot be compared with other measures of intensity.

29. Reference has been made in the Introduction to the recent book by Haining (ref. 2). This works lends colour even to the contemporary newspaper accounts and, notwithstanding his apparent access to family records and oral history, judgments of intensity based upon this book would be grossly exaggerated.

DAMAGE TO SECULAR BUILDINGS

30. The vulnerability of buildings is an important factor in relating the reported damage to the intensity of the earthquake. A study of the contemporary photographs, (particularly those by Damant, ref. 7), sketches and descriptions shows the damage suffered by each building type described above to have been distinctive and to reflect differing responses of the construction materials to the seismic shock, that is differing vulnerability. Chimneys, however, are highly vulnerable and damage to them appears to be governed more by their free-standing height, slenderness and state of repair than the character of the buildings of which they are a part.

31. Unmodified timber frame buildings (Type X) suffered the cracking and loss of plaster, and sometimes the laths to which it was fixed, together with the bulging and cracking of wattle and daub infill. The buildings themselves did not fall although they give the impression of having been very severely shaken and disrupted. No doubt some of the joints would have been damaged but their construction ensured that this did not cause collapse. The Rose Inn, Peldon, in which the walls were "cracked and bulged" (Meldola and White, ref. 1, p 68) and cottages in Wivenhoe, Abberton and Peldon, (photographs in ref. 2, and sketches in Illustrated London News, 3 May) illustrate this type of damage. Highly vulnerable bulging plaster is clearly visible in many Type X buildings at the present day.

32. Veneered timber frame buildings (Type Y) produced some of the most photogenic damage due to the poor connection between the brick veneer and the timber frame and to the slenderness of the brickwork (ref. 8). Seismic shaking caused displacement of the timber frame beyond that which the brickwork could withstand producing large cracks, often at the corners of the building, and sometimes the collapse of large areas of brickwork. Many of Damant's photographs are of damage to this type of building. Tie bars can be seen restraining the brick veneers of a number of Type Y buildings at the present day. Some of these may have been inserted as a consequence of the earthquake damage, for example at East Donyland Hall where the "front of house was left insecure and had to be shored up" (Meldola and White, ref. 1, p 80).

33. Brick built (Type Z) buildings are more enigmatic. No pictorial record has been found of a positively identified example of serious damage to this type although detailed reports concerning three such buildings indicate that they were so badly damaged that they required rebuilding or very major repairs. These are Stroud Villa, Peldon (built 1860), The Rectory, Abberton (nearing completion in 1884) and Ballast Quay Farm, Fingringhoe (built about 1870). The damage to these buildings comprised extensive cracking of the brickwork, displacement of doors and stonework and the ubiquitous chimney and roof damage. Very little evidence of the damage to these Type Z buildings can be seen today.

34. Ballast Quay Farm has the date 1887 above the east door which may perhaps be taken as confirmation that it required "virtual rebuilding" (ref. 2, p 76), although no evidence was found of the "numerous" cracks mentioned by Meldola and White (ref. 1, p 80). The Rectory at Abberton also lacks visible cracks, and indeed the present Rector told the authors that the building was "virtually pulled down and started again". Documentary evidence of the rebuilding of Strood Villa, Peldon is not available. The building has visible cracks but these are on a scale which could be explained by settlement since 1884. There is no evidence or displacement of the building on its foundations.

35. The amount of cracking which may be produced by settlement is illustrated by the New Rectory at Salcott. "Meldola and White (ref. 1, p 73) report that no injury (was) done" to this building and yet a wing in the front elevation now displays fairly extensive cracking and is restrained by tie bars. This demonstrates the extent to which the reserve strength of a building can be depleted by settlement; the New Rectory is no longer in pristine condition and is much more vulnerable than in 1884.

THE ASSESSED MAXIMUM INTENSITY

36. The maximum intensity of VIII for the Colchester earthquake is dependent upon the correct assessment of both the degree of damage sustained and the structural class of the buildings. There is no doubt that damage in the epicentral area was generally of Grade 3 together with a few cases of Grade 4. The structural class or vulnerability of the buildings, however, is more equivocal.

37. In the rural epicentral area it appears that a significant proportion of the few modern brick buildings (Type Z) were severely damaged. This provides a strong case for an epicentral intensity VIII as these buildings are certainly MSK Type B Structures (Table 1). This may not be the case for Types X and Y. These were old and possibly in a poor state of repair at the time of the earthquake and, especially in the case of Type Y, had vulnerable additions to the original structure. Some of them may have been approaching MSK Type A Structures, although they can hardly be classified as Type A which is reserved for rural possibly non-domestic structures.

Timber frame buildings, that is our Type X, are Type B Structures in the MSK scale but brick veneers, our Type Y, are not specifically mentioned. The Modified Mercalli (MM) scale used in New Zealand (ref. 9) refers to veneers, stating that "some" brick veneers are damaged at MM intensity VIII and that "brick veneers fall and expose frames" at MM intensity IX. It seems reasonable therefore to assess an intensity of at least MM intensity VIII and hence about MSK intensity VIII for the damage sustained by Type Y buildings in the epicentral area.

38. In conclusion, our assessed maximum intensity of VIII, which relies heavily on the damage to load-bearing brick buildings (Type Z), is in accord with the extensive damage to other building types, notwithstanding their enhanced vulnerability due to construction method and state of repair.

DAMAGE TO CHURCHES AND LARGE HOUSES

39. Although the assessment of intensity discussed above avoided the use of damage to churches it is worthwhile considering both churches and large older secular buildings within isoseismal VII. Such buildings tend to survive the centuries with additions and restorations and cannot be regarded as a class with uniform vulnerability (ref. 10).

40. Churches have been classified as: substantially mediaeval; mediaeval, restored in the 19th century; and 'modern', built in the 19th century. An analagous classification of older great houses, manors and large farms has been adopted, comparable to that used for smaller secular buildings, as follows: timber frame; timber frame with additions; and load-bearing brick. The damage 'profile' of each set in terms of assigned local intensity is shown in Table 2 for churches, both in the rural area and in Colchester itself, and Table 3 for the rural large houses.

Table 2. Damage to churches within isoseismal VII

CLASS	TOTAL	NO REPORTED DAMAGE	INTENSITY ASSIGNED				
			VI	VI/VII	VII	VII/VIII	VIII
Mediaeval	7(3)	2(1)	1	1(1)	2(1)	-	1
Mediaeval, restored in 19th century	12(2)	2(1)	2	2(1)	2	2	2
19th century	2(14)	1(11)	-	-	(1)	(2)	1

Note - Numbers in brackets from Colchester town.

Table 3. Damage to large houses within isoseismal VII

CLASS	TOTAL	NO REPORTED DAMAGE	INTENSITY ASSIGNED				
			VI	VI/VII	VII	VII/VIII	VIII
Timber frame	5	5	-	-	-	-	-
Timber frame with additions	11	5	-	3	1	2	-
Brick	4	3	-	-	-	1	-

41. The samples are small but there are indications of a wider spread of damage and a higher proportion of intensities of VII or more for restored or modified mediaeval than unrestored or modern buildings. More striking is the contrast between the proportion of newer churches in Colchester as compared with the rural area. Probably as a consequence there is a higher proportion of undamaged churches in the town itself, notwithstanding the course of the isoseismal VII. Intensity is however a combination of shaking severity and building vulnerability and the damage sustained. The procedures for evaluation downgrade the intensity for a given damage indicator when the buildings are judged to be more vulnerable so there is a tendency towards a uniform intensity even though a set of buildings have wide disparities of style and condition.

LOCAL GEOLOGY AND TOPOGRAPHY

42. The depth to Palaeozoic basement is conjectural in this region and the bases of the Chalk and London Clay are deduced with uncertainty of local detail from quite widely spaced wells. Near surface soil conditions have been evaluated from maps of the British Geological Survey and a very simplified outline of the exposed London Clay, stiff Glacial and Post-glacial Clays, Gravel and Alluvial Clay covered areas is shown on Fig. 5.

43. The distribution of damage indicated by the present study (and indeed by Meldola and White, ref. 1) points to local concentrations of damage at Wivenhoe and The Hythe, Colchester outside the immediate apparent epicentre. These locations are in part on alluvium and had a high proportion of old and probably vulnerable buildings. The reasons why alluvial areas may show high damage cannot be ascribed solely to dynamic effects. Old buildings on alluvium are likely to have suffered differential settlement which reduces the reserve strength of the structures and indeed this is evident from present day inspection. Conversely the low intensities in the gravel area between Rowhedge and Colchester may arise from the compounded effects of newer brick dwellings and less differential settlement.

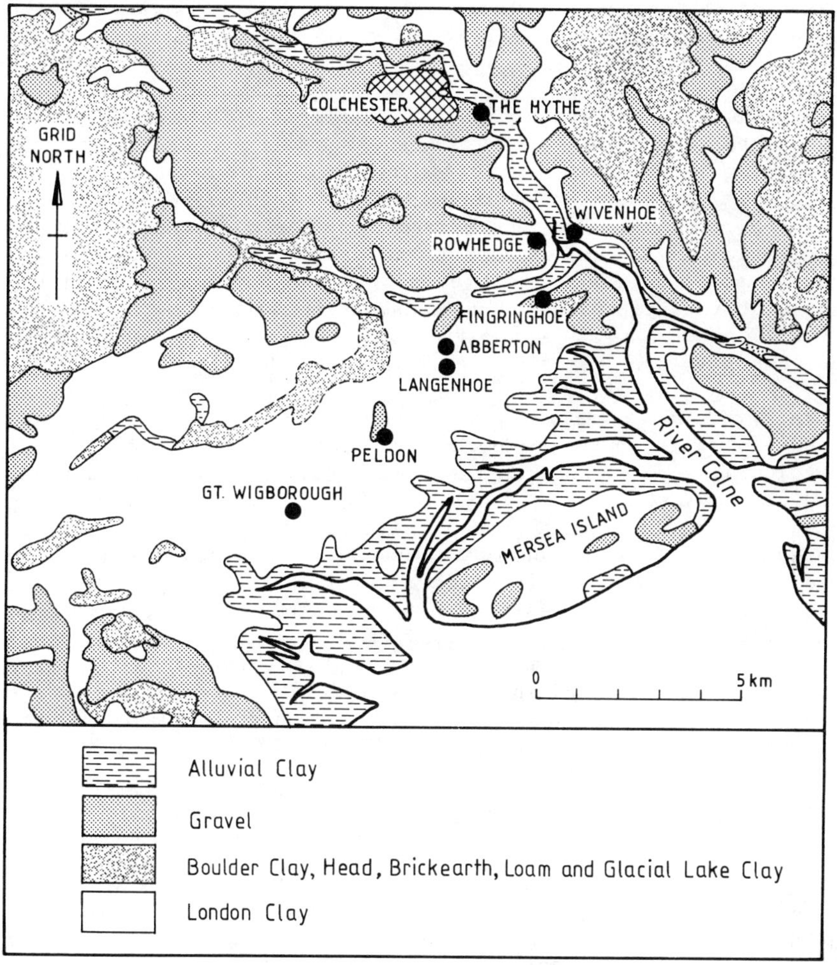

Fig. 5. Colchester earthquake, 1884: Simplified geology of the epicentral area

44. The effects of topography are similarly difficult to disentangle from other determinants. Undoubtedly Polden with its modest topographic high was severely shaken. This is not surprising as asperities such as this, with heights in the range 10-50 m and lateral dimensions of a few hundred metres, could respond to probably predominantly SH wave energy of comparable wave lengths. Locally steep topography could also have contributed to the severe damage to Ballast Quay Farm, Fingringhoe. But elsewhere the evidence is less clear cut. Strood Villa which was newly built of brick and severely damaged was on a low lying alluvial site, whereas the unscathed terrace of brick houses bounding the railway cutting at

Wivenhoe were elevated with respect to the river and on the edge of the artificial defile.

CONCLUSIONS

45. The detailed reappraisal of contemporary evidence for the Colchester earthquake has exposed some of the limitations of assessing intensity from diverse kinds of primary and secondary sources. It illustrates the benefits of intensity assessment informed by historical understanding. The differences between technical and newspaper reports are significant at intensities based upon the effects felt by people but perhaps surprisingly small at intensities based upon damage to buildings.

46. The region as a whole is rich in old and vulnerable buildings with a relative dearth of load-bearing brick houses in the epicentral area. The population of churches in the rural area was significantly different in structural style and condition from that in Colchester itself. These two factors precluded over-reliance on the damage suffered by both churches and old houses.

47. In the epicentral area itself the assessment of maximum intensity VIII was strongly influenced by reports of damage to the few substantial load-bearing brick dwellings therein. Confidence in this assessment was increased when long-postponed historical and field studies confirmed that the small number of these buildings so damaged was a high proportion of the dwellings of that kind in the area.

48. The significance of local ground conditions and topography is equivocal but probably more geotechnical in nature than dynamical.

ACKNOWLEDGEMENTS

The work upon which this paper is based was carried out mainly between 1981 and 1982 by Soil Mechanics Limited on behalf of the Central Electricity Generating Board as part of a programme to re-assess the seismicity of the United Kingdom (ref. 11). The authors would like to thank the CEGB for permission to publish this paper.

REFERENCES

1. MELDOLA R. AND WHITE W. Report on the East Anglian Earthquake of April 22nd 1884. Essex Field Club Special Memoirs, vol 1, 1885.

2. HAINING P. The Great English Earthquake. Robert Hale, London, 1976.

3. SCOTT R. F. The Essex Earthquake of 1884. Earthquake Engineering and Structural Dynamics, 1977, vol 5, 145-155.

4. BROWN A. F. J. Colchester 1815-1914. Colchester Record Office, 1979.

5. PEVSNER N. Buildings of England, Essex. Revised Edn., Penguin, 1965.

6. SORBY H. C. The Earthquake. Letter to Nature, May 29, 1884, p 101.

7. WESTCLIFFE LITHO. Earthquake in Essex, 1884, a unique record. Westcliffe Litho Company, Essex, 1974.
8. KINAHAN G. H. Notes on the earthquake that took place in Essex on the morning of April 22 1884. Scientific Proc. Royal Dublin Soc., 1884, vol 4, 318-325.
9. EIBY G. A. The Modified Mercalli scale of earthquake intensity and its use in New Zealand. New Zealand Journal of Geology and Geophysics 1966, vol 9, 122-129.
10. ROYAL COMMISSION ON HISTORICAL MONUMENTS. An Inventory of the Historical Monuments in Essex, Vol 3, 1922.
11. SOIL MECHANICS LTD. Re-assessment of UK Seismicity Data, Report 7984, vol 1-4, 1982.

Discussion

MR J. IRVING, CEGB
After 100 years have elapsed and intensive investigations have been carried out recently, are there any better ideas on the causes of the Colchester earthquake e.g. the source location or the causative fault?

With the investigative techniques available today would Colchester be rejected as a site for a potentially hazardous installation?

MR C. R. SMITH, National Nuclear Corporation
Is there any information which can be obtained from the data to explain why the Colchester event was so damaging for an event of its magnitude?

MR J. C. GUTMANIS, Soil Mechanics Ltd
The short answer to Mr Irving's question on the causes of the Colchester event is 'no'. Our recent intensity reappraisals have suggested a focal depth of around 5 km but very little is known about the sub-surface structural geology of the area and it is not possible to ascribe accurately the earthquake to a particular fault. Dr Muir Wood, in his paper in this volume, tentatively associates the Colchester earthquake with a NW-SE trending fault in the Palaeozoic basement (his Fig. 4). This is one of a suite of strike-slip faults in SE England and NW France. To put such speculation on a firmer footing would require deep seismic profiling.

MR SHILSTON
An understanding of the Colchester earthquake will make an important contribution to the assessment of seismic hazard for a nearby site. We do not feel that such a site should be rejected simply because of the earthquake and are sure that an adequately safe design can be produced using a hazard assessment of which the Colchester earthquake forms a part.

DR SKIPP
Mr Smith has asked if any information can be obtained from the data explaining the damaging nature of the Colchester earthquake. From the fall-off in isoseismals the focal depth was probably quite shallow so concentrated damage was to be expected. The extent of such damage reflects the vulnerable style of construction and the decrepitude of many medieval dwellings. We are accustomed to seeing pictures of devastation from small shallow earthquakes which, as it were, score a direct hit on fragile dwellings. Perhaps the Colchester earthquake shook more modified medieval timber framed dwellings than any other event on record. That there were no deaths and so few casualties is testimony to the underlying soundness of this kind of building. Deficiencies in the fixing of brick skins to timber frames in modern dwellings would probably be revealed in a repeat performance.

MR E. D. BOOTH, Ove Arup and Partners
Some seismologists recommend that intensity estimates should be based on the mode, rather than the extreme, of damage observed. May there not be a danger that historical records dwell on the extremes of damage, and neglect cases of low damage, so that the intensity estimates based on these accounts become exaggerated?

DR R. D. ADAMS, International Seismological Centre, Newbury
In assessing intensity there is a tendency to look at the worst damage. This is not necessarily bad, for by definition each area of given intensity must contain indications of lower intensity. It is not always good to determine the intensity of a region by strictly taking the mode of the individual intensity assessments.

16. The 8 November earthquake at Liege, Belgium

W. P. ASPINALL, Principia Mechanica Ltd

ABSTRACT

An earthquake of magnitude Mb 4.9 caused damage of intensity IV to VIII in Liege, Belgium on 8 November 1983. Damage was sustained by older, low quality domestic buildings whilst modern engineered structures and facilities were unaffected. The effects of topography, near-surface geology and mining operations were evident in the distribution of damage in the town.

A short temporary microseismic array study detected seven probable aftershocks, seven mine-related events and seven near events in the six days following the main shock. Locations of the aftershocks suggest a focal plane directly below the area of maximum intensity, in contrast with the hypocentre of the main shock obtained using fixed European stations which was offset from peak damage by about 3 km.

INTRODUCTION

The Significance of the Earthquake

At 1.49 am (local time) on 8 November 1983 the town of Liege in eastern Belgium was struck by an earthquake of body-wave magnitude M_b4.9 which caused damage and some casualties.

Western Europe is a region of low seismicity (Fig 1 shows recent Belgian earthquakes) and the seismic hazard for modern, sensitive industrial facilities is a subject of current investigation. This earthquake occurred within the boundaries of a long-established industrial and coal-mining town and close to a number of new industrial complexes and a nuclear power plant. Its effects are therefore of interest to designers, operators and regulators of similar plant elsewhere in Europe.

EARTHQUAKE DAMAGE

Liege - the Town

Liege lies on the confluence of the River Meuse with the River Ourthe and the area has been inhabited since Palaeolithic times. The Meuse can accommodate Rhenish barges of more than 2000 tonnes and some sea-going ships up to Liege,

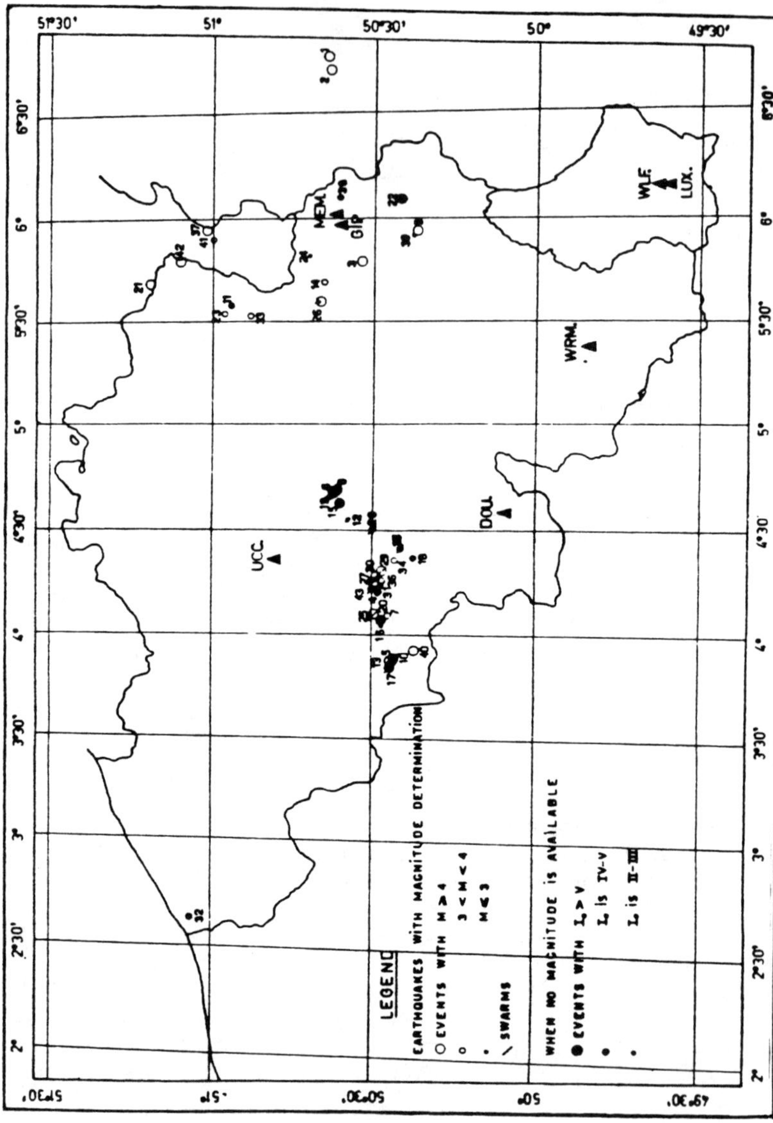

Fig. 1. Seismicity of Belgium 1950-1982 (Courtesy UCCLE Observatory)
Event 26 was the 1965 Mb4.3 Earthquake near Liege

the third most important river port of Europe; it is also the centre for mining the eastern basin of the Sambre-Meuse coalfield and enjoys accessibility by road and rail from most parts of Europe. The town has a population of about 150,000 and has been industrial since mediaeval times.

Liege comprises several fairly distinct quarters: the west bank is predominantly residential and commercial (it would be termed "downtown" by Americans). On the river at either end of the town are industrial facilities, principal amongst these being an oil terminal and petroleum products facility at Souverain Wandre (to the north-east) and old steel and coke-works at Seraing and Jemeppe on the bend of the river as it enters Liege in the south-west. Within this bend (Fig. 2) and above the commercial centre of the town, lie old and now defunct coal mines; in this area are to be found most of the oldest, decaying houses of Liege. (It is said that some illegal mining continues). Further out from the town centre, to the north-west there are new industrial parks and housing estates, reaching to a small airport at Bierset.

Damage - General Comments

Damage was concentrated in the district of St. Nicholas, and its suburbs of Glain, Montagnee and St. Gilles. This area is predominantly middle-aged (between the wars) and old (pre-1914) houses, and remnants of the declined coal-mining industry. There is very little current industrial activity although there are some small-scale operations such as small-arms and battery manufacture.

As a consequence of the prolonged coal-mining operations there is continuing subsidence in many parts of St. Nicholas and some properties are founded on ancient spoil heaps. In addition, many suffered bomb damage during the last war. Thus many houses, especially those built around the turn-of-the-century had already suffered some degree of distortion in settling before the earthquake and should be categorized as the poorest grade of masonry building on the modified Mercalli scale.

The following are the principal types of damage observed in Liege:

a) failure of domestic chimney stacks, most often by crumbling and loss of individual bricks rather than failure as whole units.

b) failure of arches and lintels over door and window openings

c) some diagonal cracking of walls in shear

d) loss of non-structural facings such as stonework ridges , blocks and decorative pinnacles

e) walls cracked or damaged by differential settling.

These effects were confined almost exclusively to old domestic and religious buildings (of which there are many in Liege) and selected examples are discussed below.

Almost without exception, modern houses, blocks of flats, schools, clinics and an electricity substation in the

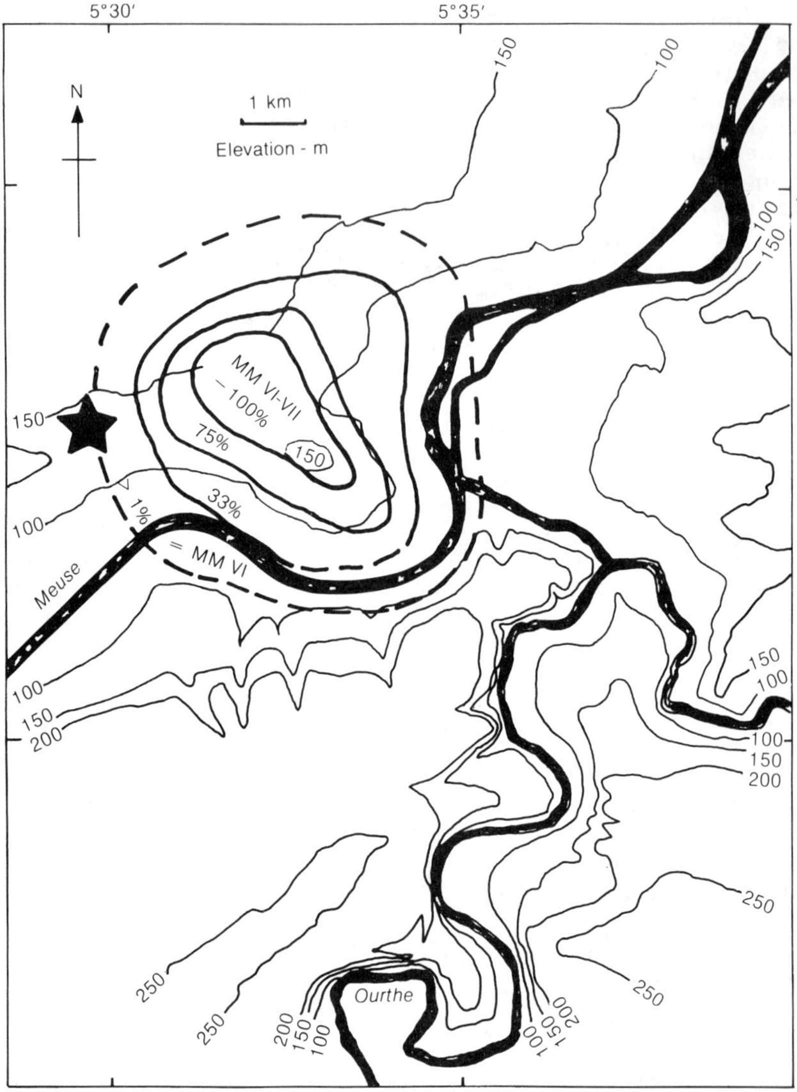

Fig. 2. Percentage of old chimney stacks damaged and preliminary instrumental epicentre

epicentral region were undamaged structurally.

Observations of Specific Damage Effects

The overriding impression of Liege following the earthquake was of damaged brick chimney stacks. As remarked above this was specifically true of older properties in which time, the elements, and the burning of coal fires had obviously taken their toll on the mortar. However- most of the damage suffered by the stacks was not that usually associated with earthquake-induced failure; whereas good quality brickwork chimneys elsewhere generally fail as a unit, sheared off at or near the roofline- in the case of Liege the majority of chimneys had the appearance of having crumbled or spalled brick by brick.

However, there was evidence that the strongest shaking had an approximate northeast-southwest orientation. It was the observation of this author that nearly all examples of shear-induced diagonal cracking occurred on walls aligned approximately NE-SW ; additionally, an example of arch failure on a NW-SE wall (Rue de Koninck) had sustained a marked lateral NE-SW displacement, and the only example of ground movement seen was consistent with easterly slippage on a slope striking NW-SE.

In terms of all wall damage the majority was in the form of vertical cracking and is consistent with accelerated differential ground settlement rather than shear failure. Several examples illustrate a feature of many Liege houses, that is, they appear to have sound brickwork but have, in fact, been refurbished with a decorative veneer. Behind these facades the original walls are often as poor in quality as some of their less resplendent neighbours and in one example what appeared to be a reinforced concrete ring-beam (which would have been ideal for anti-seismic purposes) turned out to be superficial plaster formwork.

For a number of different causes there were localized pockets of higher intensity damage throughout the epicentral area but during the survey of density of failed chimneys it was noticed in St Gilles that in three adjoining parallel streets (rue Schmerling, rue Laurent de Koninck and rue d'Omalius) maximum chimney damage had occurred in rows of houses aligned NW-SE; that is their axes of minimum stiffness were in the direction of maximum ground shaking inferred above and the observed intensification of damage, compared with their neighbours aligned in other directions, is thus explained.

Mapping Chimney Stack Damage and Considerations of Damage-Related Intensities

In order to make a map of the damage effects of this earthquake it was considered that a survey of the extent of domestic chimney stack damage in old houses (the most prevalent type of damage observed) would provide the best quantitative basis. Accordingly the author visited as much of

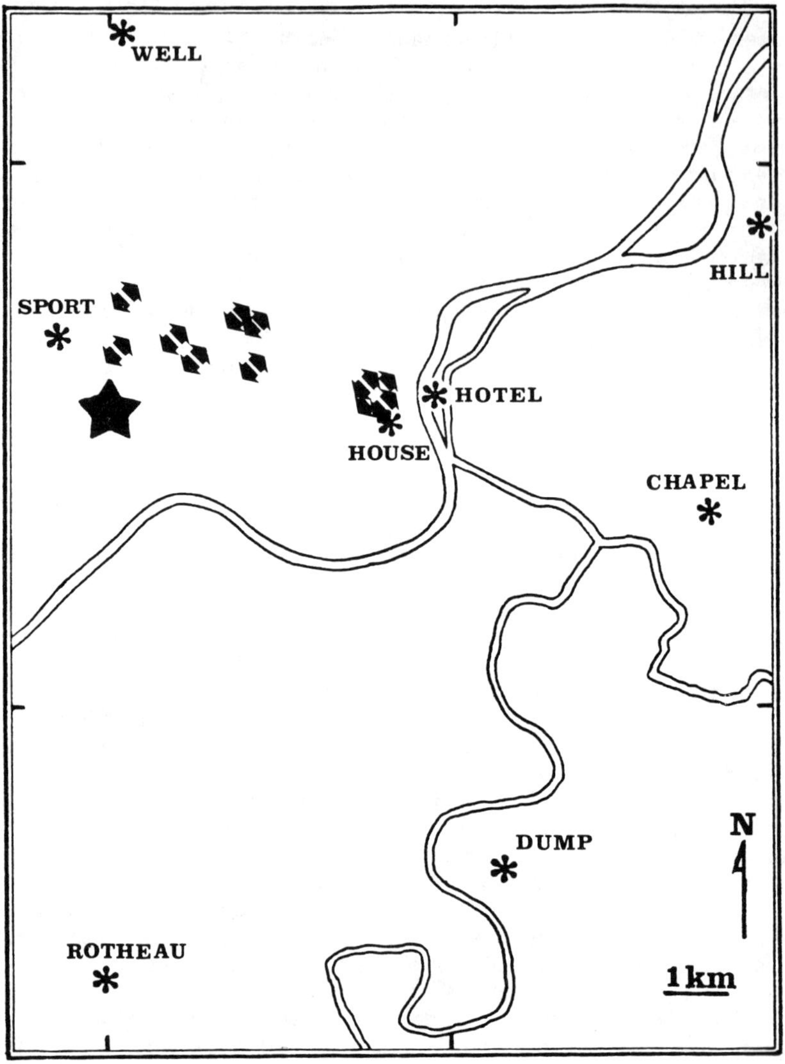

FIG. 3 Locations where NE-SW strong shaking is inferred in Liege.

the epicentral region as possible on foot and by car in the course of five days. Four levels of chimney stack failure are plotted in Fig. 2; the contours corresponding to damage to:

1) at least one stack in a hundred
2) about one-third of old stacks
3) about three-quarters of old stacks

and 4) almost complete failure of all old stacks in the given area. The preliminary instrumental epicentre given by Uccle Observatory is marked together with approximate contours of ground elevation. It will be observed that the maximum effective damage is elongated approximately NW-SE reflecting the topography of the ridge which deflects the course of the Meuse whereas the outer bound of observable damage is more circular in shape. It is inferred from this that the ridge topography had played some part in dictating damage effects.

Using the criteria of the modified Mercalli Intensity scale the >1% contour of Fig 2 encloses the area in which Intensity VI was exceeded. Although locally there was considerable damage to some of the poorest of the old buildings within the - 100% contour there was however no real evidence of damage to better quality buildings; therefore Intensity VII is only partially or locally satisfied in the peak epicentral zone in terms of structural damage effects.

It should be borne in mind that there is not a uniform density of identical old properties throughout Liege and the contours have been smoothed and interpolated.

Effects at Industrial Facilities and Public Utilities

Within the area of maximum damage there is little significant modern industrial activity; there is brokendown pithead plant and there are a few small manufacturing concerns producing small arms. transistor radio batteries and the like. It was not possible for the author to enter any of these buildings (there being two public holidays in the period of the visit and a need for official bona fides to satisfy suspicious owners) but newspapers and verbal reports spoke of overturned equipment and stores fallen from shelves.

At the Clinique de l'Esperance. a modern building which was structurally undamaged in the centre of the epicentral zone. sufficient damage was done to equipment in the radiological laboratory to render that part of the facility unserviceable The chief petrochemical plant at Souverain Wandre, about 5km from the epicentre, comprises an assortment of storage units of varying sizes. including cylindrical and elevated spherical LNG tanks. As far as could be ascertained informally no damage was sustained there.

The older industries, steel-making, coke and chemical plants on the bend of the Meuse at Jemeppe and Seraing about 2km from the epicentre, were also reported undamaged after cursory inspections despite there being many structures of old and virtually unmaintained plant. (The author observed some broken stringers on a long pipe-run in the Cockerill-Sambre steel plant, however the breaks were well rusted and therefore pre-date the earthquake.)

A modern electricity supply substation close to the epicentral area had suffered no damage and had continued to function perfectly during and after the earthquake. In the streets of Saint Nicholas many iron stanchions, carrying power lines along the upper levels of houses, had broken loose but nearly all the breaks in the lines themselves were caused when falling masonry severed the wires; the weight of such masonry can be judged by the severe damage to motor cars parked below.

Particular but brief interest was expressed in the Belgian media on the subject of Tihange nuclear power plant, situated about 25km from the epicentre. Because a visit had already been made to this facility by a specialist from the U.K.A.E.A.'s Safety and Reliability Inspectorate (Dr. O. W. Phillips) this author refrained from seeking admission. Initial reports suggested that a strong-motion accelerograph record had been obtained there but. following film processing, it was reported that there had been no seismic triggering. Dr. Phillips has reported that a peak response recorder had registered 0.02g at Tihange 2 (Phillips, 1984).

Closer to the epicentre, at Croixhe, an old coal-fired generating station continued to function without interruption.

Three or four days after the earthquake of 8 November three broken water mains manifested themselves in Jonfosse, Glain and Fragnee . It is probable that they were damaged by the earthquake but some time elapsed before the leaks became obvious by breaking the road surfaces.

Damage Intensities Summary

The 1983 November 8 Liege earthquake owes its significance (in terms of damage and intensity) to being shallow and directly under a city with a high proportion of susceptible old buildings rather than to its magnitude or other seismological parameters. There is circumstancial evidence to indicate that the local topography and ground conditions may have contributed to the damage effect but by far the most influential factor was the poor and delapidated condition of many properties. In contrast modern structures, and those with some engineering input. withstood the earthquake without effect.

This author attributes a modified Mercalli intensity VI to the epicentral area with local or partial examples of intensity VII.

INSTRUMENTAL OBSERVATIONS OF AFTERSHOCKS USING A TEMPORARY MICROSEISMIC ARRAY

Equipment and Station Locations

Eight temporary microseismic stations were deployed in and around Liege between 10th and 16th November 1983. Sprengnether MEQ-800 recorders, writing on smoked paper, were used in conjunction with Mark Products L4C 1 Hz vertical seismometers - giving a peak system response at about 20 Hz.

Individual station gain settings were adjusted for site conditions, selections of which were severely hampered by having to operate mostly within the boundaries of a town. The final array of stations is shown in Fig. 4 from which it can be seen that an attempt was made to encircle the probable epicentral area with distant stations whilst two central detectors could be used to control hypocentral depth determination; this is in line with current thinking on the best configuration for a small array (e.g. Duschenes et al., 1983).

A number of the stations were deleteriously affected by wind noise from time-to-time (Well, Chapel) and the two central stations (House,Hotel) were affected by cultural noise in the town to varying degrees. The fact that the weekend over which the array was operating was preceded by a public holiday on the Friday greatly reduced the level of cultural noise that might have otherwise occurred. The most sensitive station (Dump). founded on Rhenish Massif bedrock in a disused quarry, suffered most from mechanical defects and operator errors (this contrariness would surprise no field seismologist). On two occasions the drum rotation motor seized at night, probably because of moisture freezing in the $-10^{o}C$ temperature.

Observations and Location of Aftershocks, Microearthquakes and Mine Events

In all 21 unambiguous events were detected in the five-day period Of these. two were recorded by 7 stations five by 6 stations, six by 5 stations and five by 4 stations; those tremors constitute the event phase file. as input to the location program, HYPOELLIPSE. Also required as input to the program is some form of crustal structure model defining the distribution of seismic velocities with depth. As a simplification the ratio of compressional wave velocity to shear-wave velocity (V_p/V_s) can be fixed and Ahorner (1983) gives V_p/V_s = 1.69 for the area containing the Rhenish massif. As a preliminary test Ahorner's (1983) values of V_p with depth (Table 1) were used. however there were indications of preferential clustering in alternate layers of the model implying artefactual solutions (Fig 6). Two forms of a simple half-space velocity model were then adopted: one using Ahorner's median velocity (V_p = 6.00 km/s) down to 31 km and the other starting at V_p = 5.70 km/s at the surface with a linear increase in velocity of 0.1 km/s/km with increasing depth (Table 1). These two produced essentially similar hypocentral solutions and, pending future refinements. the simple half-space model results are presented here. In Fig. 4 are given the epicentral locations of 14 events located by the array together with the Uccle Observatory solutions for the main shock (X) and two locatable aftershocks (x's) using fixed European stations. (Recall that there was some disparity between Uccle's main event epicentre and the area of maximum damage intensity noted in Fig. 2 above). Seven other

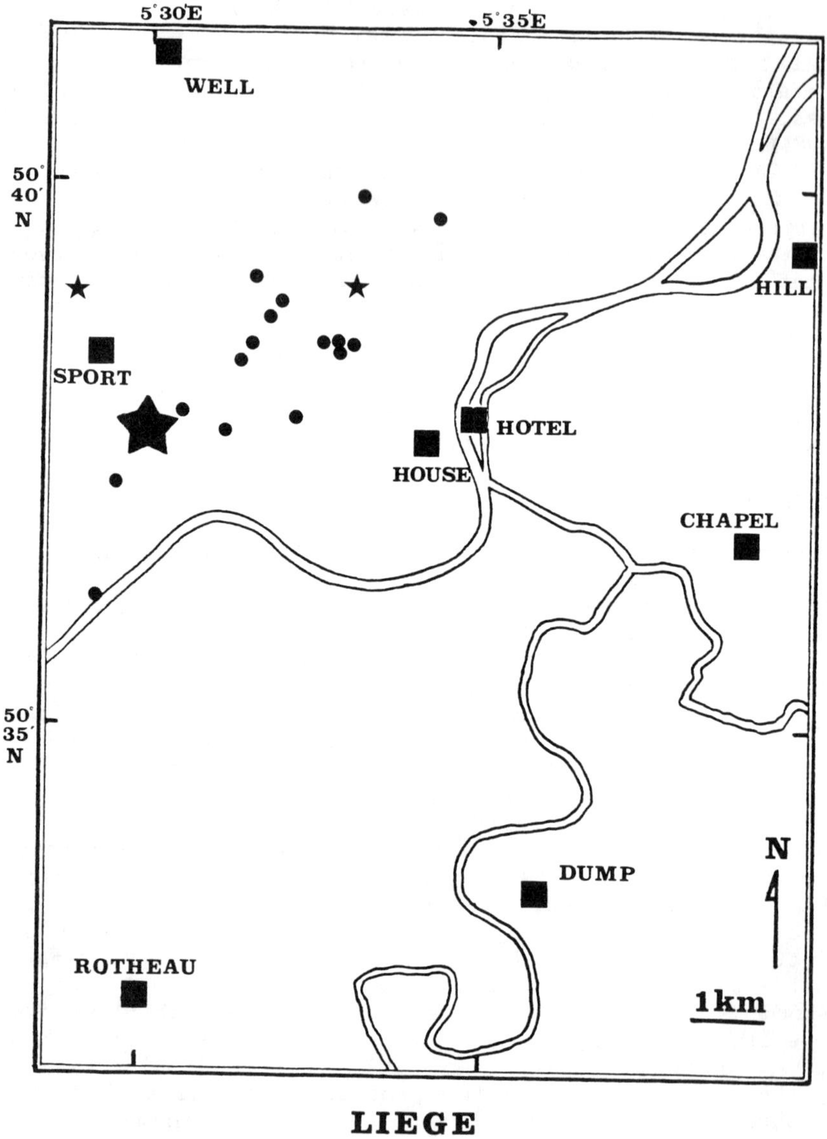

Fig. 4. Epicentres (●) of events detected by temporary array stations. Stars are epicentres of 8 November main shock and two aftershocks given by Uccle Observatory

TABLE 1

CRUSTAL STRUCTURE MODELS

SIMPLE HALF-SPACE	AFTER A. HORNER (1983)	HALF-SPACE WITH LINEAR INCREASE
Layer 1: V_p = 6.0	Layer 1: 0-1km: V_p = 5.0	Layer 1: V_p = 5.7 at surface
	Layer 2: 1-2km: V_p = 5.5	
V_p/V_s = 1.78	Layer 3: 2-3km: V_p = 5.8	+ 0.1km . s^{-1} . km^{-1}
	Layer 4: 3-10km: V_p = 6.0	to 31km
	Layer 5: 10-19km: V_p = 6.25	V_p/V_s = 1.78
	Layer 6: 19-31km: V_p = 6.9	
thickness	V_p/V_s = 1.78	Layer 2: >31km: V_p = 31km = 8.1
Layer 2: V_p = 8.1	Layer 7: >31km: V_p = 8.1	

All velocities (V_p) are in km.s^{-1}

microearthquakes picked up by the array fell outside the boundaries of the map of Fig. 4. The microearthquakes found fall quite neatly within the Uccle solutions and suggest a NE-SW trend. It is of interest to project the depths of the events onto an orthogonal vertical plane, that is, one with a NW-SW strike; this has been done in Fig. 5. Two features of Fig. 5 may be remarked: 1) there is a cluster of events of shallow depth (<1 km) which can be accounted for as probable mining-related events (Uccle Observatory recorded a fourth 'aftershock' on 8 November, felt in north Liege, which was interpreted, from considerations of its wave-form, as a mining collapse), and 2) the remaining seven events which, whilst acknowledging their sparseness, could be imagined to define a vertical lineament. This is not inconsistent with a preliminary focal mechanism solution given by Uccle Observatory, that is: predominantly strike-slip motion on one plane striking N42^{O}E with vertical dip (Fig. 7).

The epicentral centroid of the seven aftershocks is at 50° 30.5'N 05° 32,0'E (about 3 km north-east of the main Uccle event epicentre at 50° 37.8'N 05° 30.0'E) and consistent with the intensity effects. It is concluded therefore that the location of the main shock is in error by about 3 km only, a very reasonable tolerance for such an event in north-western Europe. The locations of the aftershocks detected by the temporary array define a focal volume at most 6 km deep striking NE-SW with a lateral extent also about 6 km, and suggest a probable mislocation for the principal earthquake of perhaps 3 km.

Focal Parameters

The four events reported for 8 November 1983 by Uccle Observatory were given local magnitudes M_L = 4.9 2.7, 2.9 and 3.5 respectively. These determinations were based on calibrations of Lg-wave amplitudes except for the principal event which was based on a coda duration-magnitude relationship for station Dourbes. The relationship given:

$$M_L = 2.5 \log t - 1.67$$

where t = coda duration in seconds, has been derived for Benioff and Johnson-Matheson seismograph systems having peak responses around 1 Hz for events in the 'near' range (<3°). The Sprengnether MEQ-800 system with L4C seismometer, on the other hand, has peak response at around 20 Hz and in this case was being used to record events at distances of order 10 km so the relationship for the total duration-local magnitude given by Bakun and Lindh (1977) was adopted thus:

$$M_L = 0.71 \log T + 0.28$$

where T is the total signature duration in seconds. With this equation local magnitude determinations ranged from M_L = 0.6 to M_L =1.1 for the aftershocks in this report and from

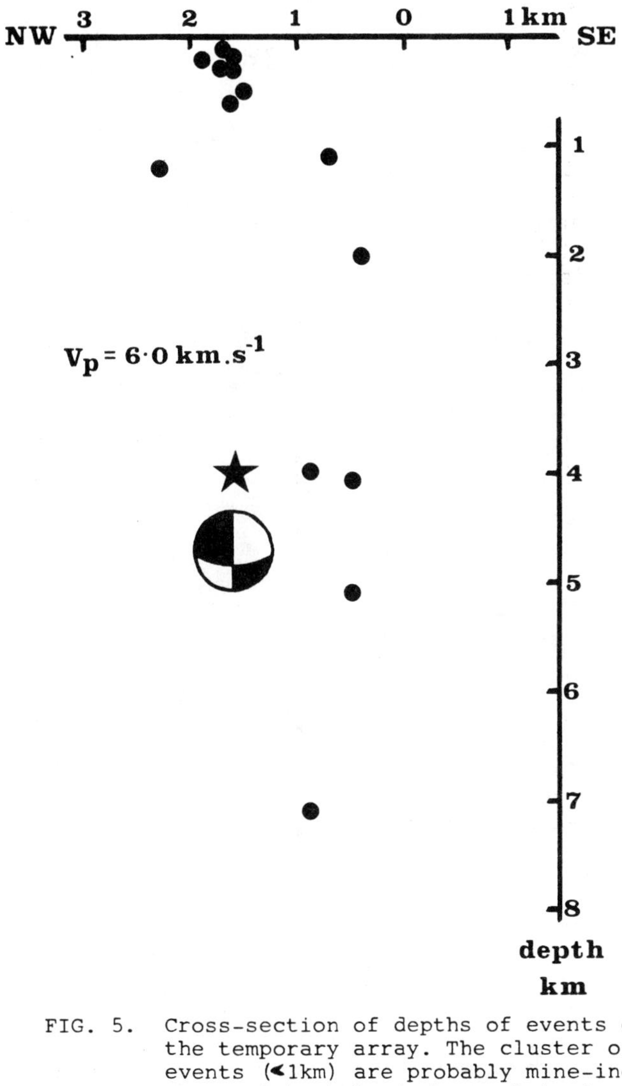

FIG. 5. Cross-section of depths of events detected by the temporary array. The cluster of shallow events (<1km) are probably mine-induced and the star and back-hemisphere plot give the Uccle Observatory main hypocentre and preliminary focal mechanism respectively.

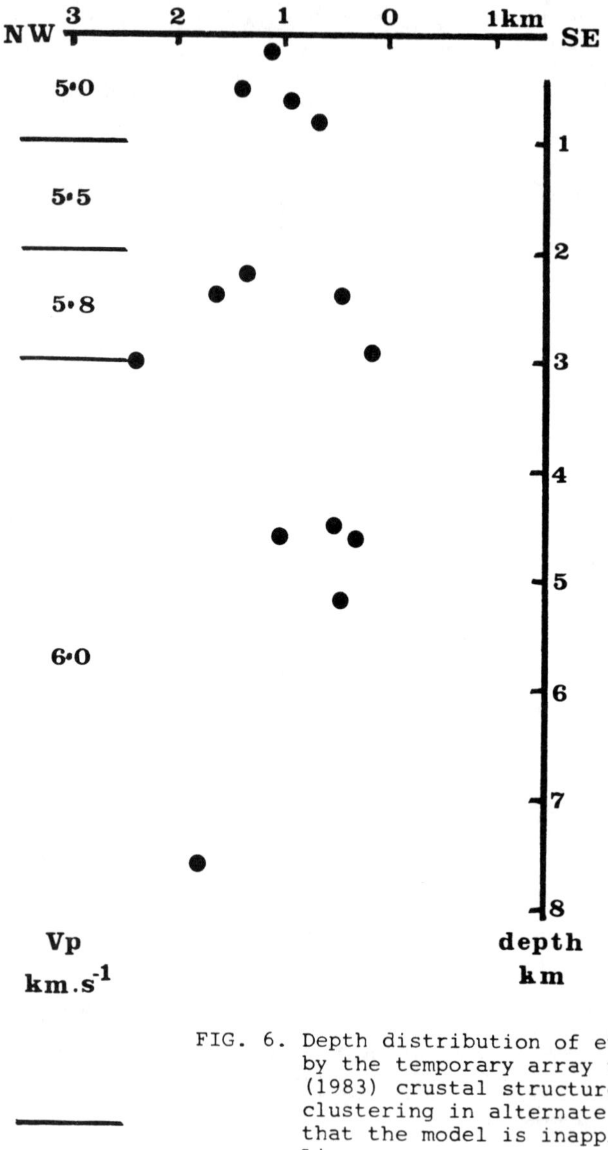

FIG. 6. Depth distribution of events detected by the temporary array using Ahorners (1983) crustal structure model. The clustering in alternate layers suggests that the model is inappropriate for Liege.

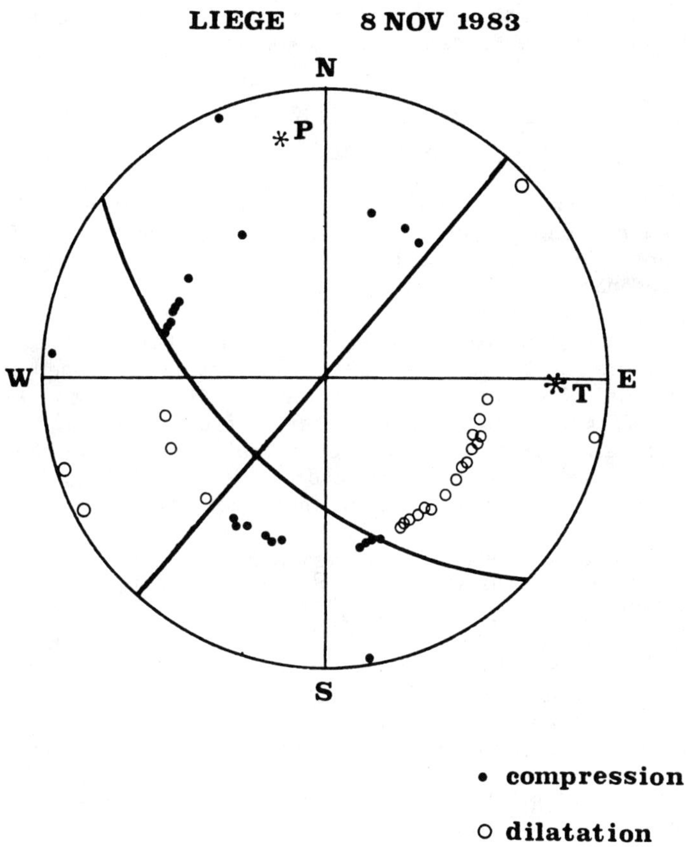

FIG. 7. A redrawn version of Uccle's preliminary focal mechanism solution suggesting a strike-slip motion with one plane (NE-SW) almost vertical in dip.

$M_L = 0.5$ to $M_L = 0.9$ for the conjectured surficial mine events. It must be stressed that these are very small seismic events and represent energy releases at source equivalent to a few pounds of buried high explosive.

The seismic moment, Mo, for the principal event of 8 November was estimated by Uccle as 1.2×10^{23} dyne.cm. If the source area is taken to have a diameter of 6 km from the extent of the aftershocks then a planar area, A, of order 3×10^{9} cm^{2} is involved. Assuming a typical value of 3×10^{11} dyne.cm^{-2} for the modulus of rigidity (μ) then the average slip D in the earthquake can be estimated from

$$\bar{D} = \frac{Mo}{\mu A} = \frac{1.2 \times 10^{-23}}{3 \times 10^{11} \times 3 \times 10^{9}} \sim 15\text{cm}$$

If this is taken to occur at a depth of about 4 km (i.e. the hypocentral depth) then from dislocation theory a surface displacement of order a few millimetres could have occurred. This is consonant with the damage observed.

The Liege earthquake of 8 November 1983 probably originated on a junction between two active tectonic zones: the Lower Rhine Graben and the Belgian strike-slip zone. This latter feature (conjectured by some to extend to the Kent coast) is estimated to have an offset of up to 6 km. A focal mechanism solution for the principal event is presented by Uccle Observatory (Camelbeeck and de Becker,1984) using 51 stations out of 62 data available. Their solution is:

	Plane 1	**Plane 2**	**Tension axis**	**Pressure axis**
Azimuth	N42^{o}	N132^{o}	N91^{o}	N351^{o}
Dip	90°	62°SW	18°	19°

This interpretation is not a unique and unambiguous one as other solutions could be made to fit the distribution of first-motions. However, taking into account all the evidence to hand, that is, local geological faulting, the inferred direction of strongest shaking, the distribution of aftershocks and the first-motion data then a strike-slip mechanism on a near-vertical fault striking NE seems the most logical result.

SUMMARY AND CONCLUSIONS

The 8 November 1983 earthquake at Liege, Belgium, had a marked impact on the population because of their lack of familiarity with felt earthquakes and because it occurred at shallow depth directly below a city containing many weakened and decaying houses. The earthquake was only of moderate magnitude (no greater than M_b 4.9) but is interesting in the context of north-western European seismicity and significant in its implications for setting a threshold below which anti-seismic design is probably not needed for modern, engineered structures.

Damage effects in Liege appeared to be due in part to local topography and ground conditions but predominantly to the poor state of domestic buildings, especially chimney stacks. In the case of the latter, the simple layered brickwork construction and absence of competent mortar contributed to their failure, not as whole units, but in brick-by-brick disintegration. This suggests that some fairly simple and cheap experimentation on a shake-table with crude replicas of Liege chimney stacks could indicate the level of shaking (and proportions of vertical and horizontal movements) needed to produce the failure observed. Making allowance for the transfer function of the supporting building, the degree of strong shaking at ground level could be estimated which would offset, in part, the absence of strong-motion accelerograph records from Liege.

In a further attempt to elucidate the nature of the 8 November earthquake, a temporary eight-station seismograph array was deployed around Liege from 10-16 November 1983. The array detected 21 local events of which seven may have been aftershocks, seven were probably local mine collapse events and the remainder were nearby microearthquakes mine events or quarry blasts. The seven conjectured aftershocks indicated the possibility of a vertical-dip source zone, not inconsistent with a preliminary regional focal mechanism solution produced by Uccle Observatory and showed a small discrepancy (3 km) from the Uccle epicentre which would put the main earthquake closer to the area of maximum intensity Unfortunately, the temporary array missed one bigger aftershock by a few hours on 10 November and was withdrawn a few hours before an aftershock on 16 November both of which were detected by the fixed Belgian seismographs. Thus no direct calibration was possible between locations obtained with the array and those with the permanent station.

REFERENCES

Ahorner, L. (1983), 'Historical seismicity and present-day microearthquake activity of the Rhenish Massif, Central Europe', Chap. 6.4 in "Plateau Uplift", eds. K. Fuchs et al., Springer - Verlag, Berlin Heidelberg 1983

Bakun, W.H. and A.G. Lindh (1977), 'Local magnitudes, seismic moments and coda durations for earthquakes near Oroville, California', Bull. Seism. Soc. Am. 67, 615-630

Camelbeeck, T. and M. de Becker (1984), 'The earthquakes of Liege of November 8, 1983 and December 21, 1965',in "Seismic activity in Western Europe",NATO ASI Series Vol. 144,ed. Paul Melchior, pub. D. Reidel, Dordrecht, pp 233-248.

Duschenes J., R.C. Lilwall and T.J.G. Francis (1983). 'The hypocentral resolution of microearthquake surveys carried out at sea'. Geophys. J R. Astr Soc. 72. 435-451

Phillips, D.W. (1984),'Macroseismic effects of the Liege earthquake with particular reference to industrial installations', in "Seismic activity in Western Europe",NATO ASI Series Vol. 144. ed. Paul Melchior. pub. D.Reidel, Dordrecht,pp 369-384.

ACKNOWLEDGEMENTS

This study was supported in part by the Central Electricity Generating Board (G.D.C.D. Barnwood). in part by British Nuclear Fuels Ltd. and in part by Principia Mechanica Ltd.

The author is grateful to Dr. G.C.P. King and Mrs. N. Threlfall for their work with the temporary array and to the Department of Earth Sciences, University of Cambridge, for the use of their equipment and van. Profs R. Mason (Dept. of Geophysics) and N.N. Ambraseys (Dept. of Civil Engineering) of Imperial College kindly loaned a seismograph recorder.

The author also acknowledges the kind courtesy, friendly discussions. and informative reports of Prof. Melchior and his staff at Uccle Observatory, Brussels.

Discussion

MR D. C. PAL, Nuclear Design Associates, NNC

The type of cracks in the very old brick or block type buildings almost conform to the classical type web shear cracks rather than the simple geological type of subsidence. The cracks also developed in potential areas like door frames and openings. Can Dr Aspinall confirm this?

Is there any evidence of measurable cracks in reinforced concrete or steel structure occurring at Liege after the earthquake?

DR ASPINALL

It did not appear to me that the cracks in the old brick buildings conformed to the conventional double-diagonal shearing failures usually encountered when a building has been subjected to severe lateral inertial loading. There was much evidence that the majority of the cracks were not symmetrical in each building and many were nearly vertical in orientation, although they did originate at openings such as windows and doors. However, this would occur in buildings distressed by settlement and hence I used the term 'accelerated subsidence' to emphasize my belief that the fundamental cause of the damage was inelastic differential displacement in the foundations rather than violent lateral shaking by the seismic waves acting on the building directly.

I was unable to find a single example of cracks in reinforced concrete or steel structures as a consequence of the Liege earthquake.

17. On the seismicity of the south-west of the British Isles

J. M. MENU, N. N. AMBRASEYS and C. P. MELVILLE, Imperial College of Science and Technology

0.0 SYNOPSIS.

The 19 July 1984 earthquake gave an opportunity to study the overall seismicity of the South-West of the British Isles. This paper describes some features of this recent shock and places it in its seismicity context.

1.0 INTRODUCTION

On Thursday 19 July 1984, at 06.56 GMT, most parts of northwest Wales were affected by a strong earthquake which was centred in the Irish Sea, near the northern coast of the Lleyn peninsula.
The shock was perceptible as far as GLASGOW,LONDON and PLYMOUTH. Felt aftershocks followed at 09.00 and 22.00 GMT the same day. The shock was felt over a large area (250,000 km2), but no important structural damage was reported from the epicentral region.

This paper presents a brief account of the effects of the earthquake, put in the context of the overall seismicity of the region.

2.0 HISTORICAL BACKGROUND AND SEISMICITY

Figure 1 shows the distribution of earthquakes reported in western and northwestern England and Wales since the 13th century.
Study of the long-term seismicity of this region during this period confirms that the largest event has not exceeded a magnitude of M_S = 5.5 and that the worst damage inflicted was never more serious than the collapse of chimneys, the cracking of walls and plaster falling off ceilings.

The largest earthquakes reported before 1800 in western and northwestern England and Wales are located approximately on fig 1.
Most of these events caused minor damage in a few locations, with maximum Intensities in the range VII-VIII (MSK).

3.0 THE RECENT WELSH EARTHQUAKE OF 19 JULY 1984

3.1 Geological setting

The Geological basement of North Wales is Precambrian as exposed in Anglesey and SW Lleyn.

The tectonic movements in the epicentral area are controlled by two old Caledonian faults: the Dinorwic fault and the Bala fault.
The Bala fault strikes NE-SW through Chester, Bala, Coris and Tywyn; it extends in the south part of the Irish Sea. The Dinorwic fault, approximately parallel, is located at the level of the Menai Strait.

Inland, only secondary faults connected with the folding of the post-carboniferious rocks overriding the Lower Paleozoic basement are surrounding the Snowdonian massif, itself intact, to give a complex pattern. The largest of them is the Lleyn Thrust in the NW of the peninsula.

Offshore, Cardigan Bay is cut across by the extension of the two faults creating a graben, covered by Mesozoic and Tertiary sediments whose thickness varies from 3 to 6 km. Like other grabens, the area is made of major parallel normal faulting associated with secondary perpendicular rupture surfaces.
This structure is abruptly stopped on the coast of the bay by the strong Cambrian Harlech Dome.

It is most uncertain, without additional evidence,

to attribute the cause of the shock to any particular fault.
However, the earthquakes of 1984, of 1852 and other events of the same order of Magnitude, and possibly the largest shock of 1247 indicate a continuing small but active tectonic element in the region.

3.2 Macroseismic Data

The earthquake was felt with an Intensity II+, over an area of at least 250,000 km2.
Figure 2 shows the distribution of Intensity in the epicentral area from Aberystwyth to Merseyside, as assessed by the authors on the basis of field observations. The map remains a preliminary assessment of the main shock and may be modified in the light of additional information.

The shock was most strongly felt along the coast of Cardigan and Tremadog bays, and in the Lleyn peninsula south of Caernarfon.

The buildings in the area are typically built out of local hard sandstones.The traditional architecture is massive and the constructions of good quality.

Most damage occured in the modern and cheaper types of housing mainly in small towns and consisted of dislodged masonry units or cracked indoor plaster.
In the epicentral area, weak chimney-pots collapsed. The observed damage was found to be very dependant of the type and quality of constructions. Houses and structures of the massive rural local type, strongly built, were not affected.

No fatality was reported but a few people were taken to hospital for treatment of shock and minor injuries.
The best assessment of Intensity would indicate a maximum observed value of VI (MSK) at certain locations and on particularly vulnerable constructions, but the mode for the area should not exceed V (MSK), indicated by the widespread alarm caused by the shock, the rushing of people into the streets, the overthrow of loose objects in some cases, and the isolated damage caused to chimneys, windows and plasterwork.

3.3 Epicentre and Magnitude Determination.

The epicentre was given soon after the shock by the European-Mediterranean Seismological Centre

(Strasbourg). It was located at 52.800 deg.N - 04.600 deg.W..
No foreshock of significant warning lėvel has been observed on any seismographic network.

The determination of the Richter magnitude was M_L = 5.7 and the body wave magnitude m_B = 5.3 based on recordings of several European stations.
In the case of the present earthquake, surface waves were recorded by many of the seismological stations in Europe. The quality of the recordings are such that reasonable assessment of M_S can be made. The periods of the surface waves vary from 7 to 17 sec depending on the distance, and computation was possible using the Prague formula.
An average value M_S = 4.5 (0.25) was obtained from 18 stations.

4.0 DISCUSSION ON THE SEISMICITY OF THE REGION

Data from field observations of damage to buildings and other man-made structures caused by earthquakes since 1957 suggest that the threshold of damage varies with the type and age of building as well as the type of foundations. For urban centres this threshold has been found to consist mainly of imaginary or "zero-intensity" damage, i.e. damage which the occupant believed occured during the earthquake, but which was actually in existence before the event. Typical examples where zero-Intensity damage overshadowed damage caused by the shock are the Derby earthquake of 11 February 1957, the Carlisle earthquake of 26 December 1979 and the Belgian earthquakes of December 1965 and November 1983.

Earthquakes of all magnitudes have occasionally triggered minor landslides and rockfalls on land, particularly after a wet period. However, neither the far-field effects from offshore earthquakes nor local shocks seem to have been responsible for failures in natural slopes on land or for submarine flow slides; all submarine cable breaks in the Irish Sea were found to be associated with other causes. In a very few cases, earthquake shocks have caused localised liquefaction of loose sands on beaches and near canals, particularly at low tide.

It is rather unlikely that earthquakes centred in the British Isles will cause serious structural damage to well designed and built engineering structures on stable foundation materials. However, considering the degradation of the rapidly ageing building stock, particularly of the urban areas, as well as the high vulnerability of some of the more recently developed, but dynamically not tested, building systems and materials, some of which are already showing signs of distress even without help from earthquakes, a magnitude 5+ earthquake occurring in a densely populated region is likely to cause widespread structural damage and most probably the collapse of substandard constructions with the loss of life.

The earthquake problem in the U.K. is not one of a large magnitude earthquake occuring but rather of a small to medium magnitude event shaking a large number of highly vulnerable structures subsisting, under normal conditions, with a factor of safety very close to one.

The overall seismicity of the region is rather subdued, typical of an intra-plate region, exhibiting a long term moment-rate of only $M_O = 8\ 10^{20}$ dynes.cm/year, five orders of magnitude smaller than at plate boundaries. Also in terms of the complementary cumulative distribution, the seismic activity of the area is low and follows a non-linear recurrence law, Figure 3, with an annual probability of 10^{-4} corresponding to a magnitude of just under 6.0.

The spatial and to a lesser extent the temporal distribution of the more important events ($M_S > 4.5$) seem to remain clustered even for the relatively long period of observation of seven centuries. This may very well be due to the relatively small numbers available of the larger events which allow the small magnitude parasitic shocks to produce a "noise" of random events devoid of any significant signal associated with regional tectonics.

The conspicuous gaps of seismic activity around areas affected by early earthquakes, along the northeast coast and central Wales, shown in figure 1, clearly support the findings of a non-random distribution of events.

5.0 CONCLUSION

Preliminary investigations of the 19 July 1984 North Wales earthquake suggest that maximum on land Intensities were in the range V-VI (MSK) and that minor damage to vulnerable structures occured over an area of 250,000 km2. The surface wave Magnitude of the event from instrumental recordings and derived from the felt area of the shock agree at a value of $M_S = 4.5$. The earthquake is thus one of a series of small to medium magnitude events that have been reported in the area since the 13th century.

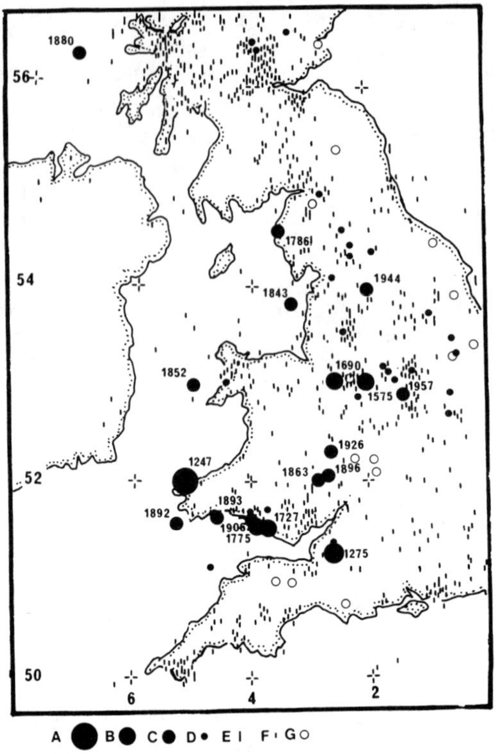

Figure 1: DISTRIBUTION OF EARTHQUAKES OF THE SOUTHWEST OF THE BRITISH ISLES (excluding aftershocks and swarms).

Earthquakes on land have a relatively good epicentral location, controlled by macroseismic observations. Offshore locations are approximate.
For most of the 20th century, earthquake Magnitudes have been re-calculated using amplitude-period readings from surface, Lg and Sg phases.
For historical events, Magnitudes have been assessed using macroseismic information calibrated against similar 20th-century shocks for which instrumental data is available.

A : 5.5 < Ms
B : 5.5 > Ms > 5.0
C : 5.0 > Ms > 4.5
D : 4.5 > Ms > 4.0
E : 4.0 > Ms > 3.5
F : 3.5 > Ms > 2.0 and shocks for which no M-determination was possible but with III < Io < V (MM).
G : Approximate location of epicentres of historical (pre-1800) earthquakes of Magnitude around Ms = 4.0.

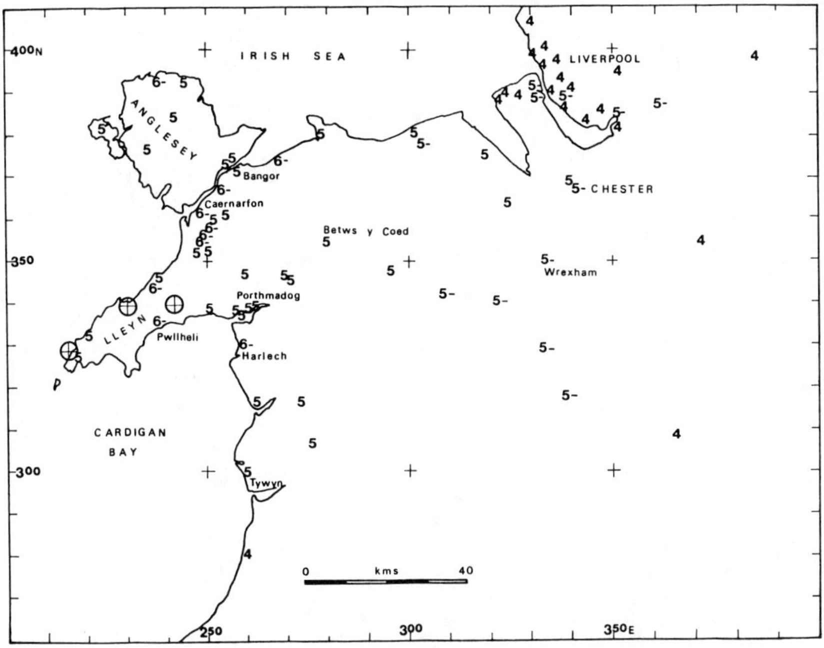

Figure 2: INTENSITY MAP IN THE EPICENTRAL AREA OF THE 19 JULY 1984 EARTHQUAKE.

Intensity points were obtained on the basis of field data and newspaper articles.
These assessments and the eventual construction of isoseismals require further additions from regional sources.
Instrumental locations of epicentres were provided by BGS, Strasbourg and LDG.

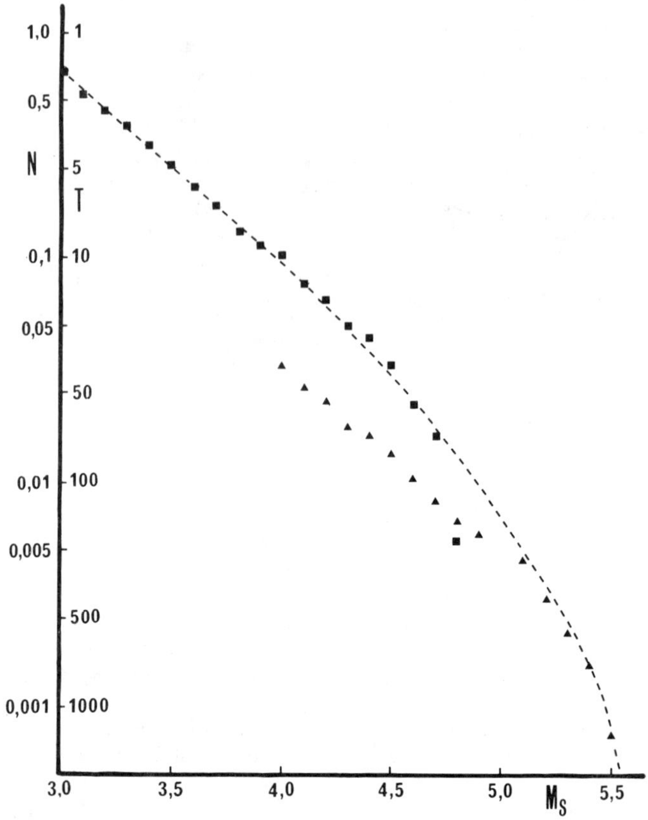

Figure 3: EARTHQUAKE RECURRENCE RELATIONS FOR THE LAND AREA (shown in fig. 1)

N = number of earthquakes per annum occurring in this area of magnitude equal to or greater than Ms. Ms = surface wave Magnitude; for 20th-century and historical events derived from intrumental or macroseismic data respectively.

Solid squares and triangles refer to the periods of 1800-1984 and 1247-1984 respectively, excluding the earthquake of July 1984. The former set of data is homogeneous for 3.0 < Ms < 4.7, and the latter for 5.1 < Ms < 5.5.

Discussion

MR J. IRVING, CEGB, Barnwood, Gloucester

I would like to illustrate strong motion recordings obtained by CEGB of one of the aftershocks of the North Wales earthquake of July 1984.

Following the main shock on 19 July 1984, the CEGB installed an SMA - 1 instrument on the floor of the security gatehouse at Trawsfynydd nuclear power station. The installation was completed by the evening of 19 July. On 27 July 1984, I was telephoned by the Station Physics Department with a message that, at 1256 hours (GMT), they had felt a strong tremor. I asked them to check the event indicator on the SMA - 1 and this was found to be positive.

The record was retrieved by Dr W. Aspinall and subsequently developed by CEGB. The trace showed an event of short duration and the film was given to Professor Ambraseys of Imperial College for processing, some of which was carried out by Dr Menu. The event with which the trace was connected was a 3.1 (M_L) event, as determined by BGS with a focal distance approximately 30 to 35 km from the instrument location. Figure 1 and Tables 1 to 3 show the raw trace, instrument factors and peak values of strong motion respectively. Figure 2 shows the digitized record of acceleration,[1] velocity and displacement. Figure 3 shows the acceleration response spectra. The peak recorded acceleration was 0.059 gravity which is higher than might be expected at this distance from a small magnitude event and there are other features of the spectra which indicate that there may have been some local ground, structural transmission path factors which were unusual.

At the time, the gatehouse guard noticed the large armour plated glass windows flexing, made a point of observing the car park standard lamps and saw them sway. No damage or malfunctions were found at the station either from the main shock nor this aftershock. Elsewhere in the area, detailed inspections were made but no structural damage was found in CEGB installations throughout the main and aftershock sequence. The most notable incident occurred on a 132 kv transformer at Four Crosses, near Pentir substation, which tripped and cut off supply for a large part of the Lleyn Peninsula. The District Engineer who investigated the fault

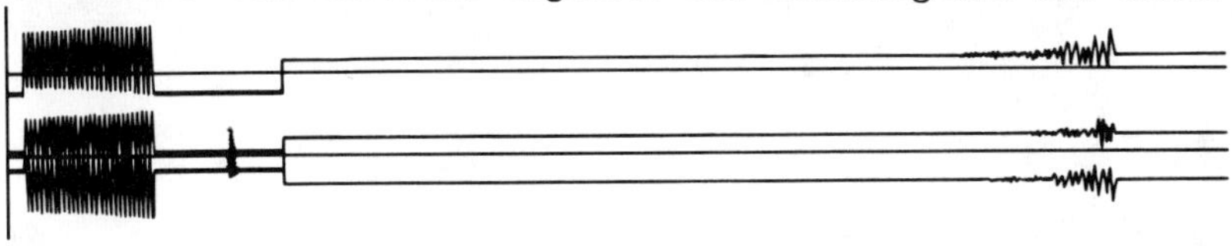

Fig. 1. Welsh earthquake - recorded traces and calibration test

Table 1. Scaling Factors

	Time	Acceleration
Longitudinal	3.9714	0.6054
Vertical	3.4025	0.5004
Transverse	3.9714	0.5875

Table 2.

Sensitivity for 1/4 g mm	Kinematrics Fo % crit.	Hz	Calibration 1 Fo % crit.	Hz	Calibration 2 Fo % crit.	Hz
16.4	0.60	13.6	0.67	37.7	0.68	13.6
17.0	0.60	13.3	0.68	13.7	0.70	13.5
16.9	0.60	13.4	0.70	13.7	0.72	13.5

Table 3. Peak values of the strong motion

		Long.	Vert.	Trans.
Acceleration	g	0.045	0.029	0.059
Velocity	cm/s	0.7	0.4	1.0

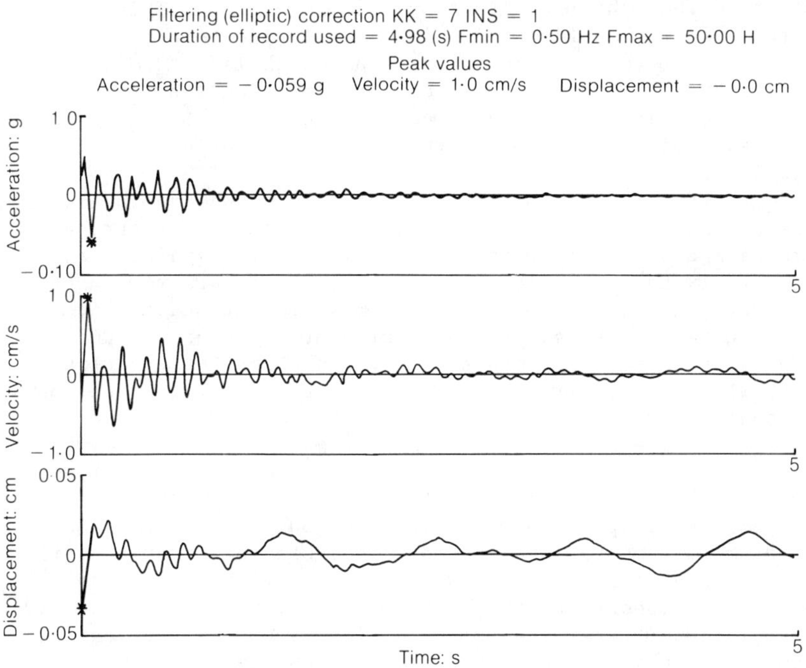

Fig. 2. Welsh earthquake (aftershock of 27 July 1984): transverse component: acceleration, velocity and displacement

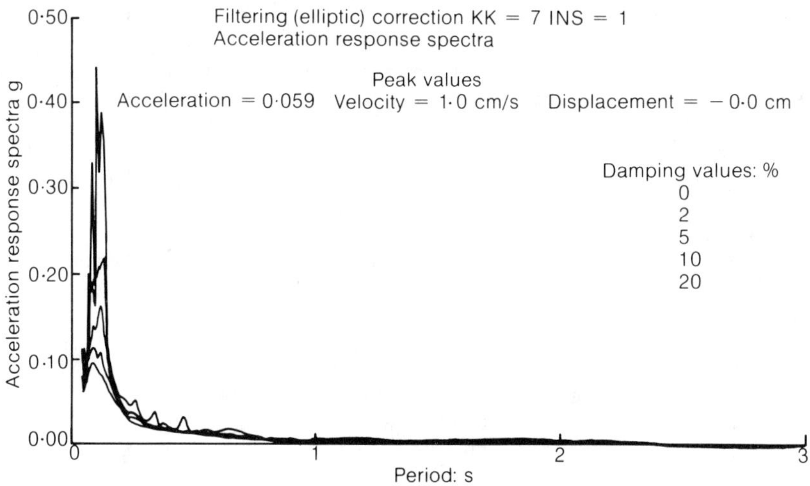

Fig. 3. Welsh earthquake (aftershock of 27 July 1984): transverse component: acceleration response spectra

reported that the Bucholtz trip relay operated. This is associated with a flap valve in the oil pipe feed from a reservoir above the transformer tank which is set to detect a sudden change in oil flow in either direction and is very sensitive. This could have been caused either by sloshing in the reservoir or perhaps by the flap valve moving under its own inertia.

In answer to Professor Finn who has asked whether he may be present at an historic occasion, to whit the first strong ground motion record ever obtained in Great Britain, I can only reply that I believe that is so. There are some unusual features in the trace which might make it dubious as an earthquake recording, but the timing of the event by the Station Physics Departments observations and the experienced installation, calibration, and prompt retrieval carried out personally by Dr Aspinall led me to the conclusion that we have a genuine strong motion recording.

DR W. P. ASPINALL, Principia Mechanica
The accelerograph record was obtained on a Kinemetrics SMA - 1 instrument temporarily located in the single-storey security gatehouse at Trawsfynydd power station during one of the larger aftershocks of the North Wales earthquake. It should be pointed out that the SMA - 1 triggers on the vertical component only and that it is quite possible that the trigger threshold was not exceeded until quite late in the event. Thus the record does not look quite like other strong motion traces but is probably only the coda of the event, with the signature dominated by transmission path scattering and by local geology at the station. Further, the hypocentral distance, about 45 Km, is great enough to make this a far-field not a near-field record. As the instrument was in perfect working order and I recovered the film within a couple of hours of the aftershock I have no doubt that this is a genuine record, albeit a very weak-motion accelerogram.

DR SKIPP
Mr Menu referred to his test for completeness. Did he use methods such as that proposed by Stepp?[2] Should distinction not be made between completeness statistically - i.e. deficiency in an M of greater than 4.5 because the record has not been long enough and the completeness or otherwise which refers to the capacity of the observatory system?

MR G. NEILSON, British Geological Survey
Mr Menu quotes 'the Richter magnitude was M_L = 5.7.' Is this a local magnitude or is it an M_{LH} surface wave magnitude? Could you also expand on the way in which this magnitude was determined?

J. M. H. MENU
Dr Aspinall's comments are essential for the qualification of the record as a genuine one. I am particularly satisified to

learn that the SMA - 1 recording instrument was situated in a single-storey type of construction and also that, to his view, the local geology at the station played a dominant role. These two points correlate reasonably with the results of the processing which was carried out without this information.[3]

Fourier amplitude spectra of the three components (Fig 4) exhibit strong predominant frequencies at 8-10 Hz and this certainly fits the natural frequency of the gatehouse or of the soil layer or both. Traces for the three components against time are shown in Fig. 5.

The two calibrations carried out before and after triggering have enabled the constants of the sensor to be recomputed. Although they show some variations with the values given by Kinemetrics, they are similar, which tends to indicate that there was not any kind of electromagnetic triggering which would have affected these characteristics.

To define the completeness of a set of data is not straightforward and depends on the model the data are referred to. I am not sure whether the approach which refers to the capacity of the observation system is applicable here. The method used was that proposed by Stepp which consists of testing the stationarity of

$$s = \frac{(\text{No of earthquakes per time interval}/T)^{1/2}}{T^{1/2}}$$

where T is a sample length measured backwards starting from the most recent year of the set.

The magnitude of M_L is a local magnitude. It was simply obtained by averaging values obtained by eleven stations in France (LDG network) and 4 stations in Germany. The final result was $M_L = 5.7$ (± 0.15).

REFERENCES

1. MENU J. M. H. A computer programme to correct the strong motion acceleration by efficient filters: ELLICOR. Imperial College Report, 1983.

2. STEPP J. C. Analysis of completeness of the earthquake sample in the Puget Sound area - contribution to seismic zoning. National Oceanic And Atmospheric Administration - TR - ERL 267, ESL 30, Boulder, Colorado.

3. MENU J. M. H. The Welsh earthquake stong motion record (aftershock of 27 July 1984). Preliminary Imperial College Report, CEGB, Jan 1985.

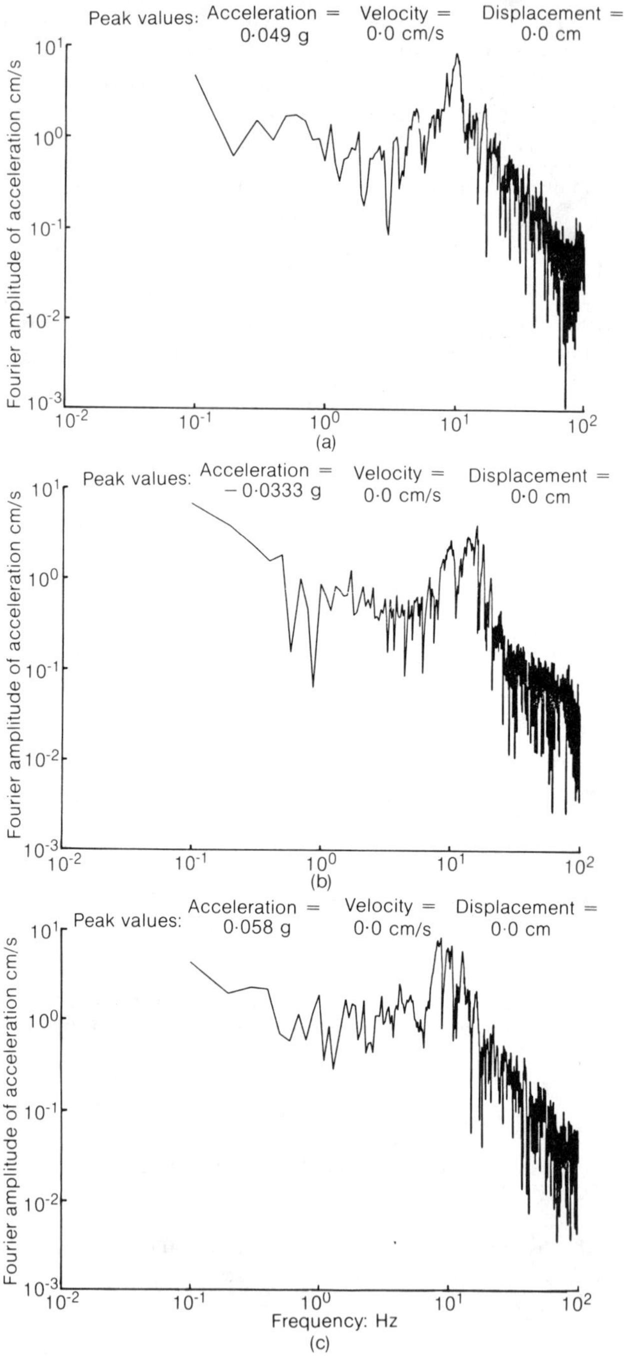

Fig. 4. Welsh earthquake (aftershock of 27 July 1984): Fourier analysis spectra, uncorrected record KK=0, frequency resolution = 0.1953 Hz, sample frequency = 200.0 Hz - components (a) longitudinal (b) vertical (c) transverse

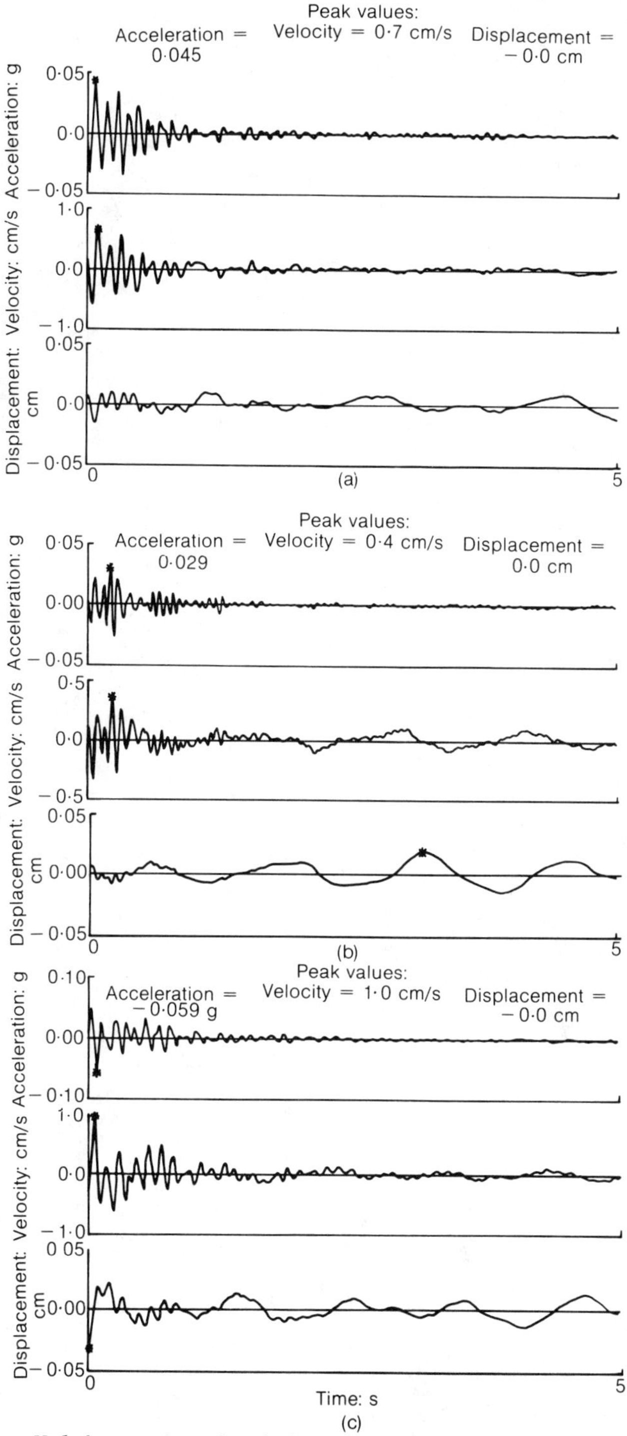

Fig. 5. Welsh earthquake (aftershock of 27 July 1984) traces: filtering (elliptic) correction KK=7, INS=1, duration of record = 4.98s, F_{min} = 0.50 Hz, F_{max} = 50.00Hz, components (a) longitudinal (b) vertical (c) transverse

SEISMIC ENGINEERING

18. Seismic engineering requirements in British industry

M. A. H. G. ALDERSON, MSc, MIMechE, United Kingdom Atomic Energy Authority

SYNOPSIS. The implications of the effect of earthquakes on and the potential consequences of failure within hazardous industrial plant have been discussed in relation to the acceptability of the risk involved. A methodology has been described which enables the risk due to structural failure as a result of earthquakes occurring at a rate typical of the United Kingdom to be quantified and thus usefully leading to a comparison with the risk due to other initiating events. The method is still developing but provides a useful way of selecting suitable levels for design.

INTRODUCTION

1. The increasing awareness of risk has precipitated various studies on the development of methodologies for quantifying the hazards and consequences from a wide range of potential phenomena. Earthquakes in Great Britain can be regarded as low probability events, especially those of magnitude 4.5 and above. However with the increasing reliance on technology, the consequence of damage to a number of installations in this country can represent a hazard which might not be acceptable.

2. At the present time, the Nuclear Industry is the only industry where there is a definite requirement to consider the occurrence of earthquakes in the design and operation of the plants. To be consistent, many other industries and installations should consider the earthquake hazard in the safety assessment of their operation and this paper is intended to outline the requirements in this direction.

3. It is the intention to discuss the characterisation of the earthquake hazard at a site and to consider the consequences in broad terms. In order to illustrate the consequences, the effects of several large recent earthquakes are briefly described. The dangers of ignoring the lessons to be learnt from studying past damage are emphasised.

4. The consideration of the acceptability of risk is prefaced by a description of the historical development of the approach. The response of the site and structure to the earthquake loading is outlined including the consideration of services and equipment.

5. The assessment of a design with respect to its acceptability in terms of safety and risk is described. It is at this stage that an iterative procedure may be found to be necessary. The methodology has the potential for quantifying the risk due to the occurrence of earthquakes and thus enable comparisons to be made with the risk due to other initiating events.

THE HAZARD

6. Because the United Kingdom is in an area of relatively low seismicity, the probability of a person at a particular point in this country experiencing a large or damaging earthquake is relatively low and corresponds to a return period in excess of his allotted time-span of three score years and ten. Hence the reluctance to consider earthquakes as a hazard in the normal course of design. However, earthquakes have occurred and continue to occur in the United Kingdom and represent a quantifiable environmental hazard.

7. The seismic hazard at a site refers to some suitable characterisation of the naturally random occurrence of earthquake intensities in the future. This topic has been covered in more detail in the first part of the Conference and the characterisation is generally expressed as a hazard function L(s) where L(s) is the probability that the intensity s will be exceeded at the site, at least once per year. The quantity s may represent peak ground acceleration, peak ground velocity, Modified Mercalli Intensity, or any other parameter related to the potential for causing damage.

8. Several methods are available for determining the hazard function L(s). These can be simply divided into two groups, the widely used 'seismic source' methods as described by Cornell (refs 1,2), Cornell and Merz (ref 3) and McGuire (ref 4) and the 'historical' methods described by Bernreuter (ref 5) and Milne and Davenport (ref 6) and extended by Veneziano et al (ref 7). These methods lead to a characterisation of the seismic hazard at the site, which can typically be represented by Fig 1.

9. The consequences of the earthquake hazard are twofold:

(a) Injury and loss of human life,
(b) Destruction of and damage to the man-made and natural environment.

10. It is only possible to reduce and not to eliminate these consequences for large earthquakes for financial and technical reasons and the basic aims of the designer are to minimise the loss of life through either structural collapse or secondary damage such as falling debris, fire or the release of toxic substances, and to reduce the damage and loss of use of the man-made environment. This naturally implies an approach towards an ideal situation or target and this will be discussed later.

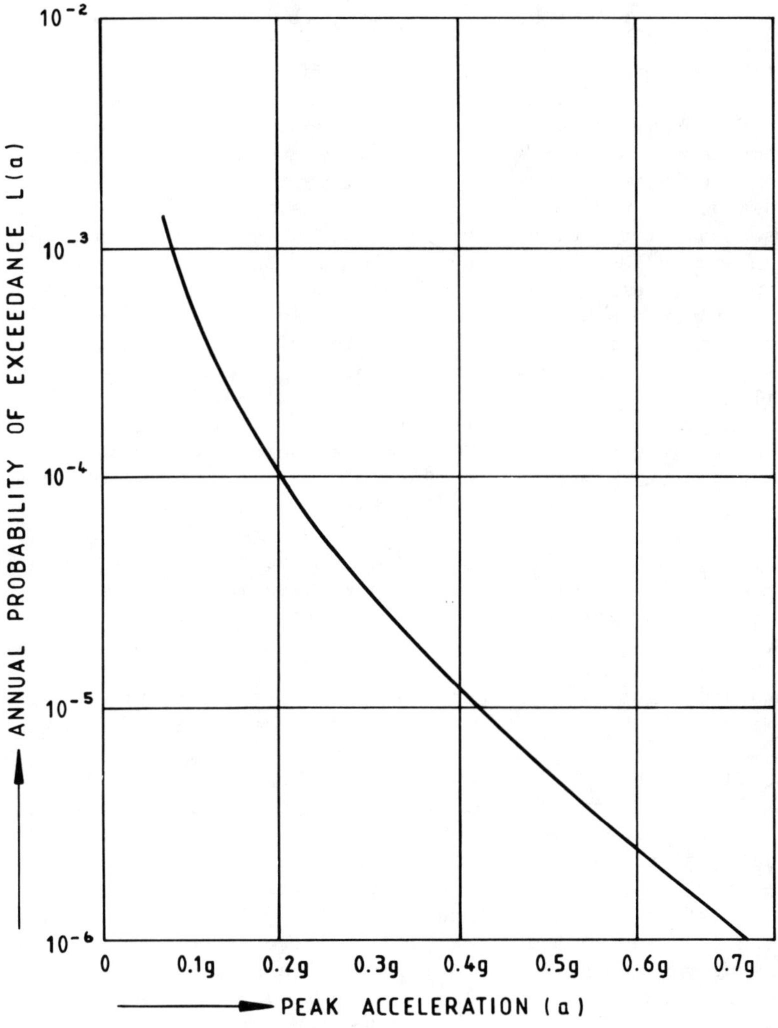

Fig 1. Typical seismic hazard curve

11. Before considering the risks involved it will be useful to be reminded of some examples of past damage from earthquakes. Several large earthquakes have occurred during the last thirty years and have been studied in some depth. These have been reviewed by Alderson (ref 8) and give an appreciation of the sort of damage that is sustained during and after the events and also the wide variation in the damage.

12. The earthquake damage described by Alderson (ref 8) has indicated several features relevant to major hazard plant. In brief, during the 1952 Kern County event, steel and reinforced concrete or concrete block structures erected on thick alluvium with a relatively deep water table, performed well. Brick construction was not so effective. The presence of flammable gases and electricity services on one site resulted in a large fire and explosion, although prompt action in closing valves reduced the damage that might have occurred. Cast iron pipes did not perform well and fluid surges in piping systems can have serious effects. Electrical installations were badly damaged and elevated water tanks of the inverted pendulum type failed in many instances. At Eureka in 1954, which was a lesser event, again, electrical installations and elevated water tanks were damaged. There was panic but only slight damage at San Francisco in 1957. A small fire broke out due to a failed gas pipe and several water lines were ruptured.

13. The Chilean earthquakes of 1960 were different in that building standards were lower than in the developed world. The damage to the steel works could have been much more serious due to the coincidental failure of electricity and water suplies. The Skopje earthquake of 1963 was of a very short duration and much of the damage was the result of falling debris. A longer duration might have produced quite a different story. The damage in Alaska in 1964 was very much the result of massive ground failure and extremely long duration. Equipment damage was extensive. The Columbian earthquake of 1967 affected a mainly urban area but again vessel support failure damaged connecting pipework. The San Fernando 1971 event caused a great deal of damage to both modern and old structures. Electrical and telephone equipment was also severely damaged. The Friuli series of earthquakes in 1976 also affected a mainly urban area and showed the effect of large aftershocks on already damaged buildings. The Santa Barbara event of 1978 demonstrated the effects of local geology on the damage pattern. The timing and short duration certainly reduced the amount of damage that might have occurred.

14. It is important to recognise and understand the features of previous earthquakes, and to take this information into account when planning and designing a plant of a sensitive nature. Berg and Hanson (ref 9) have pointed out for example, that the geological hazard in and around Anchorage was known and documented some five years previous to the earthquake, but that knowledge went unheeded. Often structures built on alluvium were damaged considerably more than those built on firm ground nearer the epicentre. Not only was there more

damage, but these sites were also more sensitive to distant earthquakes. In some cases, man has altered geological hazards and possibly created new ones. We have come to recognise that seismic activity in the vicinity of dams may increase as the reservoir is filled. Koyna in India is situated in a seismically stable region. Earthquakes were rare until Koyna Dam was built, but tremors were felt frequently after filling began. In 1967, a destructive earthquake occurred and the dam was destroyed. The reservoir was apparently the trigger, although not the underlying cause. Many lessons have been learnt from destructive earthquakes, but too often the lessons are forgotten or unheeded. Kletz (ref 10) has pointed out that there are several examples of history repeating itself.

15. An important feature which distinguishes the earthquake from other external hazards is that an earthquake affects everything on the particular site in question. That is, the plant or plants, the control buildings and control systems of the plants, the operators, the connections to the electricity grid, the telephone network, mains water supply etc. Damage or failure at one particular location can have a 'knock-on' or 'domino' effect on other items of plant already weakened or affected by the same earthquake. From this it can be seen that this results in a very difficult situation to analyse. A practical example of such an escalation was at the Palomar Recycling Plant where flammable gas leaking from a damaged tank was ignited by sparks from a damaged transformer resulting in a serious fire and explosion. This happened during the Kern County earthquake in California where the operators were aware of the occurrence of earthquakes. Prompt action on their part in closing valves resulted in the isolation of various parts of the plant and reduced the amount of damage that might have occurred.

16. Two lessons are immediately apparent from this incident. It is reasonable to assume that the occurrence of such an earthquake in an area of less common seismicity, acting on a similar plant, would have resulted in much more damage, primarily through lack of awareness or understanding of earthquakes. The second lesson is that this incident occurred in one plant under central control. Had there been more than one plant each under separate control, then, in order to minimise the interaction, communication between control centres would be essential and this communication link would have to survive the earthquake. Although the dynamic behaviour of large components containing hazardous materials is important, it is the attention to details in understanding previous events that will make the difference between minimal and serious loss prevention.

17. An attempt has been made to illustrate and identify the possible features of the interaction of a severe earthquake with structures of the type where damage could lead to high

consequences. So far this particular combination has not occurred to any large extent anywhere in the world. However, two further points emerge which should be emphasised:

(a) trivial failures of equipment in conventional buildings do occur and continue to occur;

(b) failure of similar equipment could have serious implications on the control and safety of a hazardous installation, even if the main structure remained intact.

THE RISK

18. In discussing the risk from the occurrence of earthquakes, it is useful to review the historical development of the approach to safety philosophy. As society developed after the start of the industrial revolution, legislation was gradually introduced to ensure certain basic minimum standards. Experience had shown that the concept of rating success as equivalent to maximising profits had led to the requirement that costs should be reduced to an absolute minimum. This in turn had led to unsatisfactory working conditions and the impingement of industrial activities on the environment. That is an almost complete disregard for the well-being of the workforce and the population in the immediate neighbourhood.

19. This unsatisfactory state of affairs was slowly recognised and a plethora of statutes introduced in this country, concerning working conditions, working hours, safety, pollution, public health etc. Some of these statutes did in fact reflect the desire of society to prevent recurring accidents, for example the Boiler Explosions Acts of 1882 and 1890.

20. The chief concern with regard to safety was directed towards improving conditions in the workplace. This was mainly because accidents involving employees were well known, but the attitude towards public safety was not usually of great interest. Many other countries adopted legislation on safety which was broadly similar to that in this country. The main facet of the legislation was the recognition or indeed acceptance that absolute safety was impossible to achieve and therefore the concept of 'reasonably practicable' was a necessary feature.

21. The phrase 'reasonably practicable' has been in use in UK safety legislation for a period of time approaching a century, and although it is not defined anywhere in the legislation, case law has clarified its meaning. The definition (ref 11) adopted by the Courts in Civil Actions, is repeated here for completeness:

> 'Reasonably practicable' is a narrower term than 'physically possible', and in the Judge's opinion would imply that a computation must be made in which the

quantum of risk is placed in one scale and the sacrifice involved in the measures necessary for averting the risk (whether in money, time or trouble) is placed in the other, and that, if it be shown that there is a gross disproportion between them - the risk being insignificant in relation to the sacrifice - the defendants discharge the onus upon them. Moreover, this computation falls to be made by the owner (of the mine, factory, etc) at a point of time anterior to the accident (ref 12).

22. The concept that 'reasonably practicable' entails a comparison between costs and risk, has not, therefore, developed from the use of quantified risk analysis, but considerably predates such techniques. It should also be noted that the above definition does not suggest a strict balance between cost and risk but implies that the risk should be reduced to insignificance in relation to the cost.

23. In recent times, a number of developments have had a significant bearing on the attitude towards safety of the public authorities, design organisations and operating companies. These have included:

(a) The commercial exploitation of the nuclear power industry, perceived by the public still as a novel technology with the potential for large accidents.
(b) An increase in the scale of many industries, due to economic factors, which has led to very much larger inventories in the chemical and petro-chemical industry, resulting in the potential for very damaging accidents to occur.
(c) The development of assessment techniques for the quantification of risk.
(d) The steady reduction of 'background' risks due to continuous efforts to reduce accidental deaths and improved medical care.
(e) The attention of the news media to serious accidents and those industries where large accidents are possible.
(f) The influence of pressure groups to provoke investigation of other situations.

24. The logical study of safety through identifying possible hazards and causes leading to a quantitative assessment of the potential risk has provided a powerful discipline in which both designers and operators can deploy effort in the areas where the most beneficial reductions in risk can be achieved. It is now standard practice to use these techniques especially where the scale of possible accidents meant that the predominantly retrospective approach of learning by experience without prior consideration of how such accidents might occur, is totally unacceptable. Although these new techniques have not yet been incorporated into the legal framework, most authorities accept and encourage their use.

25. The availability of such methodology has, of course, raised the philosophical question of what levels of public risk, as estimated by these techniques, may be regarded as acceptable, and how legal phraseology such as 'reasonably practicable' should be interpreted when such methods are used.

26. A full discussion on the acceptability of risk is beyond the scope of this paper. However, on the assumption that some target level of risk exists and which is related to the consequences of the various hazards under consideration in the design process, it is clear that the occurrence of earthquakes is one of the hazards which the designer must recognise when structural damage or loss of function due to ground movement can lead to public safety being jeopardised. The question of acceptability of seismic risk will be returned to in the Chapter on Assessment

THE RESPONSE

27. Having discussed the earthquake hazard and the risks involved, it is time to consider the response to earthquakes. The fundamental feature of an earthquake is the fact that it affects the total installation being considered including the site. Although the response to a given earthquake will vary enormously within a plant, no part can be neglected in the design process. Before the design process can begin however, it is necessary to select or define the design earthquake or earthquakes. It can immediately be seen that this is the start of a possibly iterative procedure since it is only by going through an initial design phase can any measure of its acceptability be made.

28. The United Kingdom is beset by a fundamental difficulty in that, to date, no strong motion earthquake records exist for any site. Even if such records did exist, it would be extremely unlikely that they would have been recorded at the site being considered for a particular design, or in fact that the identical earthquake would be repeated. Hence, the selection of the design earthquake is of necessity a matter of judgement involving the study of past events in other locations.

29. The local geology and soil conditions have a profound effect on the site response. If one considers an area in which distances are small compared with the focal distance, local differences in the geology and soil can produce quite different surface ground motions within this area from the same event. Dowrick (ref 13) has discussed the more important features in some detail.

30. Experience in the examination of earthquake damage has indicated that any weakness in the design will manifest itself during the vibration and that the simplest structures have the best chance of survival. Therefore, when embarking upon a design, the guiding principles should be:

(a) to keep it simple, since a simple design is better understood,

(b) to maintain symmetry, since asymmetry produces additional torsional loadings,

(c) to keep overall size down to minimise effects of differential settlement and phase of loadings,

(d) to maintain a reasonable uniformity of strength distribution and thus avoid unnecessary stress concentrations,

(e) ensure failure of horizontal members before vertical members to postpone complete collapse, and

(f) relate the rigidity of the structure to the local subsoil conditions and benefit from the filter action of the ground.

The choice of material is obviously important but often dictated by other considerations.

31. Returning to the design earthquake, it is necessary to incorporate the characteristics of earthquake ground motion into some form which can be used for a design analysis. The simplest approach is to assume a completely rigid structure and that all points move together in response to the ground disturbance. When this is combined with the peak ground acceleration we have the simplest procedure of all corresponding to the base shear coefficient method as recommended by the various building codes.

32. However, for structures where failure can have high consequences, this type of approach is no longer considered to be suitable. The design earthquake can be specified in the form of a response spectrum or a time history generated from a response spectrum, which is compatible with United Kingdom conditions. A methodology has been developed in the nuclear industry which follows this approach and which will be the subject of the paper by Irving and Smith (ref 14) in this Conference. It is considered that this technology should be transferred to the non-nuclear industry where the consequences of failure can be of the same order, or more, than the nuclear industry.

33. In addition to the main task of structural design is the consideration of various items of equipment and services. It is important to recognise that all are affected by the earthquake and that many failures in the past have been attributed to lack of attention to details such as sufficient anchorage. Often equipment itself is quite robust, but failures can arise for trivial reasons such as sloshing oil in transformers for example. Thorough consideration of all these things can make all the difference between superficial damage and complete loss of plant coupled with loss of life.

THE ASSESSMENT

34. The actual choice of the design level of earthquake is of course determined by the consequences of failure or loss of function, and the acceptability of such failure or loss of function when extended to the effects on the population and the environment. Hence the need for some form of assessment at an early stage to ascertain the acceptability of the first choice of design level.

35. A typical hazard curve has been presented in Fig 1 and it is important to appreciate that this represents the hazard corresponding to the occurrence of earthquakes, not the hazard posed by one event. Thus, for these purposes we are electing to use peak ground acceleration as the independent variable representing the loading.

36. The calculation of the overall risk due to earthquakes is dependent upon the probability of failure of the structure due to the occurrence of earthquakes. This probability is determined by combining the conditional failure probability of the structure with the seismic hazard curve referred to earlier.

37. The conditional failure probability of the structure or fragility curve as it is known, is best represented by a log-normal distribution based on peak ground acceleration as the independent variable and was originally proposed by Ang and Newmark (ref 15). This can be expressed as:

$$P(a) = \Phi\left(\frac{\ln(a/m)}{\xi}\right) \tag{1}$$

in which $\Phi(.)$ is the standard normal cumulative distribution function

a is the peak ground acceleration

m is the median level of peak ground acceleration, that is the level of peak ground acceleration for 50% probability of failure

ξ is the variance given by $\xi^2 = \ln(1+\Omega^2)$

and Ω is the coefficient-of-variation representing the uncertainty in the structure

38. Several areas of uncertainty surround the prediction of the structural response. The effects of the site dependent factors on the definition of ground motion, which includes the effects of local soil conditions and geology, are to a large extent based on scarce information but the uncertainty has been assessed by Newmark (ref 16) to be the order of 20%, expressed as a coefficient-of-variation.

39. The effect of variability in the amplification factors in the response spectra, based on the work of Mohraz, Hall and Newmark (ref 17) has been assumed to be the order of 30%

expressed as a coefficient-of-variation. The uncertainty associated with the damping coefficients, used in the response calculation of a structure, has been suggested by Newmark (ref 16) to be of the order of 30% expressed as a coefficient-of-variation.

40. Because of the complexity of a typical hazardous plant, there is a significant amount of uncertainty in the mathematical modelling of such a structure and the underlying soil and this could be of the order of 30%, expressed as a coefficient-of-variation, although this value may be a little high.

41. Ang (ref 18) and Ang and Cornell (ref 19) have shown that by first order approximations, the total uncertainty associated with the prediction of the response of a structure is the square root of the sum of the squares of the individual uncertainties. that is:

$$\Omega_1 = (0.2^2 + 0.3^2 + 0.3^2 + 0.3^2)^{1/2}$$

$$= 0.557$$

42. The uncertainty associated with the estimation of the capacity of the structure is considered to be comparable to that associated with the prediction of the resistance to blast forces as determined by Ang, Hall and Hendron (ref 20), and this is taken to be 30% expressed as a coefficient-of-variation.

43. The total uncertainty covering both response and capacity is therefore:

$$\Omega = (0.557^2 + 0.3^2)^{1/2}$$

$$= 0.632$$

from which

$$\xi = (\ln(1 + 0.632^2))^{1/2}$$

$$= 0.58$$

44. As stated earlier, the probability of failure due to earthquakes is determined by combining the conditional failure probability of the structure with the hazard curve and is thus given by:

$$F(g) = \int_0^g L(a)\left(\frac{dP}{da}\right) da \qquad (2)$$

where L(a) is the seismic hazard curve
and P(a) is the conditional probability of failure as before.

45. Equation (2) can be evaluated as a function of m the median value by integrating over a range to a very large value of g and the results shown in Figure 2. Also on this figure is shown the effects of increasing and decreasing all the uncertainties by $\pm 5\%$

46. Now to return to the design process. A design level of peak acceleration (a_d) is chosen and this is used to characterise the input response spectrum or time history. It is at this level that the structure, according to some ground rules, code or guide is considered to be safe or satisfactory. Now there will be some higher level of loading of the structure at which 'failure' or 'loss of function' will occur and the level at which this just occurs is the median level m, ie the level at which the structure might or might not fail, or 50% probability of failure.

47. The ratio of m the median level of peak acceleration to the design level a_d is known as the Factor of Safety (FOS) (not to be confused with the more conventional definitions of factor of safety). Figure 3 illustrates a generalised plot covering a wide range of variables relating the design level of peak acceleration to the annual probability of failure through the uncertainty and the factor of safety FOS.

48. It is from this figure that the acceptability of the initial choice of design level can be judged. It is quite possible that 'failure' or 'loss of function' will not lead directly to the release of toxic substances for example and due allowance must be made for this in the form of an additional factor. However, it does illustrate the relative sensitivity of the risk to the factor of safety FOS and the relative insensitivity to the uncertainty parameter.

49. This approach can of course be extended to other parts of a typical plant and a complete characterisation of the response to earthquake loading obtained. It is thus possible, using event tree logic for example, to determine for a plant the risk due to the occurrence of earthquakes which will enable this to be compared with the many other risks normally considered in the design and operation of a hazardous installation.

50. By this form of comparison the suitability of the choice of design level can be ascertained by considering the acceptability or otherwise of the overall seismic risk and hence the possibility of an iterative procedure. It is only by this type of exercise can the relative importance of seismic design be determined.

SUMMARY AND CONCLUSIONS

51. The occurrence of earthquakes in the United Kingdom, although at a lower frequency than other parts of the world, nevertheless represents a hazard which has the potential for causing considerable damage. It is not justifiable to design for the largest ever earthquake that has ever occurred and this results in a choice having to be made of the appropriate level for design commensurate with acceptable safety.

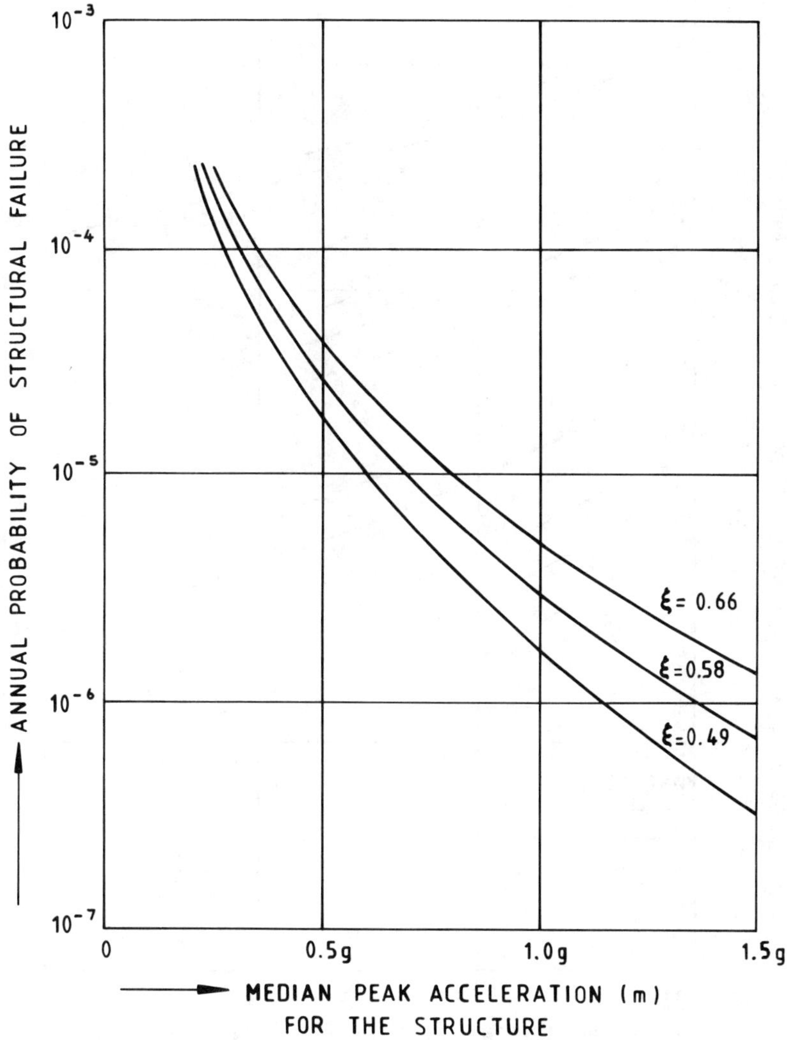

Fig 2. Variation of annual probability of structural failure with median resistance and uncertainty

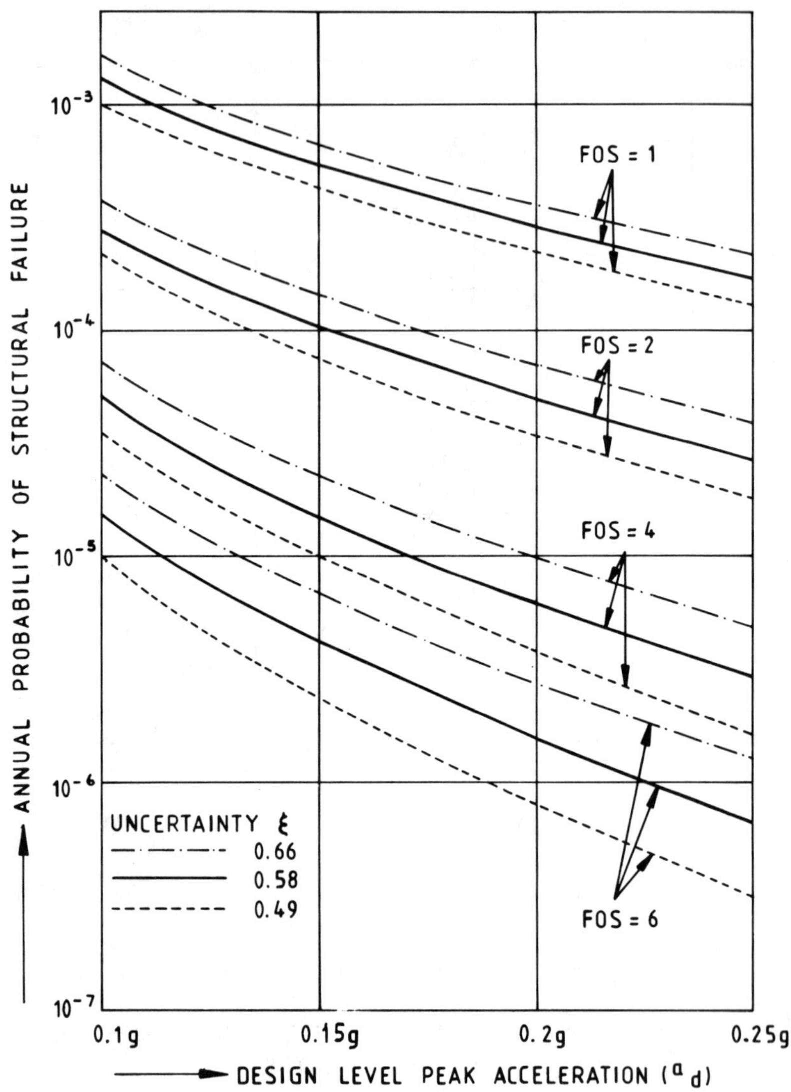

Fig 3. Variation of annual probability of structural failure with design level, factor of safety and uncertainty

52. A methodology has been described which is still developing and which has the potential for quantifying the seismic risk in order that this can be compared with the risk due to other initiating events. It has been used with varying levels of complexity on existing nuclear power plants in the United States and is believed to be the only way that a comparative risk study can be made.

53. As more data becomes available it will be possible to refine the uncertainties in the evaluation of risk although it will still be necessary to ensure a reasonably large FOS unless design levels are allowed to increase, to achieve acceptable overall risk levels.

ACKNOWLEDGEMENT

54. The views expressed in this paper are those of the author and do not necessarily reflect the views or policy of the UKAEA. The author wishes to acknowledge many stimulating discussions with his colleagues, in particular Dr D W Phillips who made several helpful suggestions during the preparation of the text.

REFERENCES

1. CORNELL C.A. Engineering Seismic Risk Analysis, Bull. Seism. Society Am. vol. 58, no. 5, pp 1583-1606, October 1968.
2. CORNELL C.A. Probabilistic Analysis of Damage to Structures under Seismic Load, chapter 27 of Dynamic Waves in Civil Engineering, D.A. Howells, I.P. Hough and C Taylor eds. Wiley Interscience, London, pp 473-488, 1971.
3. CORNELL C.A. and MERZ H.A. Seismic Risk Analysis of Boston, J. of the Structural Division, ASCE, vol. 101, no. ST10, pp2027-2043, October 1975.
4. McGUIRE R.K. FORTRAN Computer Program for Seismic Risk Analysis, US Geological Survey, Open-File Report 76-67, January 1976.
5. BERNREUTER D.L. Seismic Hazard Analysis, NUREG/CR-1582, 1981.
6. MILNE W.G. and DAVENPORT A.G. Distribution of Earthquake Risk in Canada, Bull. Seis. Soc. Am, vol. 59, no. 2, pp729-754, 1969.
7. VENEZIANO D., CORNELL C.A. and O'HARA T. Historical Method of Seismic Hazard Analysis, EPRI NP-3438, 1984.
8. ALDERSON M.A.H.G. Seismic Design Criteria and their Applicability to Major Hazard Plant Within the United Kingdom, UKAEA Report SRD R246, 1982.
9. BERG G.V. and HANSON R.D. Engineering Lessons Taught by Earthquakes. Proc. of the Fifth World Conference in Earthquake Engineering, Rome, Italy, 1973.

10. KLETZ T.A. Organisations Have No Memory. Loss Prevention, vol. 13, A.I.Ch.E. Chemical Engineering Progress, 1980, Paper presented at 86th Annual Meeting of the A.I.Ch.E., Houston, Texas, 1979.
11. THE ROYAL SOCIETY. Risk Assessment. A Study Group Report. The Royal Society, 1983.
12. ASQUITH, LORD JUSTICE. In: Edwards vs National Coal Board (1949 1KB 704; 1949 1 All ER 743). Also Marshall vs Gotham Co Ltd, (1954 AC 360; 1954 1 All ER 937). Quoted in Redgrave's Health and Safety in Factories, by I. Fife and E.A. Mechin (1976), London, Butterworth.
13. DOWRICK D.J. Earthquake Resistant Design, John Wiley and Sons, 1977.
14. IRVING J. and SMITH C.R. Engineering Safety Aspects Related to Sizewell. Proc. Conf. on Earthquakes and Earthquake Engineering in Britain. S.E.C.E.D. University of East Anglia, April 1985.
15. ANG A. H-S. and NEWMARK N.M. A Probabilistic Seismic Safety Assessment of the Diablo Canyon Nuclear Power Plant. Report to the USNRC, September 1975.
16. NEWMARK N.M. Probability of Predicted Seismic Damage in Relation to Nuclear Reactor Facility Design. Draft Report to USNRC, September 1975.
17. MOHRAZ B., HALL W.J. and NEWMARK N.M. A Study of Vertical and Horizontal Earthquake Spectra. Report to USAEC, December 1972.
18. ANG, A. H-S. Structural Risk Analysis and Reliability Based Design. ASCE J. of Struct. Div., vol. 99, no ST9, pp1891-1910, September 1973.
19. ANG, A. H-S. and CORNELL C.A. Reliability Bases of Structural Safety and Design. ASCE J. of Struct. Div., vol. 100, no. ST9, pp1755-1769, September 1974.
20. ANG, A. H-S., HALL W.H. and HENDRON A.J. Uncertainty and Survivability Evaluations of Design to Air-Blast and Ground Shock. Final Report to US Army Construction Engineering Research Laboratory, 1974.

Discussion

MR E. BOOTH

A criticism of the type of probabilistic risk assessment methods described in the paper is that the methods do not account for gross human error. Could Mr Alderson comment?

DR KUNAR

Mr Alderson presented the procedure of using probabilistic methods for assessing the safety of hazardous structures, to include all uncertainties. Is this a realistic proposition as uncertainties may not be quantifiable at this time?

MR ALDERSON
The main problem with including human error in probabilistic risk assessment is one of lack of accurate data. This is gradually being rectified and data is being collected now from which an estimate can be made of the contribution to risk that human error can make. A difficulty is caused by the fact that all errors might not be reported. Design and construction errors are very difficult to quantify for the same reason. The way in which the information is incorporated in probabilistic risk assessment varies according to the type but is usually considered to be part of the uncertainty propagated through the analysis.

Probabilistic risk analysis has the potential for determining the overall risk of a particular operation. It is developing technology with the major contribution emanating from within the nuclear industry. The question of uncertainties is a difficult one especially in the area of external hazards as initiating events, since there is of course uncertainty associated with the hazard occurrence. This can lead to the probability of a particular plant damage state lying in a range of up to three orders of magnitude. However, it is still possible to use the methodology to determine the relative importance of various aspects of the design and this will become more quantitative as the uncertainties are reduced with the natural gain in experience.

19. Earthquake hazard in Britain

J. IRVING, BSc, MICE, Central Electricity Generating Board

SYNOPSIS. The paper describes how seismic design specifications have evolved from studies of earthquake hazard and risk carried out by CEGB with the assistance of specialist consultants engaged to acquire and assess historical and seismological data. The results of hazard analyses are described and examples are given of associated topics including long period ground motions, the reliability of electrical grid systems, and the quantification of the statistical uncertainties associated with probabilistic estimates of earthquake hazard in Britain.

INTRODUCTION

1. Unlike other natural phenomena such as tempest and flood, British earthquakes have aroused interest more as curiosities than for their destructive force. Even the famous events, for example at Colchester in 1884, caused little damage by world-wide earthquake standards. Were this not the case, the public perception of risk to life and property would have demanded long ago that the majority of buildings were designed to withstand earthquakes. As far as the author can determine, the only structures built prior to 1970 allowing for significant earthquake forces were a number of dams in an area of Scotland that had suffered lengthy and repetitive swarms of moderately sized earthquakes.

2. The overall safety standards applied by the CEGB to the design, construction and operation of nuclear power stations are several orders higher than those applied to conventional buildings. For example, non-nuclear design practice considers wind speeds with return periods of 50 to 100 years, whereas it is current nuclear practice to design safety related buildings for wind speeds extrapolated to a 10,000 year return period. In order to be consistent with these standards, which have become more stringent as time has passed, earthquakes and other rare phenomena must be included within the design basis loadings of the plant.

3. The need to specify seismic design loadings for nuclear plant has stimulated a new interest in the history, causes and effects of earthquakes in Britain and this paper describes the work that has been carried out by, and on behalf of, the CEGB in this specialist field.

4. While the studies have been primarily concerned with seismic loading specifications for the design of nuclear power plant, the methods and results may well find more general applications. For this reason, examples are given of solutions to problems which have arisen as side issues to the main objective.

5. Definition of terms

Hazard - The annual probability that a given ground motion level will be equalled or exceeded at the site. The ground motion may be defined as peak acceleration or velocity, intensity etc.

Risk - The annual probability that a specified limiting plant condition will be equalled or exceeded at the site. The limiting condition may be structural collapse, failure to perform a safety function etc.

Design Level - The ground motion level used for design purposes. This may be further defined as the SSE, DBE or OSE as described below.

BACKGROUND TO CEGB SEISMIC DESIGN SPECIFICATIONS

6. Design and safety guidelines for external hazards were required in the early 1970's for application to all future nuclear power stations. The design requirements included external environmental events such as floods, wind and earthquakes.

7. In 1973 a small working group was set up within CEGB with the objective of defining Safe Shutdown Earthquake (SSE) ground motions for design purposes. Whilst it was tempting to copy US-NRC specifications such as Reg. Guide 1.60, the group concluded that US-NRC regulations may be suitable for a seismically active region such as California but they were not appropriate in Britain, where earthquakes occur much less frequently and are of a less damaging nature both in energy release and duration of strong shaking.

8. All natural hazards were to be specified as those levels of flood, wind speed, or ground motion corresponding to a return period at the site of 10,000 years (annual probability of exceedance of 10 to-the-minus 4). In order to meet these requirements, hazard analyses were undertaken based upon seismicity data published by the Institute of Geological Sciences (now the British Geological Survey - BGS).

9. On the assumption that the seismicity of Great Britain was relatively uniform in all the areas of interest, a hazard analysis was performed which showed that at the 10 to-the-minus 4 exceedance level, the SSE ground motions should correspond to peak horizontal ground accelerations of 0.15 to 0.25 gravity and to peak velocities of 10 to 26cm/sec, depending upon two alternative interpretations of the seismicity data available from BGS.

10. A suite of US accelerograms was searched to find a set which corresponded most closely to the indicated peak velocity, which was thought to be the parameter most likely to correspond to the damage potential of each earthquake record. The Station No. 5 and Temblor records of the 1966 Parkfield event were found to be the most suitable on this basis (max. velocity of 27cm/sec) and were also representative of the magnitude and shallow focal depth characteristic of British conditions (up to 5.7 Richter magnitude and 5 to 20 km focal depth). No other records available at that time were found to fit the selection criteria established in the hazard analysis. As a result the Parkfield 5 and Temblor records were chosen for SSE design purposes and designers were supplied with digitized input data in time history format. The hazard analysis and choice of accelerograms are described in references 1 and 2.

11. An approximate probabilistic risk analysis (PRA) was undertaken in order to estimate the probability of failure to meet safety criteria given the probability distribution of peak ground accelerations and the likely fragility distribution of major plant components. It was assumed in the PRA that the seismic qualification process for the plant would be deterministic and would make use of a defined SSE peak ground acceleration of 0.25 gravity and corresponding time histories or spectra. The analysis followed the method described in reference 3. The results of the PRA indicated that a target figure of 10 to-the-minus 7 annual probability of "failure" could be achieved if the 10 to-the-minus 4 peak ground acceleration hazard levels were taken to define the SSE.

Design Ground Motion Specification for Heysham II AGR Power Station

12. The Parkfield/Temblor records, while corresponding to the required peak velocity value, contained high acceleration peaks (0.47g and 0.42g). These accelerations were well in excess of the 0.25g level which was indicated to be required for the SSE on a 10 to-the-minus 4 per annum probability basis and, furthermore, presented problems of non-linear response which would be difficult to analyse at the initial design stage.

13. At the outset of the Heysham II project, modifications were made to the accelerograms supplied for SSE design purposes. In the absence of more suitable records, the Parkfield 5 and Temblor accelerograms were modified within a narrow band to suppress the existing high peak accelerations to 0.25g. The modification technique was developed by CEGB's Berkeley Nuclear Laboratories and it allowed the peak accelerations to be decreased to the required level while leaving the peak velocity and spectral energy virtually unchanged. In this way a set of "modified" accelerograms was produced which was consistent with the hazard analysis in both peak acceleration and velocity.

14. Discussions with the designers, following the issue of seismic specifications, led to the adoption of Design Basis Earthquake (DBE) ground motions to be used for initial design purposes with elastic limits on stress and deformation. The DBE motions were linearly scaled to a level of one half of the above-mentioned "modified" accelerograms. The DBE was used as an intermediate design tool; the final design was validated against the full SSE motions.

Design Ground Motion Specification for Sizewell 'B' PWR Power Station

15. The start of the PWR project coincided with a search for more comprehensive ground motion data to cover a range of possible site conditions for generic design purposes. A study of the characteristics of ground motion spectra corresponding to British conditions was commissioned in collaboration with British Nuclear Fuels plc. This study included a survey of relevant world-wide strong motion records to establish a set of real and synthesized time history accelerograms and spectra for hard, medium and soft site ground conditions. The study (ref. 4) was used as the basis for the design ground motion spectra with the SSE peak horizontal ground acceleration specified as 0.25 gravity and the peak vertical ground acceleration specified as two thirds of the horizontal acceleration. The DBE concept is not included in the PWR specification. The design and qualification procedures for the PWR are described in reference 10.

16. Special consideration was given to the effects of small amplitude, long period, long duration surface waves generated by large magnitude earthquakes at great distance from the site (e.g. 1000 kms). These waves are not recorded by the strong motion instruments used to define the design spectra.

17. A study was made of European seismograms from large events in order to derive long period spectra corresponding to the 10 to-the-minus 4 hazard level. This showed that the strong motion design response spectra generally envelop the derived long period spectra at frequencies below 1 Hz. However, special attention may have to be given to critical items with natural frequencies close to 0.1 Hz and with low damping (0.5% critical), since the major peak in the long period spectra exceeds the existing strong motion design envelope at about this frequency.

UK SEISMICITY RE-ASSESSMENT STUDIES

18. Discussions during the late 1970's, with other organisations having an interest in British seismic criteria, indicated that there were doubts about the validity of the intensities, magnitudes and focal depths of earthquakes used

in the earlier hazard and risk studies. It appeared likely that a re-assessment of the historical data would lead to reductions in the uncertainties associated with the peak acceleration specified for SSE design purposes.

19. Two separate studies were commissioned from independent consultants in July 1981 and their reports (refs. 5, 6) were completed at the end of August 1982. Taken together these two reports contain a comprehensive assessment of all significant seismic events which have occurred on the land mass of Great Britain together with the felt effects of events which have occurred off-shore and in the neighbouring European land mass. Both consultants carried out detailed searches for the original sources of information (e.g. ecclesiastical records, personal letters, technical and popular publications, newspapers and instrumental recordings). The search covered the period from 600 AD to 1982 and was widely spread geographically. Primary source material was collected in catalogued archives. The primary data were then processed to produce isoseismal maps together with assessments of epicentral intensity, magnitude and focal depth for events which provided sufficient data. Spurious or apocryphal events were eliminated on the basis of the collected data. One of the consultants provided a peak acceleration attenuation scaling law appropriate for British conditions (ref. 5).

20. The detailed processing and assessment of the large volume of new data provided in the reports are described in references 7 and 8 and a summary of the method used for hazard analysis is given in paragraphs 31 to 33 herein. The results were used to update previous work and to evaluate the Sizewell 'B' site. The new hazard analysis showed that, for a 10 to-the-minus 4 per annum site exceedance probability, the SSE level based on alternative assumptions would be as follows:-

(i) Average national uniformly distributed seismic activity (see curve 2 of Fig. 1):
0.20g peak instrumental ground acceleration (p.i.g.a.)

(ii) Sizewell site and its surrounding area treated as part of the uniformly distributed activity for the whole East Anglian region (see curve 1 of Fig. 1):
0.15g p.i.g.a.

(iii) Sizewell site with historical epicentres and major faults treated as source zones. A range of seismic activity rates was used in this analysis:
0.10-0.15g p.i.g.a.

21. The major contribution to seismic hazard at the 0.25g SSE design level was found to be contained within a circle of radius 30km from the site of interest and was due to events with relatively low magnitudes between $4.0M_S$ and $5.0M_S$.

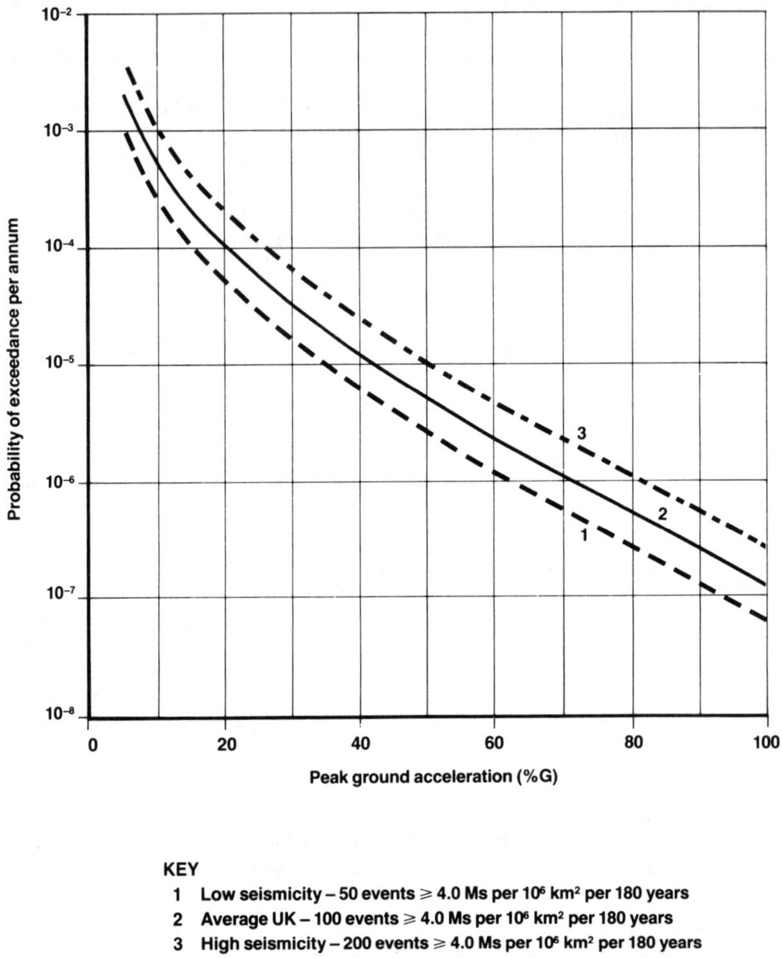

Fig 1 Effect on hazard of variations in the average historical seismic activity rate (ref. 8).

22. Estimates of a suitable Operational Shut-down Earthquake (OSE) level showed that, for sites in regions of up to average seismicity, the probability of experiencing 0.05g or greater in a 50 year period would be 0.1.

23. The results demonstrated that the overall national SSE level of 0.25g estimated in earlier hazard studies was conservative for the average site. The Sizewell site hazard was below the national average in all the cases examined. However, in the interests of standardisation, PWR design is proceeding on the basis of an SSE with 0.25g peak ground acceleration although the option to use a lower figure, if appropriate, has not been abandoned at the time of writing.

ANALYSIS OF GEOGRAPHICAL NETWORKS SUBJECTED TO EARTHQUAKE HAZARD

24. The hazard analysis can be extended to incorporate the calculation of the reliability of geographically dispersed networks, for example electrical grid systems. The likelihood of loss of off-site power supplies can be determined by considering failure of the insulators at each sub-station in the grid system to be a potential cause of loss of supply, assuming that the intervening transmission lines have greater resistance to seismic loads and remain intact. A computer program (EGRID) was developed by the author to solve this problem taking account of random spatial and temporal distributions of earthquakes, the geographical configuration of the sub-stations and the logic of their connections within the transmission line network.

25. The insulator fragility distribution for each sub-station was represented by a lognormal probability of failure distribution for earthquake induced ground motions covering the range of interest. A different fragility distribution could be specified for each sub-station site.

26. The program sets up a regional array of closely spaced epicentres and then computes the peak acceleration at each of the sub-station sites within the region due to an earthquake occurring at one epicentral location. The corresponding probability of failure at each site is found from the discretised insulator fragility distributions. The corresponding probability of failure of power supplies is then found using a Boolean logical model of the grid system network. This calculation is repeated for all epicentres in the region and for each of a set of earthquake magnitudes from a lower threshold level to an upper bound magnitude.

27. Integration of conditional system failure probabilities over the magnitude range combined with the probability of occurrence of an earthquake of the given magnitude at each epicentre completes the calculation of total annual probability of loss of power supplies.

28. The conditional probabilities of system failure, given the occurrence of an event of a set magnitude at each epicentre, can be mapped in numerical contour form; each digit representing the probability of system failure due to one or more epicentral locations.

29. Different magnitude/frequency seismicity distributions can be used and peak acceleration attenuation can be chosen from a range of scaling laws. Allowance was made for uncertainties in attenuation by total probability methods in order to determine the mean annual probability of loss of off-site power.

30. Calculations carried out using the EGRID program for a typical CEGB grid system have shown that, given reasonable geographic dispersion and redundancy of the sub-station network, no common mode failure due to a single earthquake is likely. As might be expected from actual experience, the risk of loss of off-site power for lengthy periods due to earthquakes is small compared to the risk arising from other more conventional causes.

UNCERTAINTIES IN EARTHQUAKE HAZARD ESTIMATES

31. Earlier studies (refs. 1,7,8) were concerned with estimating the annual probability of exceedance for a range of peak ground accelerations. These estimates were called hazard curves and each hazard curve represented the mean annual probability of exceedance versus peak instrumental ground acceleration as shown in Fig. 1. The mean hazard curves were obtained by treating seismic activity rates (number of earthquakes exceeding a given magnitude per year) and attenuation (peak ground acceleration for a given magnitude and source to site distance) as fundamental uncertainties; the former modelled with data-based Poisson/exponential distributions and the latter modelled with a lognormal distribution. The parameters of the seismic activity rate distributions were point estimates obtained by fitting the historical seismicity data in a conservative manner.

32. The parameters of the attenuation distribution were point estimates obtained from multiple regression analysis of recordings from European and US earthquakes. In this case, the well-known large scatter on attenuation data was treated as a fundamental uncertainty lognormally distributed about the point estimate of the median regression value.

33. The effect of uncertainties on the mean value of the annual probability of exceedance of a given peak ground motion level was addressed in reference 7, in which it was shown by Bayesian methods that statistical and model uncertainties have only a small effect on the mean hazard computed from point estimates of the parameters describing seismic activity rate and the magnitude frequency distribution.

34. The objective of the work described below was to quantify the uncertainties in the probability distribution of the hazard by treating the underlying parameters of seismicity as random variables with probability distributions determined from data-based Bayesian prior-posterior analysis. It should be noted that subjective Bayesian distributions were not used.
35. Two fundamental sources of uncertainty were considered; seismicity and attenuation, which were treated as follows.

Seismicity Uncertainties

36. The seismicity parameters of the exponential distribution of magnitude versus frequency of occurrence were treated as random variables which are Gamma distributed. The derivation of their distributions followed the Bayesian methods outlined in Appendix IV of reference 7. This allows for the statistical uncertainties associated with the number of events (data points) in the historical record.
37. The cumulative frequency distribution of magnitudes can be assumed to be exponential (ref. 5,7) with,

$$N = \alpha \exp(-\beta(M-M_o))$$

where $M_o \leq M < M_u$

and N = Average number of events greater than magnitude M per unit geographical area per unit time period

M_u = Upper threshold magnitude

M_o = Lower threshold magnitude

α = Activity rate - number of events greater than or equal to M_o

β = Exponential distribution parameter

38. The probability distributions of α and β were derived by Bayesian prior-posterior analysis. The probability density function (PDF) of α, assuming a Poisson process, is given by

$$f_A(\alpha) = \text{Gamma}(r+1,1) = \alpha^r.\exp(-\alpha)/\Gamma(r+1)$$

where r is evaluated from the historical data.

39. For example, r is the number of events that have occurred in the geographical area a_o with magnitude greater than M_o in the past t_o years. The mean value of α is given by the mean of the Gamma distribution, that is

$$\bar{\alpha} = (r+1)/t_o a_o$$

and the maximum likelihood (mode) is

$$\dot{\alpha} = r/t_o a_o$$

A similar derivation applies to the distribution of β.

40. Clearly the sensitivity of results to assumptions about the uncertainty distributions can be investigated by varying the number of events r, the past time period t_o or the area a_o. The distribution for 2 past events is shown in Figure 2.

41. The parametric uncertainties were incorporated into the cumulative magnitude-frequency distribution using the convolution integral solution (ref. 9) for the functional dependence of the number of earthquakes N on the two random parameters α and β for a range of values of earthquake magnitude.

42. In this way, the cumulative frequency distribution (CDF) of uncertainties in the number of earthquakes of each magnitude can be found.

Attenuation Uncertainties

43. Peak instrumental acceleration attenuation scaling laws are generally of the form:-

$$\dot{G} = \text{function of } (M,R) = H(M,R)$$

where $\dot{G}$ = Median value of peak instrumental ground acceleration
M = Magnitude of event
R = Hypocentral distance from source to site (assumed to be fault distance in this analysis).

44. The large data scatter is represented by a lognormal distribution, $LN(1.0,\sigma)$ about the median; that is, the distribution of uncertainties in the peak ground acceleration (G) given the magnitude and epicentral (or fault) distance correspond to a random variable multiplying term, say E

where $E = G/\dot{G}$

and $T = \ln(E)/\sigma$

with T normally distributed $N(0.0,1.0)$.

45. Treating R and T as random variables with the following functional relationship

$$G = H(M,R).\ \exp(T\sigma)$$

This relationship can be inverted to give

$$R = J(M, G \exp(-T\sigma))$$

for all M,G,T where $J \geq 0.0$.

46. This defines the radius from the site within which G will be exceeded given an event of magnitude M. Since R is monotonic in T for a given G value, the cumulative distribution function (CDF) of R can be found directly from the normal distribution of T.

47. Similarly the circular area A within which G is exceeded is a random variable with a CDF of uncertainties found directly from the monotonic functional relationship.

$$A = \Pi R^2 = \Pi J^2 (M, G \exp(-T\sigma))$$

for all M, G, T where $J \geq 0.0$.

Hazard Uncertainties

48. The assumption underlying the derivation of the Gamma distributions of uncertainties in seismicity is that earthquakes within the area under consideration are Poisson distributed in time and space (i.e. "random" arrivals). Within certain restrictions, reference 7 shows this to be a reasonable assumption based on statistical testing of the historical data available in Britain and it is also used in final stages of the analysis.

49. The algorithm used in the computer program (UNCLE) developed by the author is summarised below:-

(i) Set up arrays of magnitude (typically in 13 equal intervals from 4.0 to 6.6 M_s) and CDF's of seismicity and attenuation (typically at 100 equally spaced levels from 0.001 to 0.999).

(ii) For a given magnitude interval compute the average number of events in the interval per unit area per unit time corresponding to each CDF level of seismicity.

(iii) For the given magnitude interval compute the area A within which G is exceeded corresponding to each CDF level of attenuation.

(iv) Compute the average number of exceedances per unit time of G at the site corresponding to all combinations of seismicity and attenuation CDF levels using the product of the average number of events per unit area per unit time and the attenuation area.

(v) Repeat stages (ii) to (iv) for each magnitude interval in the range considered.

(vi) Compute the probability of one or more exceedances of G at the site using the Poisson distribution with the average rate defined by the sum of the products in stage (iv) for the same seismicity CDF level over all the magnitude intervals in the range of magnitudes considered. Store values corresponding to each combination of seismicity and attenuation CDF level.

(vii) The computation is completed by enumeration of the distribution of uncertainties for the probability of exceedance of each G level. This is achieved by integrating interpolated values of the CDF of attenuation for each seismicity CDF level at a set of chosen probability of exceedance levels and normalising to obtain a proper probability distribution of uncertainties.

Results of Analysis

50. The results obtained using the seismic activity rate distribution given in Fig. 2 and the attenuation scaling law of reference 5 are illustrated in Fig. 3. The hazard distribution shown in this figure corresponds to the East Anglia region with 2 events greater than 4.0Ms in the last 180 years, within one twentieth of a million square kilometres.

51. The most notable feature of the intermediate stages of the computations is the marked dominance of uncertainties in attenuation over those associated with seismicity. While the probability of exceedance is dependent upon seismicity, the dispersion represented by the CDF, is swamped by attenuation scatter. This implies that there is more benefit to be gained in terms of reductions in uncertainty and in the mean exceedance probability by studying the mechanisms of attenuation and in particular by developing more reliable and less dispersed measures of actual damage than provided by peak instrumental ground acceleration.

52. The final results demonstrate that, even where historical seismicity data are sparse, the mean hazard curves used in earlier studies (ref. 7) to define the SSE are conservatively placed within the distribution of uncertainties. In the case examined here the mean hazard at the SSE level of peak ground acceleration (pga) is close to 80 percentile of uncertainty and at higher pga's exceeds the 90 percentile. It can be seen that the median pga hazard curve (50 percentile) drops rapidly, so that higher levels of pga than the SSE are unlikely to have annual exceedance probabilities greater than 10 to-the-minus 6 or 7. It might also be reasonable to consider the use of the 67 percentile hazard curve to determine design levels on the grounds that the balance of uncertainty is weighted two to one in favour of this curve; so it provides conservative values of the probability of exceedance of events associated with very rare natural phenomena.

53. The uncertainty analysis has been extended to cover seismic sources associated with geological fault planes and diffuse, three dimensionally bounded source volumes, as well as the uniform seismic source density covered by the illustrations given in this paper.

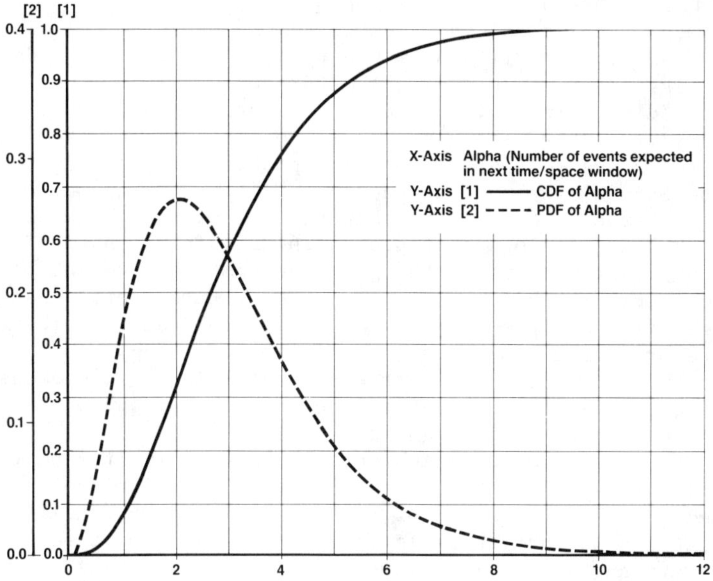

Fig 2 Posterior Gamma distribution of number of events expected in the next time-space window given the observation of 2 events in prior window.

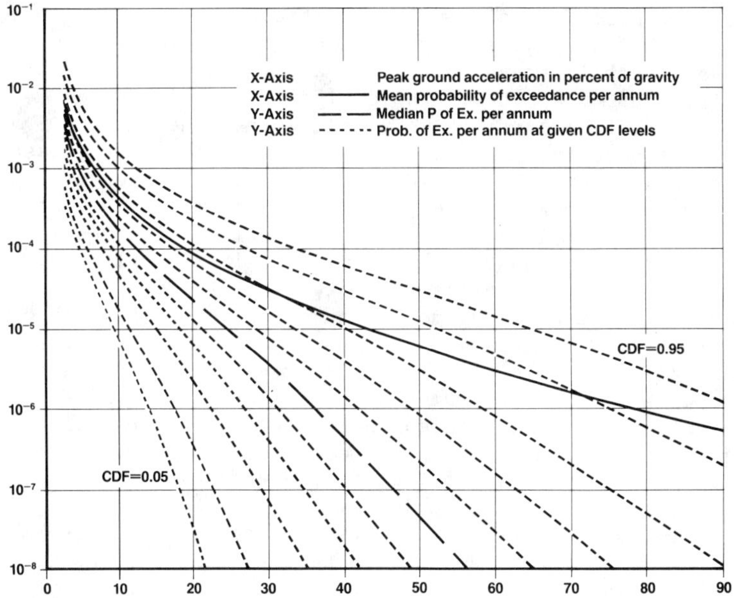

Fig 3 Effect of uncertainties on hazard given 2 events greater than 4.0 Ms observed in 0.05 million sq. km. in prior 180 years. Uncertainties are plotted at the 0.05, 8 intervals of 0.10, and 0.95 CDF levels.

FUTURE DEVELOPMENTS

54. One of the problems encountered in assessing seismic hazard at a site in Britain is the paucity of historical data on the likelihood of exceeding the SSE design level. While this is a direct indication that the hazard is low, the discovery of either geological faults or past earthquake epicentres close to the site present problems in proving that the SSE design level is adequate.

55. One reason for this is that the association between earthquakes and geological features in Britain is poorly understood, although it is thought that past and present seismic events are better indications of future trends than is given solely by the presence of geological faults. There is therefore a need for more research in seismology and neo-tectonics to provide a better understanding of the causes of earthquakes in Britain, given the enigma that geological faults are common but earthquakes are rare.

56. The other reason, which applies both to faults and to past epicentres, is that there are large uncertainties attached to the attenuation of peak instrumental ground acceleration, so that there is a high probability, based on current attenuation scaling laws that the SSE might be exceeded given even a moderate magnitude event on a fault or at a past epicentre close to the site. There is therefore a need to reduce uncertainties in site response by collecting and assessing attenuation data, particularly in the near field of moderate magnitude earthquakes, paying attention to source mechanisms and site soil conditions.

57. Furthermore, design ground motion specifications are based upon peak instrumental ground acceleration coupled with time histories or spectra. The design and qualification procedures (ref. 10) effectively assume that the plant response is sustained at the peak SSE level and, in the case of test validation, the applied motions are sustained at full amplitude for at least 8 seconds preceded by multiple tests at a lower level. Thus, tested items are typically required to withstand up to about 100 cycles at full SSE amplitude. This is much more severe than expected in a British earthquake when the strong motion duration is unlikely to be more than 2 to 3 seconds with only one or two early cycles reaching the peak level for the SSE derived from mean hazard curves based on peak instrumental acceleration. Non-linear analytical studies and data collected from actual experience of the performance of power plant components in earthquakes support the conclusion that little damage results from high peak but short duration ground motions.

58. Studies of better parameters have begun, with the objective of redefining both the hazard and the SSE design level and it is hoped that these may lead to more rational and less costly seismic provisions.

ACKNOWLEDGEMENTS
This paper is published with the permission of the Central Electricity Generating Board. Figure 1 is reproduced from reference 8 of this paper by permission of the Council of the Institution of Mechanical Engineers.

REFERENCES

1. IRVING J. An analysis of seismic risk in Great Britain. CEGB Report GD/PE-N/279(B), Re-printed April, 1982, originally issued November, 1979.
2. MALLARD D.J., IRVING J. and CORKERTON P.A. The assessment of seismic design criteria for nuclear power stations in England and Wales. Proc. C.S.N.I. Specialists Meeting on Anti-Seismic Design for Nuclear Installations. NEA-OECD, Paris, December, 1975.
3. ALDERSON M.A.H.G. and IRVING J. Probabilistic methodology for seismic risk assessment for nuclear power plants within the United Kingdom. Proc. Specialists Meeting on Probabilistic Methods in Seismic Risk Assessment for Nuclear Power Plants, NEA-OECD, Lisbon, 1980.
4. PRINCIPIA MECHANICA LTD. Seismic ground motions for UK design. Report prepared for BNFL and CEGB, April 1981.
5. PRINCIPIA MECHANICA LTD. British earthquakes. Report for CEGB, Ref. No. 115/82, 1982.
6. SOIL MECHANICS LTD. Re-assessment of UK seismicity data. Report for CEGB Ref. No. 7984, 1982.
7. IRVING J. Earthquake hazard. CEGB/GDCD Report C/JI/SD/152.0/R019, December 1982.
8. IRVING J. Seismic hazard in the United Kingdom. Proc. Inst. Mech. Eng. Seminar on Seismic Qualification of Safety Related Nuclear Plant and Equipment, Inst. Mech. Eng., London, April 1984.
9. BENJAMIN J.R. and CORNELL C.A. Probability, statistics and decision for civil engineers. McGraw-Hill, New York, 1970.
10. SMITH C.R. Seismic design practice for UK nuclear power plant. Ibid., 1985.

Discussion

PROFESSOR FINN
What attenuation relationship did Mr Irving use? On the basis of his intensity data does he find that attenuation is relatively slow in Britain?

MR E. BOOTH
Mr Irving stated that even if the maximum magnitude earthquake in the UK could be shown to be 5.5, there would still be an earthquake hazard problem, since near-field events of

magnitude 4 to 5.5 contributed most to the peak ground acceleration with a 10^{-4} annual probability. Could Mr Irving comment on the following?

PGA is not the same as damaging potential. What field evidence is there, if any, that near-field events up to magnitude 5.5 can damage installations such as those proposed for Sizewell?

From the CEGB's evidence submitted to the Sizewell inquiry, it appears that the uncertainty in attenuation laws contributed a large amount to the risk of high accelerations at long return periods. Might this be more of a mathematical anomaly than of real physical significance, as far as damage potential is concerned?

Could Mr Irving offer a personal opinion as to whether it might be possible to neglect design against earthquakes for structues designed to the other general standards applied to Sizewell B, if it could be conclusively shown that magnitudes greater than 5.5 were not possible near a given site?

Could Mr Irving comment on the relative fragilities shown in his paper for structures and pipelines on the one hand, and equipment on the other hand, since a very large difference is apparent.

MR IRVING

The peak ground acceleration relationship was derived by Principia Mechanica Ltd (Ref. 5 of my paper) and was based upon USA and Alpine Belt data. It is of the Campbell type with a magnitude term in the divisor.

Intensity isoseismals in Britain do show markedly slower attenuation than, for example, in California. This is illustrated in reference 5 and discussed in reference 6 of my paper.

I agree with the sentiments expressed by Mr Booth and would welcome a paper from him on this topic. I shall answer his three points in turn.

It is well known that the peak ground acceleration measured by strong motion instruments is not necessarily a good measure of damage to structures, particularly in the case of the moderate magnitude events that can be expected in Britain. Allowance has been made for this effect in the probabilistic risk assessment mentioned by Mr C. R. Smith. However, in a power station there are components which may be relatively fragile when exposed to short duration shocks typical of moderate magnitude events, so that 'damage-effective' acceleration (e.g. root mean square) is not a total panacea and, in general, for the design and qualification of safety related components including structures we have adopted the conservative assumption that peak ground acceleration is the damage measure of an earthquake. We are engaged in studies to find better (true) damage related parameters since this may lead to economies in the design and qualification of plant. As Mr Booth mentions, there has been little evidence of actual damage to power station structures in the many moderate magnitude earthquakes that have occurred in California for example.

Uncertainties, in the form of attenuation data scatter, contribute a great deal to the high peak ground accelerations that might be expected to be exceeded with low probability (e.g. less than 10^{-4} per annum). This is in part due to the assumptions made about the shape of the uncertainty distributions (e.g. untruncated lognormal distributions). As mentioned in my previous answer, a better parameter that more truly reflects damage may have to be found that also shows less scatter, thereby reducing uncertainties. A power station is not just a structure; there are many other components, some of which may well be affected by short sharp shocks which leave the structures undamaged, and this has to be taken into proper account before the existing design levels of acceleration can be abandoned.

The design of a power station therefore involves many other components than the structures. In addition, it is often better and more economic to design generically for any site, as the French are doing with their PWR programme rather than produce a one-off design for a particular site (apart from foundations). If, and it is a big if, magnitudes greater than 5.5 could be proven beyond doubt to be impossible in a particular region, say the UK, then this would make the deterministically derived seismic design level much less onerous. In this respect, I very much welcomed Professor Ambrasey's remarks, in the earlier sessions of this conference, on the small likelihood of earthquakes greater than 5.5 M_s in the UK.

The paper did not include comments on the relative fragility of power station components. However, I showed a slide while presenting the paper to illustrate the different risks of 'failure' associated with different classes of components i.e. structures, pipework and equipment.

This illustration was drawn from the approximate PRA mentioned in paragraph 11 of my paper. The work was carried out in 1978 to 1979 and was based on the fragility data given by Ang and Newmark in their study of the Diablo Canyon plant in California. The fragility data for equipment could, with hindsight, be criticized since it was based upon non-earthquake loadings and may well be very conservative when applied to the properly qualified equipment now being provided for nuclear plants. I believe this is a subject which warrants further study as it is particularly important when realistic PRA's are required in support of the safety case for external hazards such as earthquakes.

20. Promises and limitations of available techniques in UK seismotectonic investigations

D. PAPASTAMATIOU, PhD, DIC, Delta Pi Associates

SYNOPSIS. The record of historical seismicity in the UK suggests a rate of crustal deformation of the order of 10^{-2} mm/yr; this estimate is sensitive to the actual capacity of the continental crust to generate large earthquakes. If crustal deformation is released by a primary set of 100 faults, each fault would exhibit a rate of co-seismic deformation of the order of 10^{-4} mm/yr. Available techniques to quantify the correlation between earthquakes and geology are reviewed within the framework of the expected crustal deformation bracket. The review indicates the level of seismotectonic correlations that would be substantiated in the UK and the level of support that would be provided to design criteria.

INTRODUCTION

1. The precision of earthquake hazard assessment depends on the availability and the applicability of techniques that probe different aspects of the seismotectonic relationship. These techniques are developed in various disciplines in the geosciences to investigate the dynamics of the geological system in space and time.

2. The engineering assessment of earthquake hazard contains a certain degree of load factorisation that anchors it to rigid procedures with varying resilience to change. As a result, engineering earthquake hazard assessments lag behind new developments, in understanding the seismotectonic relationship, that follow the development of new techniques. The rigid element implicit in the engineering approach is expressed in terms of load criteria that have been intimately related to a particular technique in the past: early versions of earthquake resistant design have been based on felt intensities, earthquake resistance of dams has relied upon deformation in the holocene, a geologically distinct recent period covering the last 10,000 yrs. Early formalised criteria in the nuclear industry bear a direct relationship to available techniques such as the 35,000 yr mark, employed by the US NRC rules, that is considered the limit of carbon dating techniques.

3. Traditionally, the set of design criteria has dictated the level of assessing the earthquake hazard. This practice

has led to inconsistencies since it implies a change in the aperture of seismotectonic investigations depending on the design criteria set for the particular structure under consideration. Thus, in the early days of the development of the nuclear industry, earthquake hazard was assessed on the basis of the partial picture used in earthquake resistance regulations. Later, the techniques employed by the nuclear industry to satisfy more strict design criteria have exposed a real hazard potential to ordinary structures (ref. 1). It therefore appears that earthquake hazard should be assessed irrespective of design criteria; in this unified approach criteria should be then used to specify loading as a function of the consequences of failure. The success of the earthquake hazard assessment depends on how close the assessment methodology follows the natural process; this process is understood and quantified through the application of techniques. Therefore, the assessment of earthquake hazard is limited by the potential of available techniques; design requirements beyond this potential can only be satisfied with safety margins.

4. The potential of available techniques in clarifying seismotectonic relationships depends, to a great extent, on the rate of crustal deformation. The techniques that formed the data base for the development of the theory of plate tectonics were effective within the time span of a few years. The theory concentrated on the upper range of rates of deformation on a global scale; for this deformation range rates of deformation inside the plates were ignored. The same techniques applied in the low seismicity environment inside the plates have resulted in a very slow progress in understanding seismotectonic relationships there. This presentation attempts an evaluation of various techniques on the basis of results obtained in low seismicity intraplate environments similar to the UK. The basis for this presentation is a more general review (ref. 2) on international experience.

5. In the first part of this report a framework is developed for the simple representation of seismotectonic relationships. A seismotectonic chart is presented on which seismological and tectonic information can be combined. The chart is adjusted to UK to facilitate presentation of the remaining sections. In the following section available techniques are discussed and their potential is rated on the seismotectonic chart in terms of their theoretical and actual performance. In the final section a position of UK is sought on the same chart and compared to the representations of design criteria and of the potential of available techniques.

SEISMOTECTONIC MODELLING

6. A basic problem in seismotectonic modelling is the establishment of the boundary conditions, i.e., choose around the site a piece of crust, with reasonably defined boundary stress or kinematic conditions, that contains the earthquake hazard

to the site. The solution to this basic problem depends on the level of understanding the regional seismotectonics.

7. Inside this piece of crust, deformation takes place following the geometry of natural formations and of pre-existing fractures as well as the combination of rates of deformation and material properties in depth. In general, permanent crustal deformations can be expressed in two terms: a slow creep deformation u_c that is not accompanied by earthquakes and a co-seismic component u_s of intermittent motion:

$$u = u_c + u_s \tag{1}$$

The two terms are not necessarily independent. The permanent crustal deformation constitutes a fracture process that can be described at different levels. At a 'macroscopic'level the fracture process appears random and can be described by a Poissonian distribution of fracture events deprived of memory. At a 'microscopic' level, where reactivation of faults is determined, concentrations and releases of crustal strains are expressed in more complex migrations of seismic activity in space and time. The 'macroscopic' image gives the long term prediction of earthquake hazard whereas the 'microscopic' image gives the intermediate and short term predictions. A design criterion compatible with the natural process would be in terms of probability of occurrence during the expected life of the particular structure that, in general, falls in the range of intermediate predictions (a notable exception here is the stability of repositories of radioactive materials with long decay times). Traditional earthquake hazard analysis is built on the assumption of Poisson distribution and therefore corresponds to long term predictions. In this analysis the intermediate range is covered by educated judgement on the variation of the parameters involved.

8. The fracturing process can only be understood through an interrelationship of seismic activity and the fractures in the crust. The co-seismic term in this process is produced by a sequence of seismic events characterised by the size of the event (seismic moment M_o), the ruptured area (A_i) and the average dislocation $u_{s,i}$ on the ruptured area:

$$M_o = \mu \, u_{s,i} \, A_i \tag{2}$$

where μ is the rigidity at the seismic source.
The expression (2) correlates seismic activity with the geometry and the deformation along a crustal fracture and can be determined from either geological or seismological evidence. After reduction of the single dislocation u_{si} from the rupture area A_i to the total area of the fault zone A and summation of M_o over a time span (ref. 3), expression (2) leads to a relationship between the total seismic moment release ΣM_o and the total crustal deformation u_s. The seismic moment has been empirically correlated to the more conventional measure of the size of an earthquake in terms of the magnitude M.

The magnitude distribution has been shown to stabilize to a linear cumulative frequency distribution:

$$\log N = a - bM \qquad (3)$$

where N is the number of events exceeding magnitude M.

9. An equation correlating crustal deformation with the geometry of the seismogenic source and the distribution of the seismic moment is derived in ref. 4. The equation is recast in terms of earthquake magnitude (see Appendix) into the form:

$$\log N = \Big[\underbrace{-6.58 + \log\mu \tfrac{1.5-b}{b}) - (1.5-b) M_{mx}}_{seismo} \underbrace{+ \log A + c_o + \log \dot{u}_s}_{tectonic} \Big] - bM \qquad (4)$$

where $\dot{u}_s$ is the rate of co-seismic deformation in mm/y
M_{mx} is the limiting magnitude characteristic of the zone
b is the magnitude distribution parameter
μ is the rigidity at the seismic source
N is the no. of yearly events exceeding magnitude M_s
c_o is a crustal deformation factor: c_o=0 for transform deformation and c_o=.4 for deformation of a volume of continental crust
A is the active area of the seismic zone in km^2

The depth D of the zone may vary from about 10km, on a nearly vertical fault, to about 15km on a 45° dipping fault. In the latter case that corresponds to c_o=.4, seismic strain is released by the horizontal component in eqn 4 and a vertical component of crustal thickening or thinning. Expression (4) is derived for long term conditions for which seismic events may be assumed Poissonian and the deformation rates continuous; continuity does not necessarily apply to a single fault but also holds for deformation of a volume of crustal material.

10. Expression (4) can be utilized as a seismotectonic chart either in terms of the number of events/yr or in terms of the mean return period T(T=1/N) for a given length L of the seismogenic source. The chart employs the same axes used in the traditional distribution (3); on this plane, different deformation rates are represented by a family of curves sloping at an angle $\tan^{-1} b$. A seismotectonic chart is illustrated on Fig 1 for mean values $\mu = 3 \times 10^{11}$ dyn-cm^{-2}, b=1.0 and seismotectonic parameters representative of UK conditions: M_{mx}=6.25(ref. 6) and active depth of the seismic zone D=10km. To cater for the uncertainty in the mode of crustal deformation equal rates are plotted as parallel bands with a width of 0.5 log units and centered around an effective average c_o=0.25. This c_o value lies between the two extreme configurations: c_o=0, D=10km for transform deformation and c_o=0.40 , D=15km for volume deformation.

This source of uncertainty indicates possible deviations from the computed rates within a factor of about 2. Increasing the limiting magnitude M_{mx} by one unit would lift the family of curves by about half a unit on the log scale as the same seismic activity rate would correspond to a higher rate of crustal deformation. All in all, co-seismic deformation is computed from eqn 4 within a factor of about 5. The return period axes on the seismotectonic chart are drawn for three lengths of the seismogenic zone: L = 10, 50 and 100km. Clearly, the deformation rates shown on the chart do not represent deformation measured in the field: the permanent displacements of geological formations u may also include a substantial amount of creep deformation u_c (eqn 1).

SEISMOTECTONIC SIGNIFICANCE AND VALUE OF AVAILABLE TECHNIQUES.

11. Techniques from various disciplines, some of them highly specialised, are employed to determine eqn 2, for a single earthquake, or eqn 4 for continuous seismic energy release accompanying crustal deformation rates. The broad spectrum of disciplines include geology, seismology, geophysics, geodesy, mineralogy, imagery, geomorphology, rock mechanics, etc. Techniques developed in these fields provide the information necessary to establish the basic components of the seismotectonic relationship:

(a) The geometry of crustal discontinuities
(b) The mode of crustal deformation
(c) The time history of crustal deformation.

12. Eqn 2 requires a fracture geometry and expresses the earthquake mechanism. The geometry is well manifested in tectonically active areas where the historical record encounters evidence of surface faulting. Even in cases of clear surface manifestation, however, uncertainities remain on the geometry of the activated fault in depth. Although geophysical techniques provide an invaluable tool in completing the three dimensional picture, in areas of intense tectonic activity, the seismotectonic picture is put together primarily on the surface using field geology techniques. Tectonic activity in the historic period is studied through evidence on ground deformation and damage. This evidence is not clearly presented on an isoseismal map,as tectonic deformation, although implicit in the upper grades of the Mercalli Scale ($I_{MM} \geqslant IX$), is not an explicit index in assigning intensities. For prehistoric events techniques have been developed to decode evidence preserved in young sediments (paleoseismicity assessments). Both historical- and paleo-seismicity provide evidence on single events above a magnitude threshold, probably around M = 5.5 to 6.0 . The effectiveness of field geology techniques is not necessarily a function of the rate of crustal deformation as there are cases of high rates (such as the Zagros Range in the Middle East) where earthquake processes are masked by decollement surfaces (ref. 7).

13. Under the general conditions of intermediate rates of crustal deformation, techniques lose power to identify individual earthquake episodes but are still effective in picking up the rates of deformation involved in eqn 4. The rates are depicted on the ground, using various imagery techniques, or from the deformation of conveniently placed deposits using a combination of deep and shallow reflection/refraction methods. In the absence of such deposits an attempt is made to identify and date in-fill material in fault zones (ref. 8);this technique is very sensitive to the geodynamics of the regional mineralogy. Imagery as a primary technique in seismotectonic investigations is used as a geomorphological tool. A powerful alternative to the analysis of satellite imagery is being developed in digital cartography. More precise information on current rates of crustal deformation is sought in repeated measurements of tilt in the water level of lakes and of triangulation networks. For small deformation rates the measurement span is of the order of tens of years and problems are encountered with the accuracy of old measurements.

14. In areas of slow deformation rates there is a break in the potential of available techniques to establish seismotectonic relationships. With the notable exception of the 1968 Meckering earthquake in Australia (ref. 9), the historical record does not include clear surface deformation accompanying earthquakes in low seismicity intraplate environments. This lack of evidence may be due to the small tectonic deformation: after the 1978 earthquake in S Germany, geodetic traverses failed to depict any ground deformation (ref. 10) whereas after the New Brunswick earthquake, secondary ground deformation was of the order of 1" (ref. 11). Under the general conditions of slow deformation environments techniques retain their capacity to x-ray the geological system but lose their capability to pick up current geologic strain. Economic restrictions impose a grading of geophysical techniques in the x-raying procedure: coarse gravity and magnetic measurements provide a blurred picture that can be refined through crustal modelling. In these measurements there is no control on focusing to a particular depth range. Such control is possible in reflection/refraction techniques that can also provide more accurate measurements. In general, geophysical measurements are indirect and, in principle, should be confirmed with deep boreholes.

15. In slow deformation environments, therefore, the link between geology and seismology is provided by seismological evidence or, at a more fundamental level, by stress/strain measurements. The acquisition of stress/strain measurements is essential for understanding the physics of the seismotectonic relationship; in terms of earthquake hazard they only indicate the possibility of reactivation of favorably oriented crustal fractures hence satisfying an 'absolute safety' criterion. The interpretation of stress/strain measurements is conducted within the framework of a stress/strain tensor relationship that is distorted by the presence of discontinuities in the crust.

Seismological techniques are as powerful in low seismicity as in high seismicity areas; their application in low seismicity areas is drastically slowed down by the infrequent occurrence of earthquakes and requires broader time windows of observations. That applies to long period observations, such as historical records, as well as short periods of microarray deployment. For example, in intense seismicity areas historical studies have identified cycles of the order of 200 yrs (ref. 12); such cycles could be much longer in low seismicity areas. The requirement of a long period of microarrays operation in low seismicity environments (typically of the order of 5 yrs) has direct implications on their use in engineering site specific investigations.

16. The calibre of available techniques can be postulated on the seismotectonic chart. Such a representation is shown of Fig. 1 for the three lengths of active fault or fault zone: L=10, 50 and 100km. The small length is representative of site specific investigations whereas the large length of 100km is in the range of regional investigations. The limitations of various techniques are represented on the time scale or the deformation scale. Techniques providing information on seismicity or paleoseismicity have a potential that relates to the time scale and the accuracy of coverage. The potential of microarrays is represented for a length of operation of about 5 years. The instrumental and historical seismicity are represented for UK conditions. Paleoseismicity covers a large span on the time scale and therefore is not sensitive to small changes of this scale: techniques dating paleoseismicity evidence appear to have a lower magnitude bound (about $M=5\frac{1}{2}$ to 6) and no clear upper bound on the time scale. The main problem with these techniques appears to be the confidence in their measurements, as such determinations may be contaminated by other geodynamic processes without direct relevance to current deformation. The bounds on rates of crustal deformation were obtained from assumptions on the detection threshholds of associated techniques: the limits for the identification of deformation within geological epochs is obtained from an assumed detection threshold of 0.50m. The limit of geodetic determinations is obtained from a threshold of 0.5cm (two orders of magnitude higher than the geological threshold) change in a time span between measurements of 50 yrs. The geomorphic classification is taken from Matsuda (ref. 13) and shows a detection upper bound of about 10^{-2} mm/yr.

17. A more realistic picture of the potential of available techniques in clarifying the seismotectonic relationship in low seismicity intraplate environments was obtained from a review of seismotectonic investigations in NW Europe and NE America (ref. 2). Out of a compilation of 30 case histories, correlation of earthquakes with faulting emerge at the Ochtendung Fault in W Germany, the New Madrid Seismic Zone and the Humbolt Fault Zone in Central US, the Ramapo Fault Zone and

the Giles County Seismic Zone in E US as well as the Charlevoix Seismic Zone in Canada. Other case histories showed association with crustal weaknesses, such as the decollement zone in Central Virginia or the early Cretaceous Gustaf-Lougheed Arch in the Sverdrup Basin in Arctic Canada. The geodynamic process that underlies these associations has not yet been satisfactorily documented. An attempt is made to place four of the above case histories on the seismotectonic graph shown on Fig. 1: the Ochtendung Fault (ref. 14), the New Madrid Seismic Zone (ref. 15), the Ramapo Fault Zone (ref. 16) and the Giles County Seismic Zone (ref. 17). The four case histories indicate large differences in intraplate conditions in terms of recurrence rates, limiting magnitude capacity and style of crustal deformation. Co-seismic rates if not available were estimated for the specific seismotectonic conditions and then transfered to the UK adjusted plot on Fig. 1. The figure also includes a representation of the Foothill Fault System, Ca (ref. 18) a slow deformation case history in intraplate environment. The case history representation on Fig. 1 is premature but indicates an upper bound in the resolution of available techniques at about 10^{-3} mm/yr. It should also be noted that reasonable associations of earthquakes with faulting in low seismicity intraplate environments have been reached after long periods of investigations, typically longer than five yrs, at sites selected on the basis of demonstrable recent activity.

THE SITUATION IN THE UK

19. The situation in the UK is discussed in terms of proposed design criteria and of the potential of available techniques. It is first attempted to place proposed criteria on the seismotectonic chart. Then the situation in the UK is discussed on the same chart and compared to the relative position of design criteria and potential techniques.

20. Design criteria are specified independent of the actual earthquake hazard and of the techniques employed to evaluate this hazard but should be imposed on parameters significant in the seismotectonic process: A criterion compatible with earthquake prediction is in terms of the probability of exceedance during the life of the structure. Engineering structures have expected lifespans, in general, within the range of intermediate earthquake predictions; an interesting exception here are monuments that have longer life expectations. Design criteria are expressed on the basis of the response deformation of a facility at two levels: a lower level linked to the operation and a higher level associated with the safety of the facility. This higher level is more critically related to the seismotectonic relationship and has been associated with the probabilities of exceedance shown on Table 1 for a broad classification of facilities:

TABLE 1. Design probabilities for a broad classification of facilities

	probability of exceedance during the expected life	mean return period for 50 yr life
Building Codes	50-10%	200-500y
Dams	5%	1,000
NPP' s/LNG's	0.5%	10,000

23. Probabilistic analysis of the earthquake hazard is built on the assumption that seismic events are statistically independent in space and time, i.e., long term seismotectonic behaviour. Although mathematical models are available for more realistic statistical distributions, their application is limited to areas where earthquakes are associated with major faults with substantial documentation of their long term behaviour. In low seismicity areas the engineering aspect of intermediate earthquake predictions is treated with a variation of the seismotectonic parameters within bounds (ref. 6). This treatment is critical with respect to possible fault activity near the site, a stumbling block in the application of probabilistic techniques in the US NRC procedure. This eventuality has been treated with a separate criterion that refers, in some form, to the rate of deformation on a fault. In updates of the UBC a special zone has been assigned to areas close to active faults in W US with deformation rates seemingly exceeding 10mm/yr. For dams the design criterion refers to holocene activity and can be placed on the postulated limit of detecting such activity shown on Fig. 1. For Nuclear Power Plants the criterion is explicit (ref. 19) although its association with fault activity is not clear. Within the framework of W US seismotectonics the US NRC criterion would be associated with limitations in measuring geological deformation. This interpretation places the above criterion inside the bounds shown on Fig. 2: 500mm/30,000yr= $1.67\ 10^{-2}$ and 500mm/500,000yr=10^{-3} mm/yr. More recent guidelines developed in Japan (ref. 10) link the fault activity criterion with the limit of geomorphic investigations potential i.e., 10^{-2} mm/yr. Comparison with Fig. 1 indicates that some areas in intraplate regions (eg, New Madrid Seismic Zone) may come close to meeting the UBC criterion.

24. A relationship of the type of eqn. 4 is ill-defined in the UK because seismicity is low and spectral analysis of seismic signals is not an established routine in seismological data processing. Thus, there is very little evidence on the mode of crustal deformation from fault plane solutions of recent good size events and no basis for correlating seismic moment with magnitude, let alone a standard instrumental

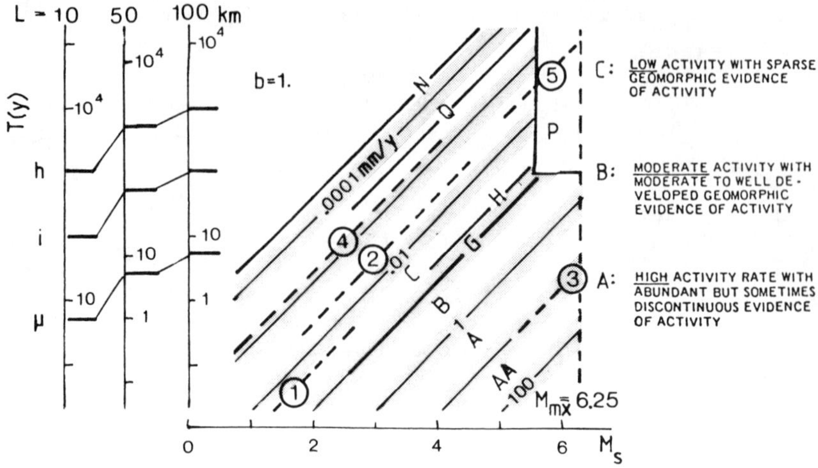

Fig 1. Seismotectonic representation of technique potential and of case histories. TECHNIQUE POTENTIAL: h historical seismicity, i instrumental seismicity, μ microseismicity, P paleoseismicity, G geodetic, H holocene, Q quaternary, N neotectonic, A,B,C geomorphic. CASE HISTORIES: 1 Ochtendung Fault(FRG), 2 Ramapo Fault Zone(E US), 3 New Madrid Seismic Zone(C US), 4 Giles County Fault Zone(E US), 5 Foothill Fault System(W US). T is the mean return period on a length L of the seismogenic zone.

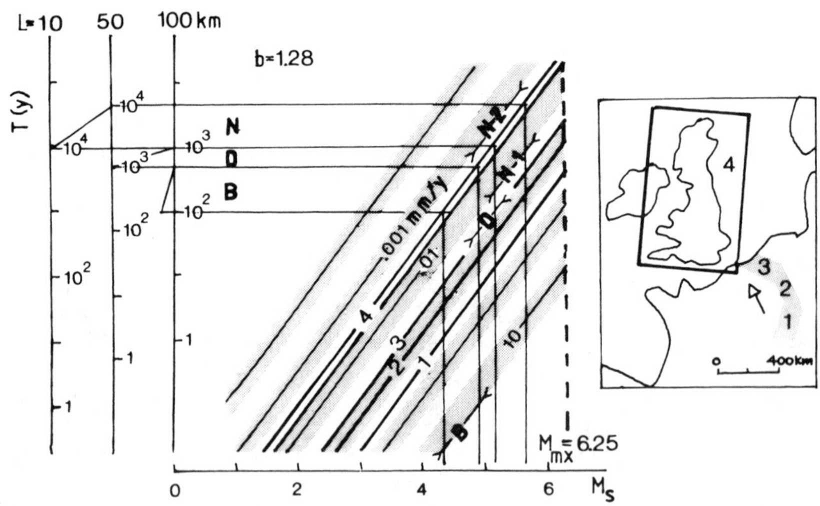

Fig 2. Seismotectonic interpretation of UK and environs and of design criteria. DESIGN CRITERIA: B buildings, D dams, N nuclear Power Plants. T is the mean return period on a length L of the seismogenic zone.

determination of the magnitude of British earthquakes. Moreover, there is no direct evidence on the slow crustal deformation thought to occur in the UK. However, an attempt can be made to represent the UK within the broad terms of the seismotectonic chart. The potential of the historical record on the occurrence of British earthquakes is being utilised (ref. 21) and quantified on the basis of calibrating felt areas of events in the 20th century with instrumentally determined surface magnitudes M_s. The magnitude rated data set has been statistically processed (ref. 6); the record appears complete for events $M_s \geqslant 5$ for about the last 1,000 yrs and for events $M_s \geqslant 4.0$ for about the last 200 yrs. The data set is best fitted by b = 1.28 and corresponds to an activity rate for the whole of the country of about 0.25 events/yr with $M_s \geqslant 4$.

25. The deformation rates shown on Fig. 2 are derived for the UK compatible values assumed in the seismotectonic representation of Fig. 1 modified to the steeper b-value obtained in the UK. The higher b-value renders the representation insensitive to the limiting magnitude M_{mx}, a result that obscures the physical meaning of this high value. Placing the UK on the seismotectonic chart involves additional assumptions on the mode of strain release inside the box shown on Fig. 2. For a uniform seismic strain field, gross seismotectonic conditions would correspond to deformation along a zone of length intermediate between the two dimensions of the block, i.e., between 700 and 1,000km. This gross estimate is shown on Fig. 2. Accommodation of the gross rate within existing crustal weaknesses will have a slight effect on this estimate: in a uniform stress field strains may incease by about 50%. A more pronounced effect is suggested by the diffused seismicity pattern in the UK (ref. 22) that may be interpreted as seismic strain release along a tortuous path inside the box with effective length much longer than the sides of the box. As a result, deformation rates across seismogenic zones in the UK may show values lower than 10^{-2} mm/yr; rates of deformation across faults inside these zones would be really small. Conversely, the gross estimate may be manifested with increased activity rates on a shorter actual fault zone. An important factor in distributing the gross rate is the time migration of tectonic activity that may shift around the UK block shown on Fig. 2, the single zone configuration deformation rate ($\sim 10^{-2}$ mm/yr) corresponding to the gross value.

26. It is interesting to place the gross UK estimate in the regional context of NW Europe. This comparison is shown on Fig. 2 along a well defined seismic zone running to the north of the Alps and along the river Rhine. The rates for the different segments along this zone are taken from refs. 14 and 23 and do not correspond to the seismic conditions on Fig. 2. Total deformation rates obtained from precision geodetic releveling are about an order of magnitude higher. Deformation along the Rhine Seismic Zone shows the accommodation of the apparently uniform stress field in pre-existing

crustal weaknesses and a consistent decrease of co-seismic rates away from the Alps. If a similar pattern extends into the UK, co-seismic rates there would be expected in the range 10^{-2} mm/yr.

27. The logic of the probabilistic analysis treats uniform seismicity as equivalent to a fault with concentrated activity going through the site. Uniform seismicity probabilistic calculations in the UK have indicated that for a design criterion of 10^{-4} mean annual probability any site is exposed to the occurrence of a magnitude 5½ within 10-30km radius. The gross estimate plotted on Fig. 2 amounts to the same physical interpretation and indicates the same extreme earthquake exposure. For a dam, the 10^{-3} criterion would increase the radius of a potential event and would probably reduce the extreme design event to a value $M \sim 5.0$. Within the recommendations of Building Codes a site will be exposed to more distant events and the extreme design event will drop probably around the M=4.50 threshold of structural damage, i.e., high risk areas in the UK would be placed to the equivalent of UBC Zone 1. In terms of fault activity criteria the gross UK estimate has a large factor of safety with regard to Building Code requirements, a smaller factor of safety with regard to dam design requirements but lies, as expected for a design event close to the site, within the more strict US NRC criteria. However, the criteria apply to fault activity whereas the UK gross estimate refers to the hypothetical case of releasing all the seismic activity in the country along a single fault. As discussed in par. 25, actual rates of fault activity in the UK are expected to be far below the 10^{-2} mm/y gross value

28. Placing the expected range of crustal deformation in the UK onto Fig 1 suggests that seismotectonic correlations may emerge through careful and apparently time consuming investigations in selected areas where seismic activity indicates rates in the upper end of the postulated bracket. Such investigations are guided by the long historical record. For events $M_s \geqslant 4$ this record would reveal patterns at a scale involving hundreds of kms whereas smaller events may reveal patterns at a scale involving tens of kms. Of particular interest is the potential of microearthquake measurements that follows the high b-value obtained from UK data.

29. The approach followed in this investigation is general and the conclusions remain tentative. Estimates of crustal deformation will improve as more experience is accumulated on the mode of crustal deformation in the UK. At this stage it appears that available techniques can support generally accepted probabilistic criteria only within safety margins. In engineering terms, the safety margins can be interpreted as factors of safety on the earthquake hazard estimates based on historical evidence and are associated with higher uncertainty in low probability design. The evaluation of the factor of safety is rationalised through comparisons to more active sites in the region where seismotectonic relationships are better understood.

ACKNOWLEDGEMENTS. The formulation of the seismotectonic relationship was developed in a site specific project under BP contract. The calibration of the relationship to UK conditions and the compilation of case histories from abroad was performed in a generic study for the Seismic Hazard Working Party (SHWP) under CEGB contract. This compilation was greatly assisted through interviews with international experts in seismotectonic investigations; space does not permit individual appreciation of the long list of consulted persons. The views and conclusions contained in this publication are those of the author.

APPENDIX. Correlation of seismic with tectonic parameters.

The equation correlating crustal deformation with the geometry of the seismogenic zone and the seismic parameters(ref 4, eqns 18 & 20) reads:

$$\log\dot{u}_s - \log n'_o = (1-B)\log M_{o,mx} - \log(\mu \frac{1-B}{B}) - c_o \qquad (A.1)$$

where u_s is the rate of crustal deformation in cm/y
$M_{o,mx}$ is the limiting seismic moment in the zone
B is the seismic moment distribution parameter
μ is the rigidity in the seismogenic zone
n'_o is the no. of events/y/source unit exceeding M_o=1
c_o is a crustal deformation factor(see par. 9)

The above relationship can be converted from seismic moment M_o to earthquake magnitude M following the relationships:

$$\log M_o = c + dM \quad , \quad B = b/d \quad , \quad \log n'_o = \log N'_o + cb/d \qquad (A.2)$$

where c,d are region dependent parameters
b is the magnitude distribution parameter
N'_o is the no. of events/y/source unit exceeding M=0

Substitution in (A.1) gives the general form:

$$\log N' = \log\dot{u}_s - (d-b)M_{mx} - (c - \log\mu \frac{d-b}{b}) + c_o - bM \qquad (A.3)$$

where N' is the no. of events/y/source unit exceeding M
To convert N' to the no. of events N in a segment of the seismogenic zone it is necessary to introduce an active area A for this segment. After conversion to km, for the dimensions of A, and to mm/y for $\dot{u}_s$ eqn (A.3) yields:

$$\log N = \left[9.o - (c - \log\mu \frac{d-b}{b}) - (d-b)M_{mx} + \log A + c_o + \log\dot{u}_s\right] - bM \qquad (A.4)$$

Substitution with the values suggested in ref. 5 for intraplate regions (c=15.58 and d=1.5) gives eqn 4 in par. 9 .

REFERENCES.

1. RYALL A S & J D VanWORMER Estimation of Maximum Magnitude and Recommended Seismic Zone Changes in the Western Great Basin Bull Seism Soc Ame, 70(5): 1573, 1980.
2. CEGB-SHWP International Experience in Seismotectonic Investigations (in preparation).
3. BRUNE J Seismic Moment, Seismicity and Rate of Slip along Major Fault Zones, J Geoph Res, 73: 777, 1968.
4. PAPASTAMATIOU D Incorporation of Crustal Deformation to Seismic Hazard Analysis, Bull Seism Soc Ame, 70(4): 1331, 1980
5. KANAMORI H & D L ANDERSON Theoretical Basis of Some Empirical Relations in Seismology, Bull Seism Soc Ame, 65(5): 1073, 1975.
6. IRVING J Earthquake Hazard, CEGB/GDCD Report C/JI/SD/152.0/R019, December 1982.
7. BERBERIAN M & D PAPASTAMATIOU Khurgu(North Bandar Abbas,Iran) Earthquake of March 21, 1977, Bull Seism Soc Ame, 68(2):411,1978
8. CURTIS G H A A Guide to Dating Methods of the Last Time of Movement of Faults, NUREG/CR-2382, 1981.
9. EVERINGHAM I B et al Thrust Fault Scarp in the Western Australian Shield, Nature, 223: 701, 1969.
10. AHORNER L pers comm, 1983.
11. WETMILLER R J et al Aftershock Sequences of the 1982 Miramichi, New Brunswick Earthquake, Bull Seism Soc Ame,74(2):621,1984
12. AMBRASEYS N Value of Historical Records of Earthquakes, Nature 232: 375, 1971.
13. MATSUDA T Active Fault Assessment for Irozaki Fault System, Izu Peninsula, Earthquake: 121, 1975.
14. AHORNER L Historical Seismicity and Present Day Microearthquake Activity of the Rhenish Massif, Central Europe, in Plateau Uplift, eds K Fucks et al, Springer-Verlag Berlin: 198, 1983.
15. NUTTLI O W Average Seismic Source-Parameter Relations for Mid-plate Earthquakes, Bull Seism Soc Ame, 73(2): 519, 1983.
16. AGGARWAL Y O & L R SYKES Earthquake Faults and Nuclear Power Plants, Science, 200: 425, 1978.
17. BOLLINGER G A & R L WHEELER The Giles County, Virginia Seismic Zone, Science 219: 1063, 1983.
18. SWAN F H Applications of Quaternary Geology to Evaluations of Earthquake Hazards, Earthq Eng Res Inst Seminar, 1983.
19. US NUCL REG COM Seismic and Geologic Siting Criteria for Nuclear Power Plants, Guidelines 10 CFR, Part 100, App A, 1973.
20. JAPAN ATOMIC ENERGY COM Regulatory Guide for Aseismic Design of Nuclear Power Reactor Facilities, 1978.
21. PRINCIPIA MECHANICA Ltd British Earthquakes, report for CEGB ref no. 7984, 1982.
22. MELVILLE C P The Geography and Intensity of Earthquakes in Britain: the Eighteenth Century, this Volume, 1985.
23. AHORNER L Present-Day Stress Field and Seismotectonic Block Movements along Major Fault Zones in Central Europe, Tectonophysics, 29(1-4): 233, 1975.

21. Methods of seismic qualifications for hazardous facilities in the UK

R. R. KUNAR, PhD, Structural Dynamics Research Corporation

SYNOPSIS. The seismic qualification of new and existing hazardous facilities has evoked considerable interest following the growing awareness of the potential earthquake hazard in the United Kingdom. This paper reviews the methods of seismic qualification that have found application in the United Kingdom. The paper also proposes an efficient procedure for the seismic qualification of existing facilities that were not originally designed or built with seismic considerations in mind.

INTRODUCTION

1. 'Earthquakes do not occur in Britain!' is often the response of engineers when questioned about seismic requirements in the design of structures. For many years this view has prevailed and many existing structures that are considered hazardous were not designed for seismic loads. These structures include nuclear power plants, petrochemical plants, LNG storage facilities, offshore platforms, and dangerous chemical storage complexes.

2. Over the last decade or so, there has been a growing awareness of the potential earthquake hazard in Britain. Recent reminders in the form of the Boxing Day Carlisle earthquake of 1979 and the North Wales earthquake of July 1984 has helped to focus further attention on the need for seismic considerations in the design of hazardous structures. In Britain, the nuclear industry has taken the lead in incorporating seismic requirements into the design of structures, plants and components. Other industries have been slower to react. The offshore and the gas industries have more recently begun to give serious consideration to the earthquake hazard though they lag considerably behind the nuclear industry. There is some indication that the increasing standards of safety demanded by our modern society will at some time in the future force the consideration of seismic loads on all structures that pose a threat to public safety.

3. The purpose of seismic qualification is to ensure that the structures, plants and components meet the pre-determined safety requirements. These requirements may vary from industry to industry and their stringency is generally decided on the severity of the hazard posed by the facility. In the design of new plant or facilities, the standard practice is to define a level of seismic loading and to design the structures, plants and components to ensure that in the event of this extreme loading, disastrous consequences are avoided. In Britain, there is an additional consideration. What of the older existing hazardous installations that were not designed seismically?

4. It would be unreasonable to expect that non-seismically designed facilities to satisfy even the 'intent' of current seismic design criteria which specifies design ground motions of up to 0.25g for some parts of the United Kingdom. The seismic qualification philosophy should be different from that of new plant. A logical approach for existing facilities is to ask the question:

> 'Given that the facility was not originally designed for seismic loads, what is the probability of 'catastrophic failure' of the facility in the event of an earthquake over its remaining period of operation?'.

It then remains to determine whether the computed probability leads to an acceptable or unacceptable level of risk. Whether new or existing facilities are being considered, there are methods of qualification that are generally applied to satisfy the requirements of seismic safety. This paper presents a review of those methods of qualification.

METHODS OF SEISMIC QUALIFICATION

5. Four basic procedures can be used for the seismic qualification of structures, plant and components:

(i) Analysis
(ii) Shake Table Testing
(iii) Low vibration modal testing in combination with analysis
(iv) Experience

6. Analysis: Analysis techniques predict responses of system models by calculation. The structure, component or system to be analysed is first modelled with discrete elements to simulate the stiffness and inertia characteristics. This model is then analysed on a computer to evaluate its behaviour to seismic excitation. The accuracy of this approach obviously depends on the accuracy of modelling the physical system.

7. Shake-Table Testing: The geometric complexity of certain components and equipment can sometimes make modelling very difficult and uncertain. Under these circumstances it is better to perform the seismic qualification by testing on a shake-table.

8. Modal testing and analysis: In cases where the facility is existing, low vibration modal testing can be performed to obtain the modes of vibration and associated frequencies of the structure or component. This modal information can be used to verify computer models or can be used directly with analysis techniques to determine the responses to the seismic loading.

9. Experience: This method of assessment relies on experience drawn from two sources:

(a) The observed behaviour of structures and components in real earthquakes.

(b) The seismic assessment of plants and components from similar items in similar environments.

10. All of these procedures have been used in Britain. The analytical and experimental procedures for seismic qualification have their own areas of superiority, and the most appropriate method for a particular system is best decided upon from practical, economical and reliability considerations. In many cases the procedures can be used in a complementary fashion.

ANALYSIS PROCEDURES

11. The items that are qualified by analysis include major structures, their foundations, some electrical and mechanical components and some sub-structures. To perform the analysis of components and sub-structures the seismic excitation at the support and attachment positions are required. These motions are usually obtained from an analysis of the overall structure to which the components are connected. It is appropriate, therefore, to consider the response of the complete building and their foundations, as the first set of analyses to be performed. (See figure 1). This is referred to as the floor spectrum generation phase or soil-structure interaction phase.

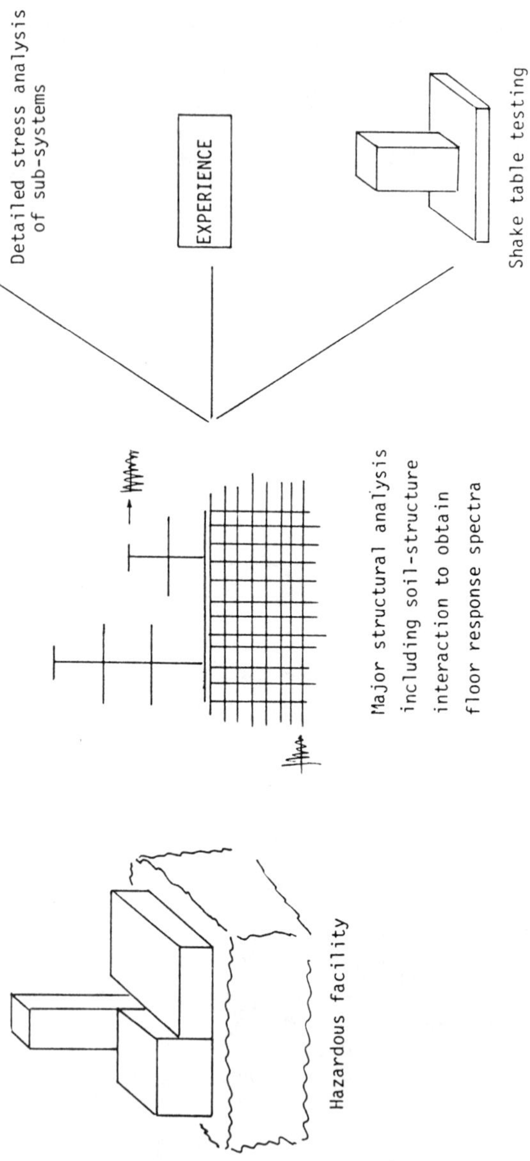

Fig. 1 : Seismic Qualification Procedures

Soil-structure interaction procedures

12. When considering the response of large structures to seismic ground motion, it is important to account for the effect of the flexibility of the subsoil on the dynamic response of the structures. The interaction effects between the soil and the structure can be very important. The deformation of the soil due to seismic ground motion induces the structures to respond, and, in so doing, the structures act as a dynamic load on the soil to modify its motion.

13. For very stiff foundations (e.g. rock), the dynamic loads resulting from the structural movements are not large enough to influence the behaviour of the foundation. The effects of soil-structure interaction are demonstrated in Ref.1.

14. Methods of analysis of soil-structure interaction systems fall into two broad categories: the direct method and the substructure method. In the direct method, the structure and the soil are modelled with discrete elements using either a finite element or finite difference formulation. The resulting dynamic equations of motion are solved directly in one step. This method has been used quite extensively in the nuclear industry in Britain. Two computer programs in particular have had substantial useage, FLUSH (Ref.2) and PRESS (Ref.3). The two programs use different numerical techniques for solving the dynamic equations of motion. FLUSH uses a complex frequency response while PRESS uses direct integration. Methods that use the substructure approach has also found application in the British nuclear industry. In this approach, the foundation stiffness is first determined for a massless foundation (Kinematic interaction) and then the dynamic analysis is performed (dynamic interaction) (see Ref.4).

15. In Britain, a number of nuclear hazardous structures are constructed on soil deposits of low or medium hardness. In these circumstances, soil-structure interaction effects are likely to be very important. Some offshore structures, and some on-shore storage facilities, are supported on piles in soft soil. The piles can be modelled as discrete beam elements in a finite element model of a soil-structure system. In the substructure approach, there are methods of evaluating the effective stiffness of pile groups taking into consideration soil-pile interactions. (See Refs. 5 and 6).

General Analysis Procedures

16. Qualification by analysis can be performed using equivalent static procedures or fully dynamic procedures. The equivalent static method (Ref.7) is used mostly in the preliminary design phases of a project but has very limited application in seismic qualification. It should only be used if justification is provided that the system can be realistically represented by a simple model and the method produces conservative results in terms of responses.

17. Most hazardous structures like refineries or nuclear plants tend to be very complex structures and it is very difficult to satisfy this condition. In general therefore, dynamic procedures are more commonly used for seismic qualification by analysis.

18. In a fully dynamic procedure, the equations of motion are solved by one of three techniques:

(a) Direct analysis in the time domain
(b) Direct analysis in the frequency domain
(c) Modal analysis

19. In the direct analysis in the time domain, the equations of motion are numerically integrated. All practical numerical integration schemes may be described as step-by-step or successive extrapolation procedures. There are implicit and explicit time integration procedures. The explicit methods express new vectors only in terms of vectors at preceding times which are already known. The implicit methods include new vectors in the expression for the new vectors and are generally solved by iteration. Mixed implicit - explicit procedures can also be used to good effect in soil-structure interaction problems (Ref. 8). The equations of motion can also be solved directly in the frequency domain by transforming the seismic input into its frequency components (using a Fourier transform) and then solving the steady state response for each component. Full descriptions of these procedures can be found in standard textbooks like Ref. 9.

20. In Britain, it is more commonplace to use modal analysis. In this procedure, the dynamic equations of motion are uncoupled by a transformation matrix obtained from an eigenvalue solution. Ref. 9 can be consulted for details. Each uncoupled equation is solved by time integration, or by application of convolution integral schemes. With the modal analysis scheme it is also possible to obtain approximate maximum responses using the response spectrum method. The maximum displacement of each mode of vibration can be determined by evaluating the product of the modal participation factor, the modal vector and the spectral displacement corresponding to the frequency of the mode. The modal displacements are then combined using an empirical rule like the square root of the sum of the squares.

21. In general, the modal analysis procedure is very efficient for linear problems, whereas the direct time integration procedure are recommended for non-linear problems.

22. The purpose of the analysis procedure is to determine the stresses and displacements in the item being qualified to compare these with appropriate allowable values determined from codes like ANSI, ASME, IEEE, API etc.

QUALIFICATION BY SHAKE-TABLE TESTING

23. Components which are too complex for numerical modelling are best qualified by shake-table testing. The components are securely bolted to a shake-table which can be subjected to single frequency or multiple frequency tests.

24. In single frequency tests the table is excited using a continuous sinewave, a sine beat, a decaying sine or a sine sweep input. This type of test is suitable under the following conditions:

(i) When the test response spectrum is fundamentally a single narrow peak response spectrum
(ii) The response of the equipment being tested has one or no resonances in the frequency range of 0-33 Hz.

25. When the conditions are not met, multiple frequency tests must be performed. These tests are more suitable for enveloping broad floor response spectra and if the equipment has more than one spaced dominant modes over the frequency range of interest. The input for such tests can take the form of a seismic time history (real or artificial), random motion, or a complex wave generated by summing a group of decaying sinesoids. The random motion input can be combined with a sine beat so as to achieve the required response spectrum without a large zero-period value. Whatever input is used (either for single or multiple frequency excitation), the required floor response spectrum must be enveloped by the test table response spectrum for the appropriate damping value.

26. Shake table tests can be carried out on single, or multi-axis machines. It is important to check if coupling exists between the different axes chosen for the test. This will be the case for example, if the centre of mass and the centre of rigidity are not coincident. Significant coupling must be accounted for by increasing the input accordingly. In the United Kingdom, most of the seismic shake table tests are carried out on multi-axis tables. The most commonly used facilities for seismic qualification are at the National Engineering Laboratories (one, two and three axis) and at Imperial College of Science and Technology (two axis). The procedures specified by the IEEE standards (Ref. 10) are generally used in the United Kingdom.

QUALIFICATION BY MODAL TESTING AND ANALYSIS

27. Equipment and structures can be qualified by a combination of testing and analysis. This method can be used on larger equipment which are impractical to test on a shake table. In this type of test, portable vibration generators are used to excite the modes of the equipment and this information is used to verify mathematical models which can be then analysed using standard dynamic procedures. In cases where all of the significant modes in the frequency of interest can

be measured, they can be used to generate the mathematical models without recourse to engineering drawings. (see Ref.11).

28. As mentioned previously, in the United Kingdom, there are a number of hazardous structures that have not been seismically designed. For these structures, the method of qualification which combines modal testing and analysis can prove extremely effective and attractive. For the older structures, the engineering drawings may well be out of date in that significant modifications to the structures could have been made during the period of operation.

29. To counter this problem of lack of 'as-built' conditions it is of course possible to prepare the 'as-built' drawings from site surveys and measurements. Instead of this expensive excercise, low vibration modal testing can be used to check the mathematical models derived from existing engineering drawings. If differences between test and analysis show up in the modal frequencies, on-site inspections can be instituted to check the as-built conditions. The mathematical models can be tuned to match the test results.

30. Verification of analytical models by practical comparison will ensure that the effects of ageing of the strcuture will be implicitly included. For example, corroded pipe work will exhibit lower stiffness than predicted on the basis of 'as new' and will show lower frequencies for its natural modes of vibration. This technique has been used in the United Kingdom in some areas of the qualification of three existing Magnox nuclear stations. Ref. 12 describes the procedures and shows how important and effective such tests can be. Typical equipment used for the modal tests are the Hewlett Packard or the GenRad Structural Dynamics Analysers together with different exciters and measurement transducers.

QUALIFICATION BY EXPERIENCE

31. Certain important safety-related components in hazardous facilities may be qualified on the basis of experience, either from assessments carried out on similar items or from behavioural data available from real earthquake observations or measurements. A data bank (Ref. 13) has been compiled recently in the United States of America of components in the nuclear and conventional power stations that have experienced earthquakes during normal operations. The components include pipes, valves and electrical relays that may be used in other hazardous industries. Qualification of equipment and components is based on a direct comparison of the behaviour and performance of the item during and after the earthquake. This method has been accepted as an adequate procedure by the USNRC. This acceptance covers only the equipment themselves and not the supports. This approach is just beginning to be used in the U.K. nuclear industries.

32. One significant concept in the qualification of existing facilities is called the 'walk-through'. This procedure, as the name suggests, consists of a site inspection by an engineer experienced in seismic matters to identify the areas of weaknesses in design and construction, particularly with regard the support and anchorages of safety-related equipment and components. This procedure, on the basis of experience, eliminates items that are robust enough not likely to be functionally affected by the earthquake, and concentrates efforts on the potentially vulnerable areas. This approach is strongly recommended as a first phase for the seismic qualification of existing hazardous facilities.

QUALIFICATION OF EXISTING FACILITIES

33. As mentioned in the introduction, there are a number of existing hazardous structures particularly in the oil, gas, nuclear and chemical industries that have not been designed seismically. It is unlikely that these facilities will satisfy the safety criteria specified for new facilities.

34. An approach developed by the author for the qualification of existing facilities proceeds in the following steps:

(i) Determine the 'limit-state' of the facility. This may be the failure to shut down a nuclear reactor or the failure of safety devices or structures leading to an explosion at a chemical factory. If the limit-state is breached, 'catastrophic failure' is assumed.

(ii) Determine those structures, components or sub-systems whose 'failure' would lead to 'catastrophic failure'. This generally requires an event-free assessment.

(iii) Use the 'qualification by experience' procedure to eliminate those items that are robust enough not to be vulnerable to earthquake loads.

(iv) Eliminate those items that can be rendered 'safe' by instituting a program of remedial action. This may mean bolting cabinets to the walls or floors, securing monitoring equipment etc.

(v) Rank the remaining items in an approximate order where the most vulnerable items are considered first. This is done on the basis of site 'walk-throughs' and experience.

(vi) Perform the seismic qualification of the most vulnerable items using on site modal testing and simplified analysis with conservative assumptions. If on this basis 'failure' is predicted at too low a level of seismic ground motion, the conservatisms are relaxed and the analysis is refined to consider non-linearities (for example strain-hardening or plastic material behaviour).

(vii) Determine the lowest level of ground motion that will lead to catastrophic failure from (vi).

(viii) Determine the probability of the site experiencing this level of ground motion in the remaining period of operation of the facility. This is done by performing a seismic hazard calcuclation for the site using available seismic, tectonic and biological information.

34. The approach described above has been applied to three existing magnox reactors in the United Kingdom. The method was developed to minimise the effort of such an exercise, while at the same time presenting a logical framework for the seismic safety case.

REFERENCES

1. KUNAR R.R. and BERESFORD P.J., The effect of site conditions on floor response spectra, Proc. of the Sixth World Cont. on Earthquake Engineering, New Delhi, India, 1977.

2. LYSMER J., SEED H.B., UDAKA T., HWANG R.N. and TSAI C.F., Efficient finite element analysis of seismic soil-structure interaction, Report EERC 75-34, Earthquake Engineering Research Centre, University of California, Berkely, 1975.

3. KUNAR R.R., An explicit method in non-linear soil-structure interaction, Invited paper to the 6th Int. Conf. on Structural Mechanics in Reactor Technology, Paper M7/1, Paris, 1981.

4. KAUSEL E. and ROESSET J.M., Soil-structure interaction for nuclear containment structures, Proc. of the ASME Power Division Specialty Conf., Colorado, 1974.

5. WOLF J.P. and ARX G., Impedance function of a group of vertical piles, Proc. of ASCE Conf. on Earthquake Engineering and Soil Dynamics, Pasadena, California, June 197298.

6. NOGAMI T. and NOVAK M, Soil-pile interaction in horizontal vibration, Jour. of Earthquake Engineering and Structural Dynamics, Vol. 5, 1977.

7. Structural Division of the ASCE, Structural Analysis and Design of nuclear plant facilities, 1982.

8. KUNAR R.R., A mixed implicit-explicit procedure for soil-structure interaction, Jour. of Nuclear Engineering and Design, Vol. 69, April 1982.

9. BATHE K. and WILSON E.L., Numerical methods in finite element analysis, Prentice-Hall Press, 1976.

10. IEEE 344-1975, IEE recommended practices for seismic qualification of class 1E equipment for nuclear power generating station, 1975.

11. SACCO N.P. and PALIE M.A., Equipment seismic qualification through the hybridization of analysis and in-situ testing, Proc. of ASCE Conf. on Structure Engineering in Nuclear Facilities, North Carolina, USA, 1984.

12. KUNAR R.R., The seismic qualification of non-seismically designed Magnox nuclear power plants, Proc. of ASCE Conf. on Structural Engineering in Nuclear Facilities, North Carolina, USA, 1984.

13. YANEV P., A pilot program for the development of an alternative approach to seismic equipment qualification, Report by EQE Inc., Jan. 1982.

Discussion

PROFESSOR R. T. SEVERN, University of Bristol
Dr Kunar refers to the validation of theoretical methods by full-scale testing. I would like to make two comments.

First, for the relatively simple type of structure which he showed, good correlation can usually be obtained, but for more complex structures, when correlation would really be useful, it is often not so easy to obtain, at least not without a lot of work. Second, something which Dr Kunar did not mention, is that in dynamic analysis there is usually insufficient information about material properties, and prototype testing combined with back-analysis is most helpful here. Admittedly such testing is usually done at low strain levels, whereas what is really needed is testing at high strains, ideally up to failure, and the chance to do this seldom arises. Material properties such as damping and elastic moduli are strain dependent, and our low-strain value is just one point on the curve, but a valuable one. Attempts to complete the picture must be carried out in the laboratory, preferably on shaking tables, but also in centrifuges and specimen testing under cyclic and transient loads.

DR KUNAR
I agree with Professor Severn's comments. It is not easy to obtain correlation for complex structures. However, with experience and a good deal of engineering intuition the task is much easier. A correlation of the modal characteristics does give confidence on the material properties for dynamic analysis, for very small strains. Ultimately, shake table or centrifuge testing is ideal but not always convenient.

22. Seismic design approach for the Sizewell 'B' nuclear power plant

C. R. SMITH, BSc, National Nuclear Corporation Ltd

SYNOPSIS. The paper describes how seismic requirements have been applied to the design of the Sizewell 'B' nuclear power plant. Topics discussed include the characterisation of the specified Safe-Shutdown Earthquake, the analysis of the response of the main structures and the qualification of plant. Additional studies to assess the possible effects of more severe earthquakes are also described. Finally, the paper discusses how the requirements for Sizewell 'B' may be developed for future nuclear power stations in the U.K.

INTRODUCTION

1. Historically the design of nuclear power plants to withstand seismic events has been achieved through the specification of a design event, the Safe Shutdown Earthquake (SSE). The designer is required to demonstrate that the plant can be safely shut down during or following this event with a degree of confidence consistent with the overall safety targets of the power station.

2. For those nuclear power plants in the United Kingdom for which a seismic capability has been required, the SSE description has been based on a free-field peak ground acceleration (pga) of 0.25g. This acceleration has been assessed to be a conservative representation of the ground motion level which will be exceeded at the site with a probability of 10 to-the-minus 4 in any year. The plants for which this SSE has been specified are the Heysham II and Torness stations presently under construction and the Sizewell 'B' station for which consent for construction is now being sought.

3. The determination of the ground motions which define the SSE is described by Irving (refs. 1 and 2). These motions are supplied to the designer in terms of horizontal and vertical free-field peak ground accelerations with the frequency content described by representative accelerograms or by elastic response spectra.

4. This paper describes the procedures being adopted for the Sizewell 'B' project by which a design is achieved which meets the safety criteria under the specified ground motion loading, and the studies which are carried out to demonstrate the adequacy of the adopted procedures.

DESIGN OBJECTIVES AND APPROACH

5. The objective of the seismic design process is to achieve a design which can be shut down during or following an SSE without exceeding permissible levels of release of activity at the site boundary. Although the SSE has an annual probability of being exceeded of 10 to-the-minus 4, the overall safety objectives are such that the probability of a release of activity due to a seismic event must be several orders of magnitude less than this. It is therefore essential that a conservative design process is employed, both to obtain a high degree of confidence that the plant will withstand the SSE and also to ensure that the plant has the capacity to survive more onerous seismic loads that would be expected to occur with an annual probability less than that of the SSE.

6. The seismic design process is divided into two broad phases. The first of these concerns the analysis of the overall response of the main structures to determine the seismic loads, in terms of accelerations, response spectra and displacements, which will occur within the structures. The second phase concerns the qualification, under the loads calculated in the first phase, of structures and equipment which are required to survive the SSE.

DESIGN GROUND MOTIONS

7. For the Sizewell 'B' PWR station the ground motion description is based on the data given in ref. 2 which recommends design response spectra for three site soil categories (hard, medium and soft). The spectrum for soft sites is used for Sizewell.

8. The response of the structures is calculated using artificially generated accelerograms whose spectra envelop the design spectra.

9. Previously seismic ground motions were defined by two accelerograms, based on the Temblor and Parkfield 5 records from the Parkfield 1966 event, considered to be representative of U.K. events (refs. 1 and 3). A rough estimate of the difference between the two specifications can be obtained by comparing the envelope of the Parkfield spectra with that given in ref. 2 for hard sites. This comparison is shown in Figure 1 which indicates that in the important 3 to 10 Hz frequency range the specifications are approximately consistent.

SELECTION OF PLANT TO BE SEISMICALLY QUALIFIED

10. There are two categories of plant for which a capability to withstand the SSE must be demonstrated. Plant in the first category is required to function so as to ensure that the safe shutdown objectives can be met. This category includes those structures which contain or support seismically qualified equipment. Plant in the second category comprises items which are not themselves required to function but whose failure could imperil other plant of the first category, which has a requirement to function.

11. It should be noted that some plant is not qualified. Firstly, since there is no requirement to ensure that the power station continues to operate during or following the SSE, plant required for operation but not for shutting down does not have to be qualified. Secondly, not all plant provided to assist in shutting down is qualified. For some safety functions an extremely high reliability is achieved through diverse provisions with the specific purpose of giving protection against faults which are more frequent than the SSE. The degree of reliability required to protect against rarer events, such as earthquakes, is less and hence it may not be necessary to qualify all diverse plant. The details of the list of structures and equipment for which qualification is not carried out depend on the station type. For Sizewell 'B', major items excluded will include turbines, the main sea water cooling systems, certain electrical systems, the electrical transmission grid and the secondary containment which is a steel clad enclosure which envelops the primary prestressed concrete containment building.

DERIVATION OF IN-STRUCTURE LOADS

12. Loads within the main structures are determined through a dynamic response analysis, referred to as a soil-structure interaction analysis, of the structure and adjacent soil. There are two types of approach to this analysis which are known as the direct and the substructure methods. In the direct approach the motion of the ground at depth (bedrock) is determined first. The response of the combined soil and structure to the bedrock motion is then calculated in a single step using a purpose-written computer program of which FLUSH (ref. 4) and PRESS (ref. 5) are examples. In the substructure approach the soil is first analysed separately to determine its compliance in forms of springs and viscous dampers which represent the reaction forces imposed on the structure due to displacement and velocity distortions of the soil relative to the free-field surface motion. These springs and dampers are then added to a finite element model of the structure for the analysis of the complete system. This approach may utilise either a standard structural analysis program or a purpose-written program, e.g. CLASSI (ref.6).

13. Despite the steady flow of enhancements to soil-structure interaction analysis techniques and computer programs over recent years, the fact remains that there is no single analysis approach that adequately models all effects which might have a significant influence on the response of the structures. For this reason the approach has been to divide the analysis into two parts. The first part consists of a soil-structure interaction analysis which uses the best available method, i.e. that available method which the analyst judges will best represent the response of his particular soil/structure system. Appropriate margins are then added to the calculated response to allow for modelling uncertainties. In the second part of the process studies are carried out to ass-

ess those aspects which might be considered not to be represented adequately in the first analysis, using whatever techniques are necessary. These studies are used to demonstrate that the responses calculated in the principal analysis are suitably representative.

14. At the start of the Sizewell 'B' project it was recognised that some of the major structures were highly asymmetric and thus the response, even to a unidirectional input, would be significantly three-dimensional. It was therefore considered that the sub-structure approach, incorporating representation of irregular basemat shapes and the variation of soil properties with depth, would be more suitable than the direct approach for which available programs were two-dimensional and hence not able to model real three-dimensional response.

15. To calculate the compliances for the Sizewell 'B' basemats, the front end of the CLASSI program was used. For this analysis the basemat plan shape is discretised to obtain a mesh such as that shown on Figure 2 for the auxiliary and control building. Soil properties, which vary with depth, were obtained from low-strain site investigation tests and then adjusted to be appropriate to the strain levels to be expected in the soil during the SSE. The output from CLASSI is the spring stiffnesses and viscous damping constants for each of the six degrees of freedom of the rigid basemat. These parameters, as calculated by CLASSI, vary with the frequency of the basemat motion. Typical variation, as calculated for the auxiliary and control building, is shown on Figure 3. For the structure response analysis discussed below a single frequency-independent value is required for each parameter. The values appropriate to the first mode natural frequency of the structure were used.

16. The soil-structure interaction analysis was carried out using the finite element structural analysis program MSC/NASTRAN (ref. 7). The details of how this analysis was performed are given by Parris (ref. 8) who describes the beam modelling of the structures, the automation of the calculation of the beam section properties using direct digitisation from drawings and the SDRC/SUPERTAB finite element modelling system, the manner in which NASTRAN was organised to allow representation of both structural and viscous damping, and the calculation of structural response and floor response spectra.

17. Figure 4 shows a typical horizontal design spectrum(at the location of 415v switchgear) derived from the above analysis. The spectra are biased towards low frequencies owing to the site being softer than average. Broadening of the spectra has been introduced following the response analysis because of uncertainties in the strain-reduced soil properties to which the response frequencies of the soil-structure system are sensitive. For firmer sites, the spectra would be shifted to higher frequencies, but, since soil-structure interaction would be less significant, it would not be necessary to broaden spectra to the same extent.

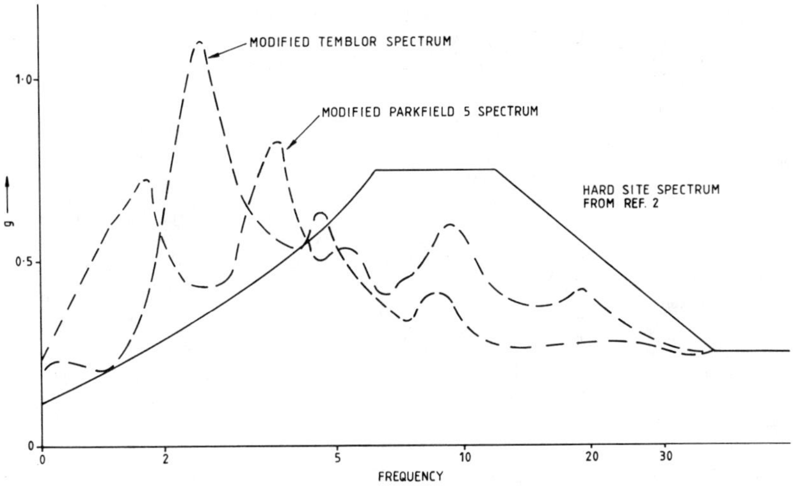

Fig. 1. Comparison of ground reponse spectra at 5% damping

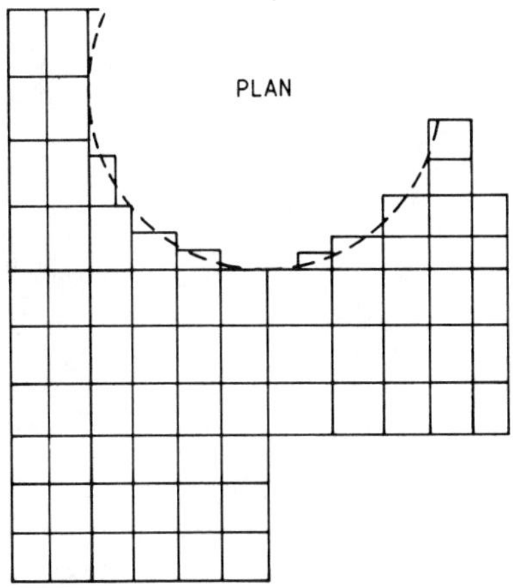

Fig. 2. Discretisation of auxiliary/control building basement

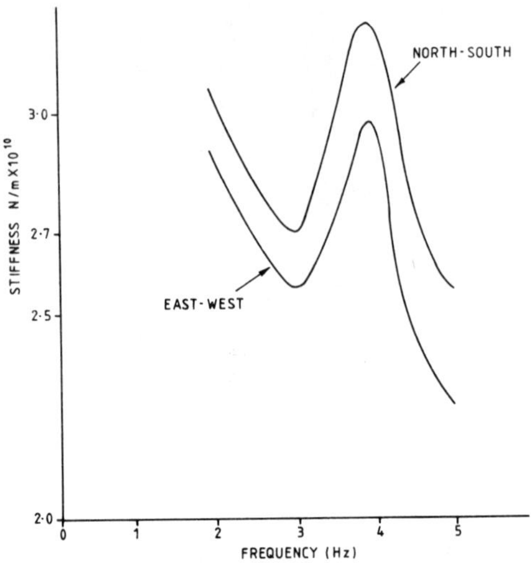

Fig. 3. Auxiliary/control building horizontal stiffness

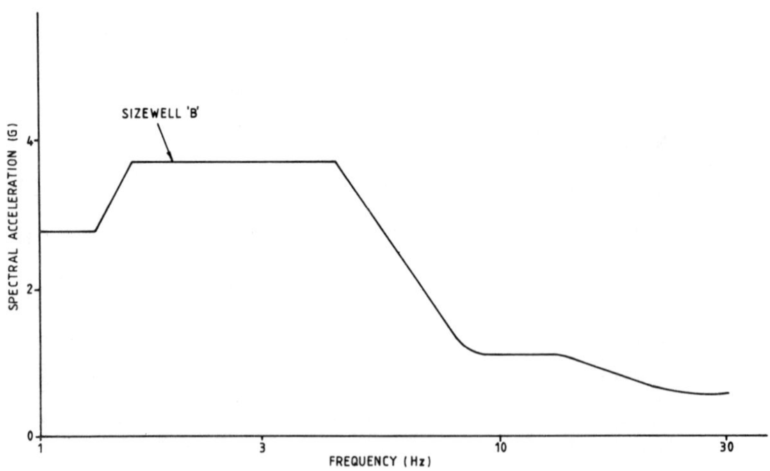

Fig. 4. Horizontal response spectrum at 5% damp envelope for 415 v switchgear

18. The use of simple beam models is generally acceptable for the calculation of the global response of the structures, particularly when the site is relatively soft, as at Sizewell, so that most deformation occurs in the soil rather than in the structures, and when the structures have a relatively regular shear wall construction so that distortion of the beam sections is minimal. However, although global response may be adequately predicted, the beam models will not predict response where the structure has local flexibility. Since many of these areas will be designed at a later stage than the global features of the structures, the most practical approach is to analyse this flexibility separately from the overall structure analysis. The most important local flexibilities occur in larger floors where they can normally be represented by a single vertical vibration mode. The vertical loading on equipment on a flexible floor can thus be calculated by passing the global vertical motion through a single-degree-of-freedom oscillator.

SUPPORTING SOIL-STRUCTURE INTERACTION STUDIES

19. The structural response analysis described above used an approach which is known not to represent a number of soil-structure interaction effects which might be considered significant. These are addressed in separate studies. At the time of writing this paper the studies for Sizewell 'B' are in progress and are therefore not reported in detail. The paragraphs below give a brief review of the most important aspects which are being addressed and the methods which are being employed.

20. In each study the analysis concentrates on a single aspect. The models in other respects (e.g. structural modelling) are simplified. Comparisons are always on a like-with-like basis in that, for a single modelling aspect, the model assumed in the main response analysis is compared with more representative models while other variables are kept constant. The studies therefore do not include comparisons between finite-element and compliance soil modelling which would not comply with this rule.

Frequency-dependency of soil compliances

21. Although the calculated compliances are frequency-dependent, a single frequency-independent value is used for the structural response analysis. This approximation is evaluated using a frequency domain analysis which can represent both cases. An internal National Nuclear Corporation program, SASSIF, has been used. Preliminary results show that the effect of the approximation is small. This is because the compliances are accurate for the lower building response modes while for higher modes, the response is less sensitive to the compliance values.

Inclined and surface waves

22. The main response analysis assumes that seismic waves

arrive vertically at the site surface so that all points on the surface move in phase. Body waves, arriving at other angles, or surface waves could lead to enhanced rocking. To analyse this effect, an additional stage is introduced into the analysis in which the motion of a rigid massless foundation is calculated. This is the same as the free-field motion in the case of the vertically incident wave. The calculated foundation motion is then input to the soil-structure interaction model. The effective filter which is applied in this stage of the analysis is dependent on the wave length of the surface motion and is thus frequency-dependent so that a frequency domain analysis is again necessary. The parameters of the filter may be approximated by the solutions for different wave types given by various papers including those by Luco and Westmann (ref. 9), Wolf (ref. 10) and Lam and Scavuzzo (ref. 11).

Adjacent Structures

23. The response of an individual structure can be influenced by the presence of an adjacent structure. To study this effect the methods given by Wong and Luco (ref. 12) are being used. Two adjacent basemats are discretised into subregions in the manner shown for a single raft in Figure 2. Coefficients which relate stresses and displacements of those subregions are then derived, from which a complex stiffness matrix linking the displacements of the two rafts and the free-field motion can be obtained. A two-structure response analysis is then carried out. Preliminary results show that the influence of one structure on an adjacent structure will not be great.

Embedment

24. In the main response analysis the effect of the embedment of the principal structures is represented by increased soil-structure interaction stiffnesses. The influences of enhanced radiation damping and reduced free-field motion at depth were neglected. Further studies will consider these effects and also review the sensitivity of the response to the effectiveness of the lateral connection between the structures and the soil.

Basemat Lift-Off

25. The normal analytical assumption that the basemat is effectively welded to the soil may be inadequate for predicting structural response if significant separation occurs when the structure rocks. A number of simple methods are available to predict whether this effect is significant. On the basis of the method proposed by Tseng and Liou (ref. 13), it has been inferred that the structural response will not be sensitive to lift-off. It is planned that this will be confirmed by a non-linear response analysis at a later date.

Non-linear Soil Properties

26. In the stress range of interest for seismic loading, the soil properties of the Sizewell site will be non-linear. For the calculation of compliances, equivalent-linear properties appropriate to the seismic stress level have been used. Typically, effective shear moduli under SSE loading are found to be of the order of 50% to 60% of the low-strain values. Non-linearity is therefore significant. It is generally recognised that in circumstances similar to those at Sizewell the equivalent-linear approach is not adequate to predict the response of a simple soil column, and comparisons between results using equivalent-linear and non-linear methods for this type of problem have confirmed that this is indeed the case at Sizewell. However, a complete soil-structure interaction problem either includes, or effectively includes, both a deconvolution analysis to determine motion at depth and a convolution analysis to obtain surface and structure motion. It is clear that to some extent the errors introduced in the first part of the process will be cancelled by those of the second part, so that the method may still give acceptable results. A preliminary check, using a one-dimensional soil model with a lumped mass to represent the structure, has shown this to be the case. However, since this model is not strictly representative, further studies are planned.

QUALIFICATION OF STRUCTURES AND EQUIPMENT

27. The demonstration that structures and equipment can withstand the loading which is imposed on them during the SSE is carried out using test and analysis procedures in line with internationally accepted standards. The principles of the approaches adopted for the Sizewell 'B' plant are discussed briefly below for structures and equipment respectively.

Structures

28. Structures are assessed by analysis. To meet a general requirement that the plant must be licensable in its country of origin, acceptance criteria are based on U.S. standards, viz. ASME III Division 2 for the containment shell and ACI 349-80 for other concrete structures. For steelwork, British standards are used with allowable stresses under seismic loading chosen to be consistent with U.S. practice. The above requirements lead to structures responding in a generally elastic manner under SSE loads.

29. The main structures have, through non-seismic design requirements (e.g. pressurisation and shielding) an inherent capacity to withstand seismic loads. This has provided a margin which has made it possible to simplify the seismic analysis procedure by neglecting the effects of soil-structure interaction when assessing the structure itself. To derive gross loads for the calculation of stresses, the ground motion is applied directly to the basemat, this procedure being shown to be conservative by comparison with the global soil-structure interaction analysis at a later stage. Civil design and

analysis work may thus proceed independently of the soil-structure interaction analysis so that the iterations which would otherwise be necessary between these two parts of the seismic design process are avoided.

30. The same global structural models are used for the civil structure assessment as for the main soil-structure interaction analysis.

31. Locally flexible areas of the concrete structures and steel structures, are analysed separately using seismic floor response spectra derived in the overall response analysis.

Equipment

32. Equipment is qualified to standards equivalent to those given by IEEE-344(1975). Mechanical plant is generally qualified by analysis using criteria given by ASME for faulted loads. Elastic analysis is normally employed, with stress limits appropriate to this type of analysis, to avoid the complexity of non-linear analysis. Since these stress limits permit a degree of inelastic movement, stresses must be limited further where it is necessary to guarantee operation of moving parts.

33. Electrical equipment is normally qualified by shaker table testing. One aspect where U.K. qualification procedures have differed from those used elsewhere concerns the required test motion duration which, at eight seconds, is relatively short, but represents better the short strong-motion duration associated with the low magnitude near-field earthquakes which contribute most to seismic hazard in the U.K. (See ref. 1).

CONSIDERATION OF EVENTS GREATER THAN THE SSE

34. It has been noted earlier in this paper that, although the SSE design ground motion has been estimated to have a nominal annual probability of being exceeded of 10 to-the-minus 4, the overall safety objectives for UK nuclear power stations require the probability of an active release due to seismic events to be considerably less than this. The safety case therefore depends upon the seismic design approach described in previous sections being conservative to ensure that the plant has a capability to withstand events greater than the SSE.

35. In recent years attempts have been made, mainly in the U.S.A., to quantify the margins in the design process through seismic probabilistic risk assessments (PRAs). These studies have followed two routes. Firstly, there have been industry-sponsored assessments, for which a relatively simple approach has been used. Secondly, the USNRC has sponsored the development of more complex procedures under the Seismic Safety Margins Research Program. Both approaches are described in a USNRC document (ref. 14).

36. For the Sizewell 'B' plant a preliminary seismic PRA assessment has been carried out using the simpler approach.

37. The industry-sponsored studies in the U.S.A. have been generally accepted as giving a reasonable estimate of the

order of magnitude of seismic risk at the plants included in the studies. However, the confidence bounds on the results are wide and there are a number of areas of uncertainty whose significance can be highly plant-dependent. For the Sizewell 'B' preliminary study, the problems have been exacerbated by uncertainties arising from the preliminary state of the design (U.S. studies have been concerned with existing plant) and doubts on the applicability to U.K. plant of sparse U.S. data on equipment seismic capacities (fragility data). In addition, the seismic risk levels which have to be achieved in the U.K. are lower than those calculated in the U.S. studies. U.K. studies will thus be more sensitive to the highly uncertain tails of the hazard and fragility probability distributions.

38. Work on the development of PRA methods is continuing in the U.S.A. but the timescales of this work are long, reflecting the general view held by the USNRC that present design procedures are known to ensure significant margins and the need to obtain accurate quantification of these margins is not urgent.

39. The Sizewell 'B' project has been asked to provide an early demonstration that there is no sudden loss of seismic capacity for events in the range immediately beyond the SSE. Because of the considerations discussed above the preliminary PRA study has not been used for this purpose. Instead, studies have been carried out to assess how the site and plant might respond to more severe events than the SSE. An important facet of these studies has been a site response assessment.

40. This study investigated the anticipated response of the Sizewell site under increasing seismic load. It was assumed that the input motion used for design was applicable at a hypothetical outcrop of the bedrock adjacent to the site but not at the site itself since the site properties were not considered in the hazard assessment which defined the SSE design level. Then, as shown on Figure 5, the motion at the bedrock depth was calculated and from this, using a non-linear model and actual Sizewell soil properties, the motion at the site surface was calculated. Figure 6 shows the results in terms of site horizontal peak ground acceleration as a function of the same parameter at the outcrop. The upper straight line shows that, if linear properties were assumed, the site would amplify the response. When the non-linear properties are represented the site response is much lower, the two curved lines showing site response for upper and lower bound soil properties. The lines of Figure 6 are generated assuming an artificial time history at the outcrop which matches the broad-band design spectrum. The points on the figure show the site response for real events assumed at the outcrop and show that the results are much the same as for the broad-band input. Similar site response has been shown to have occurred in the 1979 Imperial Valley earthquake by Mohammadioun and Pecker (ref. 15). It has been concluded that the Sizewell site has a limited capability to transmit horizontal motion.

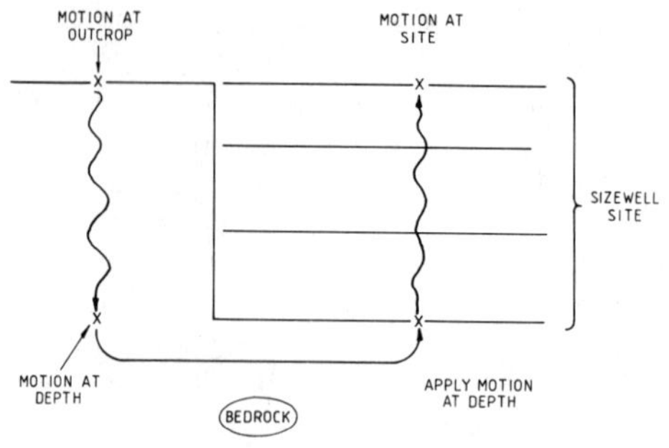

Fig. 5. Model for site response analysis

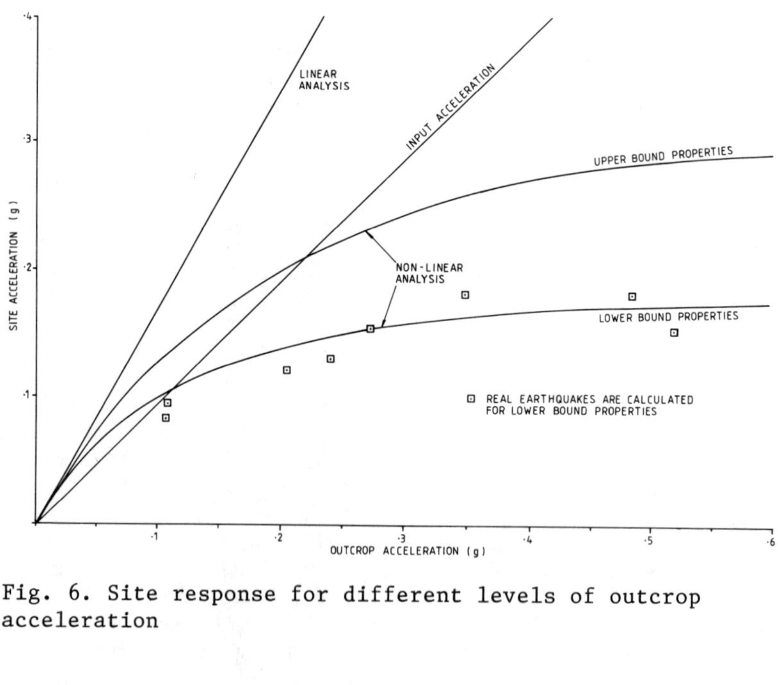

Fig. 6. Site response for different levels of outcrop acceleration

DISCUSSION

41. The procedures which are being adopted for the seismic design of the Sizewell 'B' PWR power station have been described. In addition short-term and long-term exercises aimed at showing the capacity of the plant to withstand events greater than the SSE have been briefly mentioned.

42. The procedures are based on state-of-the-art methods and are judged to ensure that the design is extremely safe with regard to seismic events. Nevertheless, as analysis methods and understanding of the physical processes develop, changes in the existing approaches are inevitable. Two areas where it is judged that developments might lead to a reassessment of seismic design requirements for future plants are the elimination of the unnecessary overconservatisms which current procedures encourage in certain areas, notably pipework design, and seismic PRA studies.

43. The difficulties in validating and carrying out nonlinear response analysis for the large quantity of piping in a nuclear power station have made it generally impractical to qualify pipework other than through the ASME elastic analysis route with its inherently restrictive criteria. This in turn has led to the use of large numbers of snubbers even though the pipework could probably withstand the applied seismic load without them. Work is in progress in the U.S.A., coordinated by the Pressure Vessel Research Committee of the Welding Research Council, to develop new criteria which allow pipework to be designed more flexibly. Work on inelastic response is also in hand in the U.K. at the Central Electricity Generating Board's Berkeley Nuclear Laboratories.

44. Concerning seismic risk, it is considered likely that seismic PRA assessments will be required for future nuclear power stations. The conclusions reached concerning the Sizewell site's limited capability to transmit motion are unlikely to be applicable on any future U.K. site. It may therefore prove necessary to use PRA studies to quantify the conservatisms in the seismic design process. Given that such studies are carried out, it is the author's view that these will focus more attention on risk-sensitive areas of the plant. As a result seismic design loads may well be required to vary from item to item depending on how critical to safety each piece of equipment is following an earthquake. This will not simplify the design process but has a clear logic if a means exists to identify plant where special treatment is warranted.

ACKNOWLEDGEMENTS

45. This paper is published with the permission of the National Nuclear Corporation Limited and the Central Electricity Generating Board.

46. The author wishes to acknowledge the assistance given to him in the preparation of this paper by Mr. J. Irving of the Central Electricity Generating Board.

REFERENCES

1. IRVING J. Earthquake hazard in Britain. Ibid.
2. IRVING J. Seismic hazard in the United Kingdom. Proceedings of the Institution of Mechanical Engineers, Seminar on seismic qualification of safety related nuclear plant and equipment, April, 1984.
3. MALLARD, D.J., IRVING J. and CORKERTON P.A. The assessment of seismic design criteria for nuclear power stations in England and Wales. Proceedings of the C.S.N.I. specialists meeting on anti-seismic design for nuclear installations, NEA-OECD, Paris, December, 1975.
4. LYSMER J. et al. FLUSH. A computer program for approximate 3-D analysis of soil-structure interaction problems. Report EERC 75-30, 1975.
5. KUNAR R.R. An explicit method in non-linear soil-structure interaction analysis. Transactions of the 6th International Conference on Structural Mechanics in Reactor Technology, Paris, August, 1981, Paper M7/1.
6. ASD INTERNATIONAL, INC. CLASSI/ASD, Computer program for three-dimensional soil/multiple-foundation interaction analysis. ASD International,Inc., San Francisco, California, January, 1984.
7. MSC/NASTRAN. Macneal-Schwendler Corporation, Los Angeles.
8. PARRIS R.A. Aspects of seismic analysis using MSC/NASTRAN. Proceedings of the MSC/NASTRAN European users' conference. Munich, June, 1983.
9. LUCO J.E. and WESTMANN R.A. Dynamic response of circular footings. Journal of the Engineering Mechanics Division, ASCE, No. EM5, Proc., Paper 8416, October, 1971, 1381-1394.
10. WOLF J.P. Seismic response due to travelling shear wave including soil-structure interaction with basemat uplift. Earthquake Engineering and Structural Dynamics, 1971, Vol.5, 337-363.
11. LAM P.C. and SCAVUZZO R.J. Lateral-torsional structure response from free-field ground motions. Nuclear Engineering and Design 65, 1981, 269-281.
12. WONG H.L. and LUCO J.E. Dynamic response of rigid foundations of arbitrary shape. Earthquake Engineering and Structural Dynamics, 1976, Vol. 4, 579-587.
13. TSENG W.S. and LIOU D.D. Simplified methods for predicting seismic basemat uplift of nuclear power plant structures. Transactions of the 6th International Conference on Structural Mechanics in Reactor Technology, Paris, August 1981, Paper K3/6.
14. USNRC. PRA Procedures Guide. NUREG/CR-2300, January, 1983.
15. MOHAMMADIOUN B. and PECKER A. Low-frequency transfer of seismic energy by superficial soil deposits and soft rocks. Earthquake Engineering and Structural Dynamics, 1984, Vol.12, 537-564.

Discussion

DR KUNAR
The seismic design of Sizewell seems to be based on the US standards and criteria. Is this because the NII have suggested that the US standards are acceptable or because the CEGB feels it is best to use the US standards?

It should be noted that while it is possible to obtain reduction of accelerations when considering non-linear soil properties, this would generally be associated with an amplification of displacements. This aspect will have to be considered in an assessment.

MR C. R. SMITH
In reply to Dr Kunar, US standards and criteria are not mandatory for the seismic design of Sizewell 'B' but are often used because of a CEGB commitment that the plant will not be manifestly unlicensable in the USA and the use of these standards and criteria for the US plant on which the Sizewell 'B' design is based.

It is acknowledged that non-linear site response can lead to an amplification of displacements. However, it should be noted that absolute displacements have no engineering significance while it has been shown in the Sizewell 'B' study that relative displacements (of the buildings relative to the ground motion) are also attenuated.

23. Response to extreme earthquakes of an offshore concrete gravity platform in the North Sea

E. D. BOOTH, MA(Cantab), MICE, MIStructE, and C. J. B. ROBERTS, BSc, MSc, Ove Arup & Partners

SYNOPSIS. A simple method is described for calculating the post elastic response of a concrete gravity platform to extreme earthquakes, and conclusions on the associated collapse mechanisms and levels of ductility are presented. The method described provides an economic and rational means of assessing the post yield behaviour of the structure early in the design process, to ensure that there is a sufficient reserve of strength to resist extreme but ill defined hazards. There are applications to extreme hazards other than earthquakes, for example ship impacts or explosions.

INTRODUCTION

1. The North Sea is an extremely harsh environment but an area of only low seismicity. Under these circumstances, it might be expected that earthquakes would not be a major factor in structural design. For loads with a reasonable chance of occurring during the lifetime of the structure, associated with return periods of the order of 100 years, a number of studies (ref 1, 2) have confirmed that wave loads are indeed generally critical, although, on somewhat conservative assumptions, seismicially induced forces may become comparable at certain critical locations, notably at the connection of the deck to support structure. For loads of this order of return period, the primary concern is to prevent any significant structural damage and hence ensure the uninterrupted operation of the installation.

2. However, the nature of offshore installations leads to the requirement that the risk of structural failure leading to collapse or impairment of escape routes and shelter areas should be very low; the Norwegian authorities (ref 3) even place a figure of 10^{-4} per platform year on the limiting risk. There are several reasons for requiring that offshore installations should not collapse under levels of extreme or accidental loads which normally are considered as unreasonable to design against. The special factors offshore include:-

- the difficulty of escape in the event of an accident,
- the possible dire environmental consequences of a disaster,

the high concentration of hazardous and high energy process plant,

and the enormous economic significance of typical installations.

3. It is therefore being recognised (explicitly, in the case of the Norwegian authorities), that in addition to a check to ensure that no damage occurs under loads with the order of 100 years return period, a check is also needed for no significant collapse under environmental loads with a much longer return period. This long return period greatly increases the potential importance of earthquakes. This is because seismic loading increases much more rapidly with return period than is the case for other environmental loads, especially in areas of low seismicity, and this is illustrated in fig 1. Therefore the fact that designing for the 100 year storm is likely to give an adequate degree of protection to North Sea platforms against earthquakes with a few hundred years return period does not mean that earthquakes can be neglected when considering extreme or survivability loadcases.

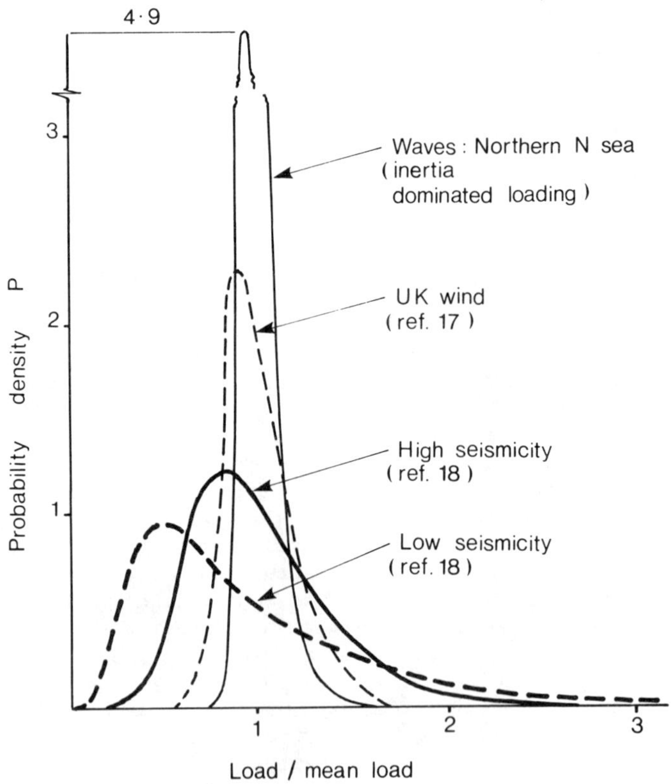

Fig. 1. Extreme value distributions for 50 year exposure, for storm and earthquake loads

4. There are several other points to bear in mind, which make it dangerous to compare earthquakes and storm loading purely on the basis of, say, induced base shears for 100 year return period. Firstly, unlike storm loads, seismic loads arise from cyclically applied displacements rather than loads. The response is therefore primarily strain and not stress controlled, and the ability to withstand repeated deformation is crucial. Hence ductility is of much greater importance in the resistance to survivability earthquakes than it is for storms. Levels of ductility are not checked when designing for storms, but for earthquakes they should be.

5. Secondly, the distribution of forces is different in the two cases. Most importantly, the high mass concentration at deck level leads to large earthquake forces there, whereas storm induced forces are concentrated lower down.

6. Thirdly, the forcing frequencies of earthquakes are likely to be much closer to the fundamental frequencies of typical offshore installations than is the case for wave loading, and this leads to high deck accelerations and displacements. This is particularly important for the process equipment and pipelines, failure of much of which may have very significant implications for safety. It is interesting to note that, in one of the few reported incidents of earthquake induced damage on offshore installations, the deck facilities of 6 platforms about 100km from the epicentre of the 1976 Tangshan, China earthquake were damaged, but the structures survived unscathed (ref 5).

BACKGROUND TO THE STUDY

7. A comprehensive evaluation of risks was to be carried out on a proposed concrete gravity platform standing in moderate water depth (See figure 2). The platform consisted of a modular topsides supported by a steel module support frame (MSF). The MSF was connected via four steel transition pieces to the four prestressed concrete shafts rising out of the concrete base caisson. The purpose of the study was to quantify the effect of residual risks from events of such low probability that it was uneconomic to design against them and to compare these residual risks with specified targets. As part of the study and in the light of the considerations outlined in the previous paragraph, it was decided to examine in detail the effects on the structure of very long return period earthquakes. Since this part of the study only formed one of the hazards being considered, and since the design of the installation was still at a preliminary stage, a very large computational effort was not appropriate. Nevertheless, the following objectives had to be met.

(a) To quantify the order of return periods associated with earthquakes causing
firstly permanent damage to the structure, and
secondly collapse of significant portions of the structure.

(b) To determine the collapse mechanisms of the structure under earthquake loading, in order to estimate its ductility and to establish if any 'weak links' existed in the initial design.
(c) To make recommendations, if necessary, for improving the seismic resistance of the structure.

The details of the analysis method chosen are described in the next section.

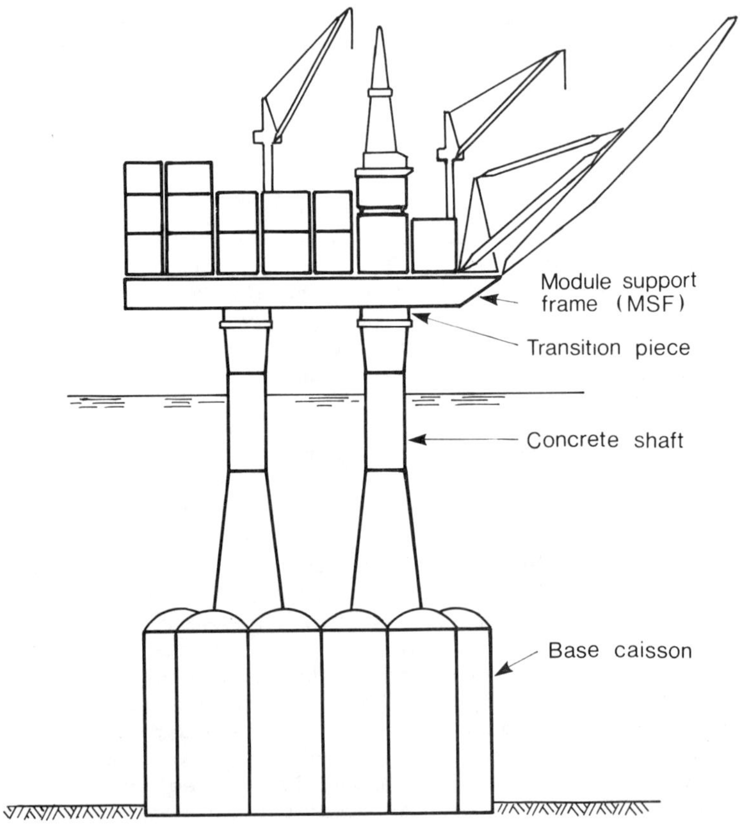

Fig. 2. Overall view of concrete gravity platform

METHOD OF ANALYSIS

8. An explicit check for the effect of very long return period earthquakes, in addition to normal design earthquake loading, is now standard practice for offshore installations in seismic areas of the United States (ref 4, 13) and API RP2A

(ref 4) gives guidance on how this check should be performed for steel jacket structures. One of the methods proposed by API RP2A is to derive a set of equivalent static forces corresponding to the normal design earthquake loads, and then to factor up these loads progressively, until the structure can no longer support an increased loading. During the course of the loading, the structure is progressively modified as members buckle, or plastic hinges form. The API RP2A requirement is that the structure should be shown by this or other methods to be capable of absorbing 4 times the energy associated with the normal design earthquake. However, it is recognised (ref 5) that this factor on energy of 4, while applicable to offshore Southern California, may be insufficient for other regions. Figure 1 illustrates this; the tail of the distribution in areas of low seismicity is significantly greater than for areas of high seismicity.

9. It was decided that this method of analysis would be particularly suitable for the present study. An elastic response spectrum analysis for the 100 year earthquake was available, which made the derivation of equivalent static forces relatively simple, (see for example Chopra (ref 15) for the mathematics involved). In addition, the fact that linear static, rather than dynamic, computer analysis was involved meant that computing costs were relatively low. Nevertheless, the analysis formed a powerful tool for investigating the post yield behaviour of the structure. It provided direct information on ductility, collapse mechanisms and the associated sequence of plastic hinge formation and drew attention to any weak links in the structure. Moreover, some estimate of the return periods of earthquakes causing damage could be made by finding the ratio of the loads corresponding to the various stages of failure to the load corresponding to the 100 year return earthquake.

10. This type of static pushover analysis has previously been described for steel jacket structures by Kallaby and Millman (ref 6). The present analysis followed similar lines, but some modifications were found necessary because a concrete gravity structure rather than a steel jacket was involved. The procedure adopted was as follows:

(a) A computer model of the structure, composed of line elements, was set up as shown in figure 3.

(b) An equivalent set of static forces, corresponding to the 100 year event was derived. A reduction in the number of loaded joints to 40 from the 113 used in the original analysis was made by combining joint masses onto a limited number of master joints. Comparison with the forces and deflections obtained in the original analysis confirmed that very little loss of accuracy resulted from this reduction.

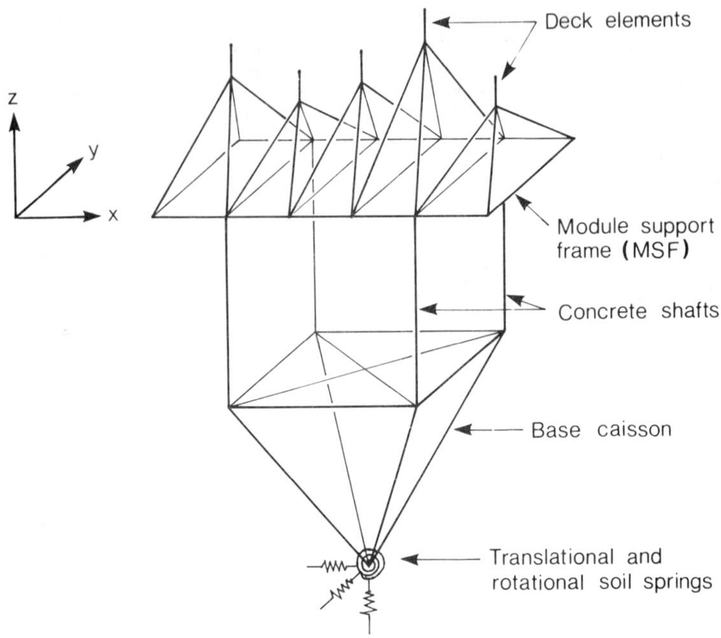

Fig. 3. Computer model of platform

(c) In Kallaby's analysis of a steel jacket, the modal forces at each joint were combined into a single force in each direction by a root sum square (RSS) method. This was found unsatisfactory in the present case, since, compared with steel jackets, concrete gravity platforms have a greater variation of mass to stiffness ratio with height, and so a greater number of modes contribute significantly to the loading. Since the RSS combination means that the direction of the higher mode forces is lost, unacceptable errors would have resulted. Separate loadsets corresponding to the first 11 modes were therefore retained throughout the analysis.

(d) All these loadsets were increased by the same factor, and applied as static loadings. The factor of increase corresponded to the load at which the first plastic hinge formed, in this case in the steel MSF (see figure 2).

(e) The resulting member forces were combined by the modified RSS method recommended by the United States NRC (ref. 14). In this method, modal responses corresponding to modes whose frequencies differ by less than 10% are added together linearly before performing the RSS addition.

(f) Before increasing the earthquake loads further, the beam in which the plastic hinge formed was released about an axis corresponding to that of the hinge. The modal components of the plastic capacity were then re-applied as external loads at the hinge - see figure 4.

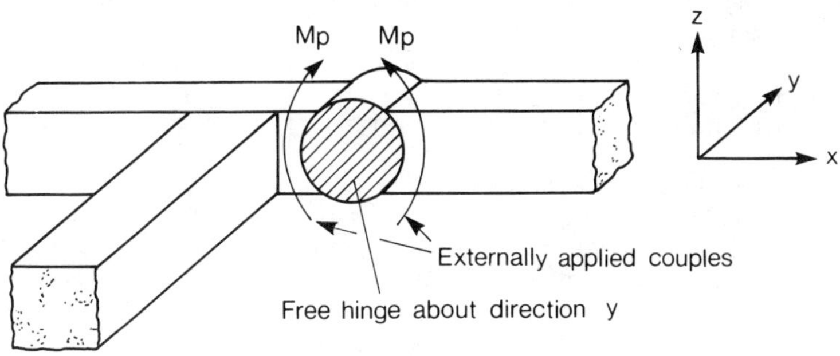

Fig. 4. Computer modelling of plastic hinge

(g) Further increments in earthquake load were then made, with progressive modifications to the structure to allow for the formation of plastic hinges. At each load increment, the rotational capacity of every hinge was checked, to see if the extreme fibre strain had reached an ultimate or fracture value. Fractured plastic hinges were assumed to have no strength.

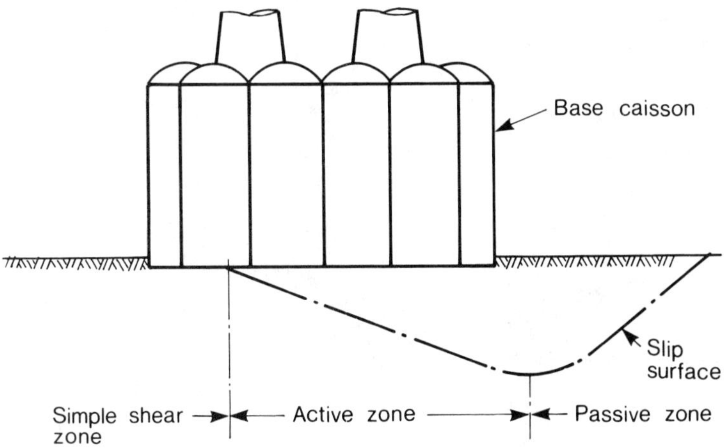

Fig. 5. Failure mode assumed in foundation analysis

(h) If necessary, the geometry could have been updated at each load increment to allow for P-delta effects (see ref 6), although in the present case, this was found to make very little difference.

(i) Linear elastic soil springs modelled the support provided by the foundation soils (see fig 3). Onset of soil plasticity was checked by performing a simple static slip surface analysis, as shown in figure 5, making some allowance for dynamic cyclical behaviour from the model tests published by Rowe (ref 7) and results published by Selnes (ref 8).

(j) Final failure was judged to occur when the structure could no longer support an increasing or constant load.

ASSESSMENT OF PLASTIC HINGE CAPACITIES

11. The plastic capacities of the concrete shaft were estimated using the OAP computer program GLADYS (ref 9). This performs a cracked section analysis, using the concrete and steel stress/strain characteristics proposed by CP110.

12. The plastic strengths of the steel members were assessed using a von Mises' criterion, to allow for the effect of shear stresses on yield, since this was found to be significant.

13. The rotational capacities of the steel beam members in the MSF (figure 2) were assessed by estimating plastic hinge lengths from the bending moment diagram, and assuming a fracture strain of 0.05. Plastic hinge lengths in the concrete shafts were estimated from standard beam data quoted by Park & Paulay (ref 16), and a fracture strain of 0.0035 was assumed in the concrete.

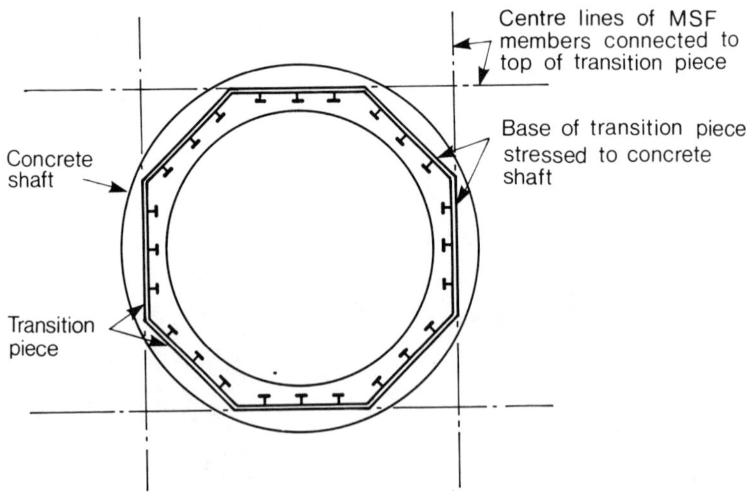

Fig. 6. Plan view of connections to transition pieces

14. The base of each transition piece (figure 2) was well connected to the concrete shaft but the top was only partially connected to the MSF - see figure 6. Failure was therefore initiated at the top, and very little opportunity for strain redistribution existed. It was assumed that none in fact took place.

ESTIMATION OF EARTHQUAKE RETURN PERIODS

15. An estimation was made of the relationship between return period and peak ground acceleration expressed as a ratio of 100 years ground acceleration. The estimate was based on a number of published sources (ref 1, 10, 11, 12) and the results are shown graphically in figure 7. The earthquake driving forces were assumed to be proportional to the peak ground acceleration, so that the factor applied in the analysis to the 100 year earthquake forces was assumed equal to the ordinate in figure 7. In reality, of course, the relationship is much more complex; major earthquakes compared with minor ones are likely to contain relatively more long period motions, and are of longer duration and the dynamic characteristics of the structure will change, due to plastic softening.

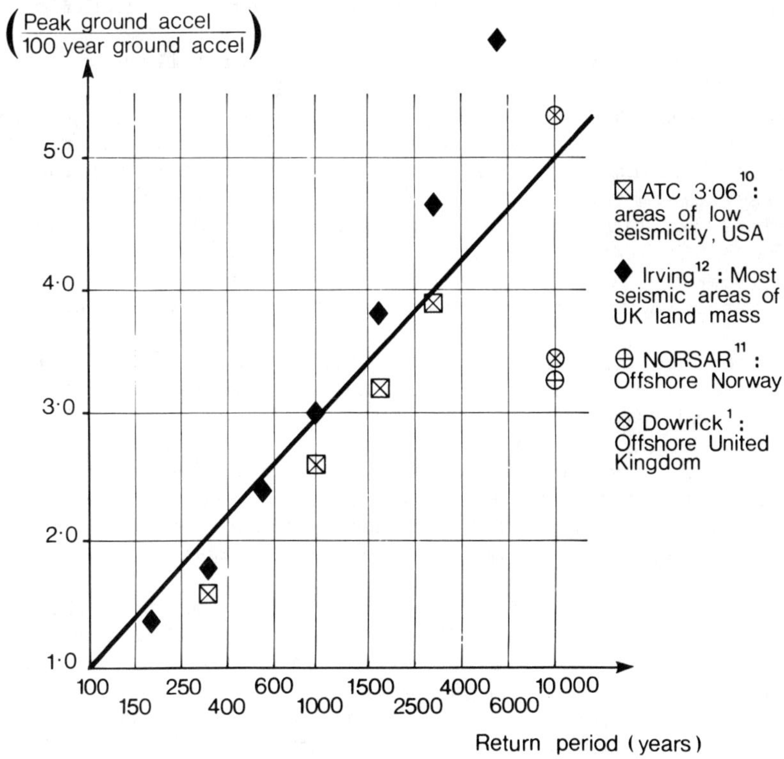

Fig. 7. Peak acceleration against return period

The approach used was therefore not rigorous, but it was commensurate with the accuracy of the rest of the analysis, and enabled some bounds to be put on the return periods involved.

RESULTS

16. The first plastic hinges were found to form in the MSF (see figure 2). Figure 7 enabled the return period associated with this onset of structural damage to be estimated. However, these plastic hinges were shown to be highly ductile, and no fracture occurred at any stage of the analysis.

17. Plastic yield occurred next in the steel transition pieces. For the reasons discussed above, these members lacked appreciable ductility and fractured soon after yielding. Failure in the transition pieces was judged to lead to overall failure of the platform.

18. In order to quantify the response of a platform with strengthened transition pieces, it was assumed that no failure occurred there, so that the applied load could be further increased. In practice, it proved simple to improve the design by increasing the coupling between the top of the transition pieces and the MSF (see figure 6). In this way, the strength of the transition pieces was significantly improved.

19. It was found that the next point of failure was in one of the concrete shafts at a load corresponding to an earthquake with a return period in excess of 10,000 years. Due to the high level of prestress in the shaft, failure arose from crushing of the concrete rather than steel yielding. It was therefore a non-ductile mode, and no significant plastic hinge rotation occurred before failure. Shaft failure was judged to lead to overall collapse of the installation. The effect of increasing ductility by reducing the level of prestress was investigated, and found to be significant.

20. Onset of significant plastic deformations in the soil system was found to occur at a loading similar to, or rather greater than, that corresponding to concrete shaft failure. The load level depended on what assumptions were made about the rate of consolidation - and hence increase in strength - of the clay soils on which the platform was founded. Based on the model tests of Rowe (ref 7), it appeared that these plastic soil movements represented a ductile mode, so that there was judged to be a large reserve against catastrophic soil failure.

21. Figure 8 shows the deflection of a point on the deck, up to a load corresponding to concrete shaft failure. The figure assumes no failure in the transition pieces, and no plastic deformations in the soil.

22. Figure 8 shows that, despite the high ductility of the MSF, where the onset of plasticity occurred early in the analysis, the overall ductility of the structure is low, being

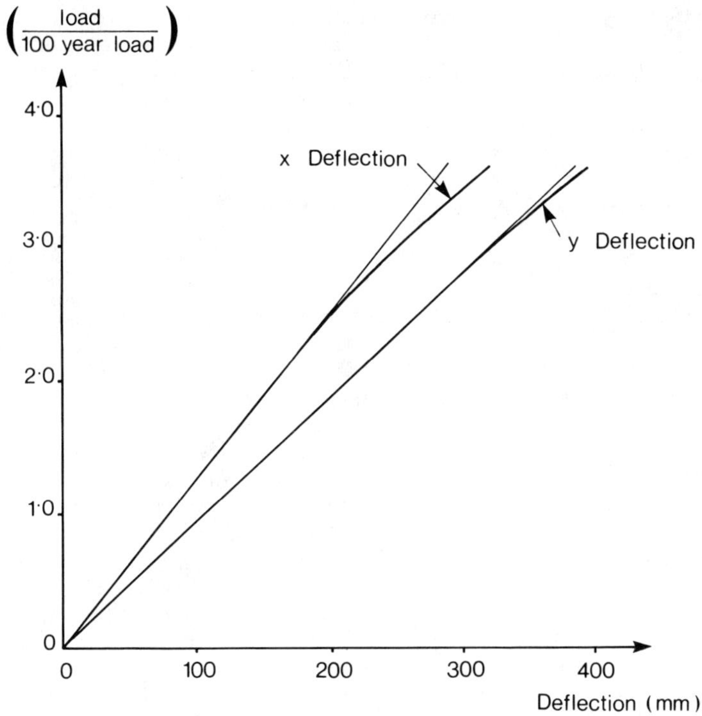

Fig. 8. Typical deflections at deck level for the concrete gravity platform

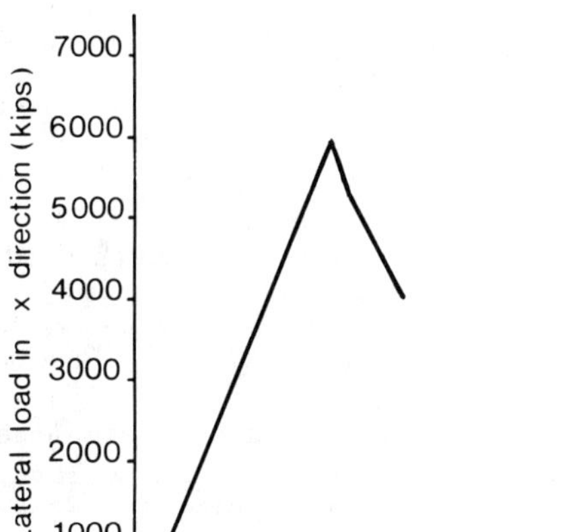

Fig. 9. Force deflection diagram at deck level for a steel jacket (taken from ref. 6, figure 9)

dominated by the concrete shafts. It is interesting to compare the shape of figure 8 with the load deflection graph (fig 9) obtained by Kallaby (ref 6) for a four legged steel jacket in a similar water depth. It can be seen that, unlike this concrete gravity platform, the steel jacket has a significant post elastic yield strength, though it appears from ref. 6 that the reserve of strength over the design (short return period) earthquake is much less than for the concrete structure.

CONCLUSIONS

23. Using the relatively simple tool of static elastic beam element computer analysis, significant conclusions were reached about the post yield behaviour of a concrete gravity platform under extreme earthquake loading. In particular, the return periods of earthquakes associated both with the onset of plasticity and with ultimate failure were estimated, and the effects of design modifications were checked.

24. The concrete gravity structure was found to have a very large reserve of strength over that required to resist a 100 year return period earthquake, largely as a result of the great strength of its concrete shafts. However, the ductility of the shafts was found to be low, due to the high level of prestress, and this limited the overall structural ductility.

25. This type of simple but indicative analysis would also be suitable to investigate other types of extreme and uncertain hazard, where the primary consideration is to prevent partial or total collapse. Examples of such extreme hazards are freak waves, ship impacts and gas cloud explosions.

26. Approximately 700 man hours was expended on this study, and the computing costs amounted to some 40% of the man hours cost.

REFERENCES

1. DOWRICK D. Earthquake effects on platforms and pipelines in the UK offshore area. Report for the UK Department of Energy, Ove Arup & Partners, 1980.
2. SELNES P et al. Earthquake risk on the Norwegian continental shelf. Norwegian Geotechnical Institute Publication No. 134, Oslo, 1981.
3. Norwegian Petroleum Directorate. Guidelines for Safety Evaluation of Platform Conceptual Design, Oslo, 1981.
4. AMERICAN PETROLEUM INSTITUTE API RP2A Recommended practice for planning, designing and constructing fixed offshore platforms. 14th edition 1984.
5. BEA R.G. et al. Seismic design of offshore platforms. EERI specialty seminar on earthquake engineering, Stanford University, July 17, 1984.
6. KALLABY J & MILLMAN D. Inelastic analysis of fixed offshore platforms for earthquake loading. OTC 2357. Offshore Technology Conference, Dallas, 1975.

7. ROWE P. Displacement and failure modes of model offshore gravity platforms founded on clay. Proceedings of the Offshore Europe Conference 1975.
8. SELNES P. Geotechnical problems in offshore earthquake engineering. NGI Publication 140. Norwegian Geotechnical Institute, Oslo, 1982.
9. CROFT D. The GLADYS computer system. The Arup Journal, March 1978.
10. APPLIED TECHNOLOGY COUNCIL ATC 3.06. Tentative provisions for the development of seismic regulations for buildings. Applied Technology Council, 1978.
11. RINGDAL F et al. Earthquake hazards offshore Norway. NORSAR contribution No 302, Oslo 1982.
12. IRVING J. Earthquake hazard. Central Electricity Generating Board, 1982.
13. AMERICAN CONCRETE INSTITUTE ACI 357R-78. Guide for the design and construction of fixed offshore concrete structures. Detroit, 1978.
14. US NUCLEAR REGULATORY COMMISSION, REGULATORY GUIDE 19.2. Combining modal responses and spatial components in seismic response analysis. Revision 1, February 1976.
15. CHOPRA A.K. Dynamics of structures : A primer. Earthquake Engineering Research Institute, California, 1981.
16. PARK R and PAULAY T. Reinforced concrete structures. John Wiley 1975.
17. CROFT D. An alternative approach to ultimate limit state design. The Arup Journal, June 1982.
18. BOOTH E. Assessment of seismic hazard. The Arup Journal, June 1984.

Discussion

MR D. W. YATES, Nuclear Design Associates

I am interested in the authors' use of plastic analysis to check the total ductility and reserve strength of the platform and supporting structure.

What degree of ductility was found to exist in the prestressed platform legs i.e. total ultimate strain/elastic limit strain?

What load factor was adopted for members liable to buckling failure such as lattice girders?

Was low temperature combined with earthquake loading and was a reduced allowable stress including fatigue effects applied to the steel?

MR BOOTH

In reply to Mr Yates, ductility was not great; approximately 1.25.

No lattice girders were present in the structure analysed,

since the module support frame was from a deep plate girder. It was assumed that sufficient restraint was present to prevent flange or web buckling. Post buckling behaviour of tubular steel struts is discussed by Marshall.[19]

The extreme air temperature assumed in design was -7°C and the transition temperatures of the streets employed were well below this, so no brittle problems were expected or allowed for. A separate analysis showed that accumulated fatigue damage was everywhere low, so again no allowance was made.

REFERENCE

19. MARSHALL P. et al. Inelastic Dynamic analysis of tubular offshore structures. Paper 2908 Offshore technology conference, Houston, May 1977.

24. The aseismic design of a reactor building for the advanced gas cooled reactor power plant

D. C. PAL, MSc, MICE, MIStructE, and J. V. PARKER, BSc(Eng), National Nuclear Corporation Ltd

SYNOPSIS. The paper describes the aseismic design procedures and criteria used for the major structures of the two advanced gas cooled reactor nuclear power stations currently being built at Heysham in Lancashire and at Torness in Scotland. In particular, the design of the reactor building which is constructed mainly of reinforced concrete is discussed. The design is unique in terms of both the size and shape of the concrete box surrounding the prestressed concrete reactor pressure vessel and may claim to be the first one of its kind built so far in the UK to a rigorous seismic safety criteria.

INTRODUCTION

1. The nuclear power stations currently under construction at Heysham and Torness are the first facilities to be built in the UK to have significant seismic capability. These advanced gas cooled reactor (AGR) power plants have been designed such that the reactors can be safely shutdown and cooled in both the short and long term in the event of an earthquake without causing an unacceptable release of radioactivity.

2. A suitable and consistent design level target for such earthquakes is considered to be an event with a probability of exceedance equal to or less than 10^{-4} per annum. This is known as the safe shutdown earthquake or SSE. The need for continued operation either during or after an earthquake is not considered necessary since the economic risk is acceptably low.

3. In order to design the complex of different types of buildings and major structural components which constitute a nuclear power station, to have the necessary high integrity and reliability, it is essential to have a thorough understanding of the dynamic behaviour of these structures when subjected to earthquakes up to the level of the SSE. However it is clear that at this SSE level of excitation, inelastic response through limited yielding and/or cracking of some of these structures can be considered acceptable. Certainly designing these components to have a dynamic response at the SSE level within the elastic limits developed for normal static loading would be extremely conservative.

4. In order to avoid unnecessary conservatism in the design and yet to maintain a relatively simple design procedure, the concept of "design basis earthquake" or "DBE" was originally

introduced. This is defined as having a free field ground motion one half as intense as that corresponding to the SSE. Any structural component designed to have an elastic response against the DBE should be capable of withstanding earthquakes up to the SSE level without prohibiting a safe shutdown of the facility. For certain components, design appraisals have been conducted in order to confirm this.

5. The application of the aseismic design procedure is discussed in this paper with particular reference to the reactor building structure. A static elastic design method was used in which the seismic coefficients were derived from the maximum dynamic response at the DBE level. The SSE capability has been confirmed through dynamic analysis studies using a single degree of freedom model with elastic-plastic stiffness characeristics.

6. The static analysis on the box shear walls of the building gave rise to a number of interesting structural responses which may be listed as follows:

(a) Horizontal shear stress contours
(b) principal stress vector plots
(c) interaction between shear and flexural stresses at corners of the box due to flange action

7. Based on the above structural responses, certain recommendations are presented in this paper as to the most effective 'form' of structure from the seismic restraint point of view. It has been felt that the aspects of shear wall design is an important consideration in a seismic restraint structure and is, therefore, discussed in some lengths in this paper.

ASEISMIC DESIGN PROCEDURE AND CRITERIA

Free field ground motion

8. The free field ground motion is the earthquake induced ground motion that would occur on firm ground at the site of the nuclear power station in the absence of any structure. Such ground motion may be expressed in the form of seismographs (eg acceleration time-history wave forms) or some form of spectra (eg ground response spectra).

9. The free field ground motion used to represent the SSE was described with two accelerograms developed from the June 1966 Parkfield earthquake and modified to have a peak ground acceleration of 0.25g horizontally and 0.17g vertically. For the DBE, the amplitudes of these acceleration time-histories were simply halved. The two accelerograms were considered at the time to be typical of a severe type of earthquake that might occur in the UK. Since then of course, UK ground spectra have been developed and these would probably be used for future designs.

Soil-structure interaction

10. The aseismic design conditions for all the plant components including the various buildings are developed through soil-structure interaction studies. These studies translate

the prescribed free-field ground motion into the dynamic response of the buildings and equipment in sufficient detail to enable the individual component seismic design loadings and criteria to be defined. These design conditions are generally in the form of static seismic coefficients and floor response spectra.

11. A typical structural idealisation used for soil-structure interaction studies of the AGR nuclear island usually consists of beam elements to model the buildings and major structures. Such an idealisation is sufficient in detail to retain the important design features that influence the dynamic response. Both modal analysis and complex frequency response methods have been used to calculate the time-history response corresponding to the prescribed free-field ground motions. The results from these analyses include natural frequencies and mode shapes, maximum relative displacements, absolute accelerations at each floor level and all the necessary floor response spectra at various levels for providing design information for plant items.

Non-linear response

12. Having designed a structure to have a linear elastic dynamic response against earthquakes up to the DBE level, it is necessary to quantify the response to earthquakes up to the SSE level. Naturally, if the whole soil-structure interaction system was to respond linearly up to the SSE level, then the SSE response would be simply twice the DBE response. Provided that the linear response of a structure to an earthquake is dominated by a particular vibration mode which is consistent also with the collapse mode, then the non-linear response may be investigated in a simple manner by considering it as a single degree of freedom system.

13. Assuming that the stiffness of the structure for these deformation modes is elastic-plastic, then the seismic response can be expressed in the non-linear regime conveniently through the ductility factor μ defined by

$$\mu = \frac{\text{maximum deformation}}{\text{deformation at yield}}$$

Thus, for $\mu \leqslant 1$, the response is elastic. The mode of vibration or collapse, together with the ductility factor completely defines the deformation of the structure and consequently the strain distribution within the structure.

14. What the DBE concept implies then is that up to the DBE level, the ductility factor will not exceed unity,

$$\text{ie} \quad \mu_{DBE} \leqslant 1$$

and that this will ensure that up to the SSE level, the ductility factor will be acceptable.

15. Non-linear seismic response analysis of a single degree of freedom system excited by the free field ground motions and with the yield or collapse load adjusted to give

$$\mu_{DBE} = 1$$

can be used to determine how the ductility factor varies with natural frequency of the structure and the amplitude of the excitation. Some of these results are shown in Fig. 1.

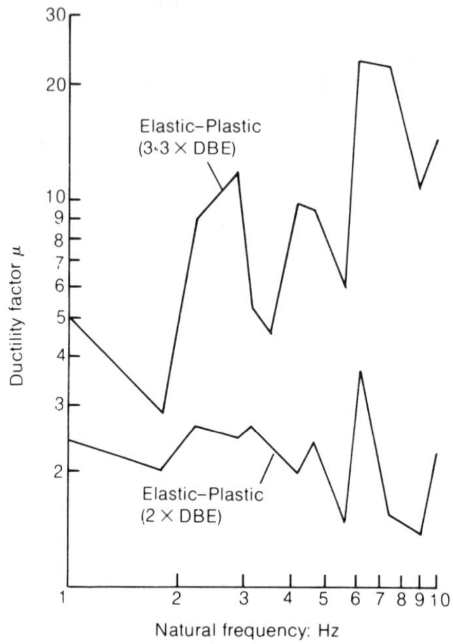

Fig. 1. Elastic-plastic seismic response

THE REACTOR BUILDING

Description of structure

16. The structure considered for presentation in this paper is an isolated box structure, separated by a movement joint from the associated nuclear island buildings as shown in Fig. 2. The reactor building is essentially a symmetric box (53m x 51m x 52m high) comprising concrete shear walls on all four sides with a central core reserved for the reactor pressure vessel as shown in Figs. 3 and 4. All floor loads either span on to walls or internal columns. Individual floors are separated from the pressure vessel structure in the centre by a circular movement joint. Each floor within the main reactor core, ie between 0.00 (ground) and +26.3m level consists of steel plated structure which in turn is supported on a system of concrete trusses between columns and perimeter walls. The foundation structure consists of a 2 metre thick raft foundation under the pressure vessel joined on either side by a 500mm thick basement slab and a perimeter strip footing under the retaining wall. The basement structure and strip footing are separated from the raft foundation by incorporating a rotational movement joint between the two. The overall basement structure and raft are founded on sound rock, both at Heysham II and Torness power stations.

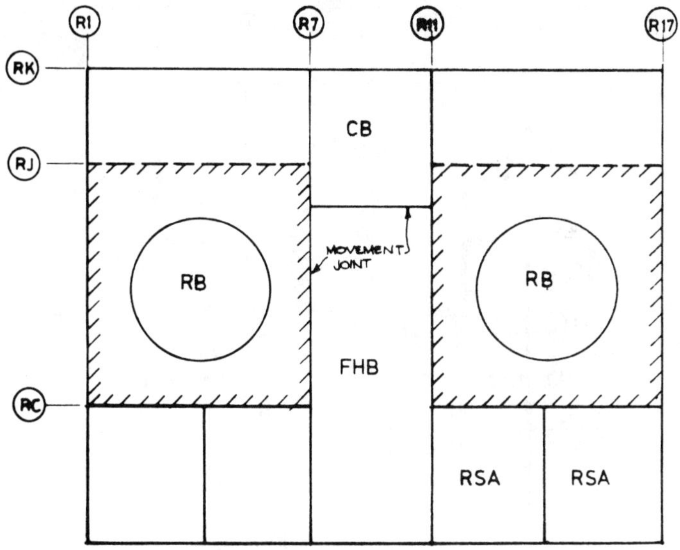

Fig. 2. AGR - Nuclear Island

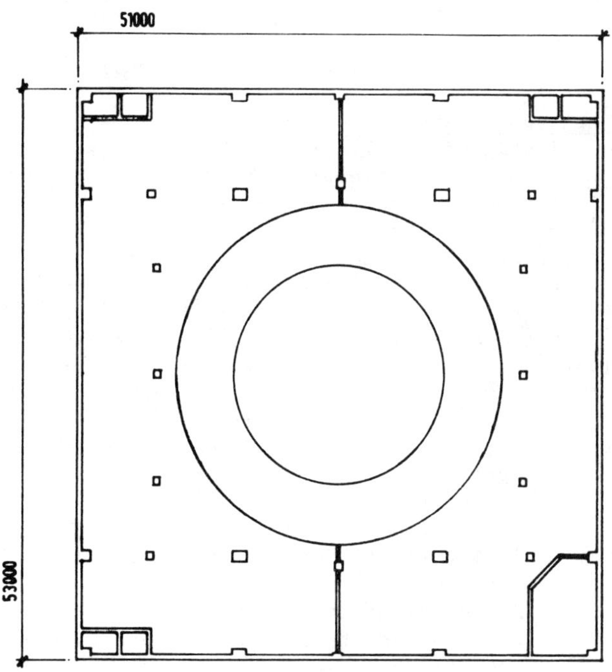

Fig. 3. Plan of Reactor Building

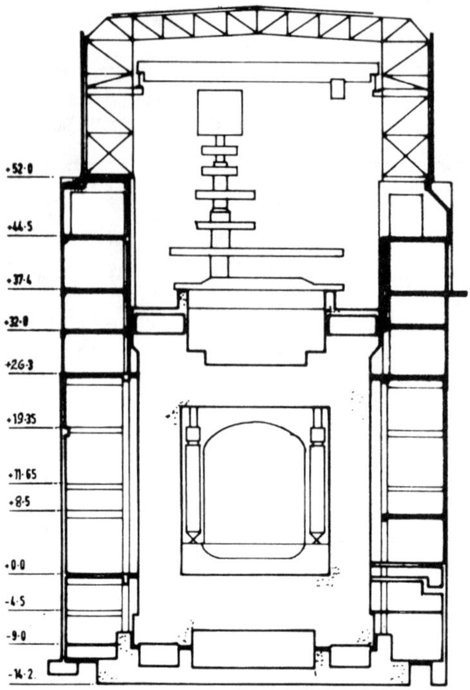

Fig. 4. Section through reactor building

Design input and assumptions

17. The criteria of civil structural design is to initiate an elastic response to the overall structure subjected to the DBE loads. This process ensures an inbuilt ductility potential in the structure, which when stretched to the target shutdown earthquake, provides adequate deformation of the structure beyond the elastic limit without loss of strength.

18. The design input was in the form of a 'plate' model using a three-dimensional finite element computer program, comprising wall, floor and columns in the form of suitable elements. A typical wall model is shown in Fig. 5. Each floor was represented by density nodes equivalent to the floor loading multiplied by static seismic coefficients. These static coefficients together with floor response spectra were derived from time-history dynamic analyses of the whole nuclear island. 'Soil-structure' interaction was taken into consideration in the static model by including the soil stiffnesses in the form of 'spring' elements.

19. An important consideration in the lateral restraint of perimeter shear walls was the choice of truss floors interposed between walls and columns. The arrangement of concrete trusses provide an effective restraint to the interstorey drift due to the increasing intensity of seismic loading from the bottom to

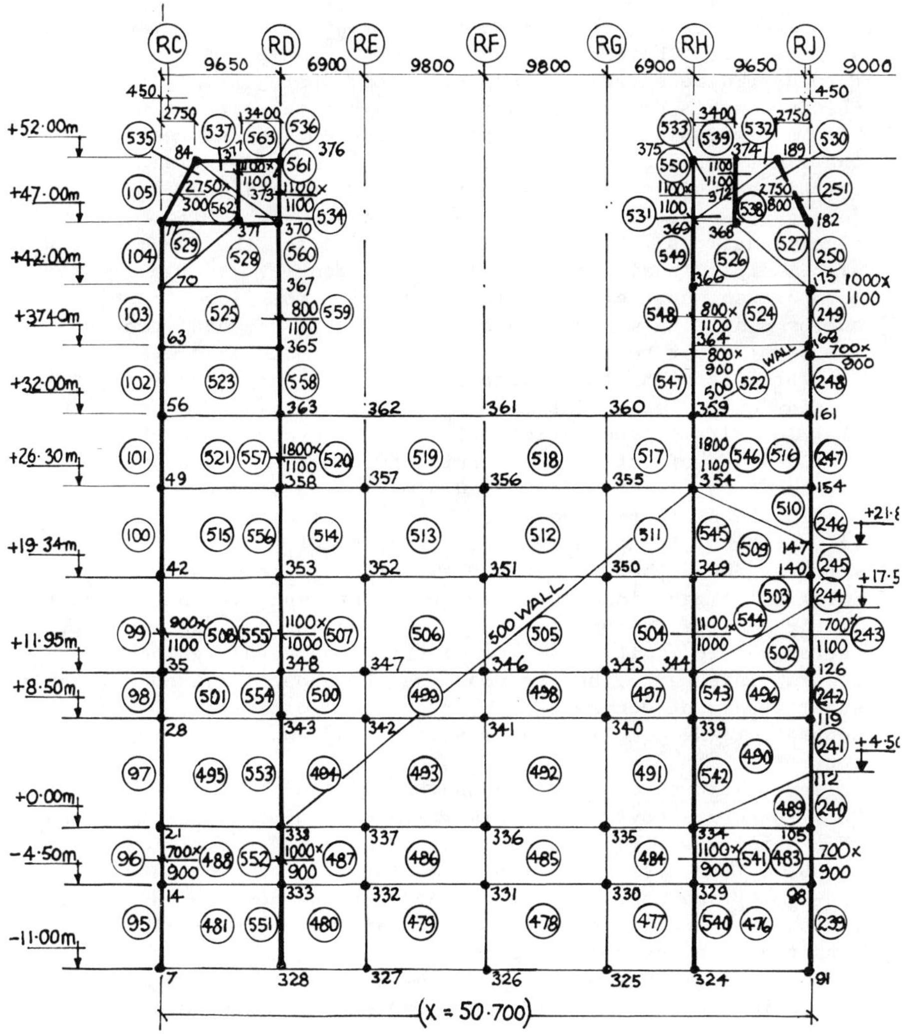

(X = as noted)

Plate elements 300, 500 Thick walls: bar elements as noted

Fig. 5. Wall on Grid Line R7

the top of structure. In contrast to the truss floors, the solid concrete floors at various levels within the reactor building act as diaphragm plates transferring horizontal shear load to the two opposite vertically held flanged walls.

Design Method

20. Shear walls provide an effective form of structural restraint in cyclic loading conditions. They contain all the three systems of inplane body forces, ie shear stresses in association with horizontal and vertical stresses which

combine to form the two principal stresses. It is the principal tensile stress which dominate the initiation of web cracking in shear walls. These 'web-shear' cracks normally inclined, once formed, progress rapidly causing catastrophic failure. Reinforcements are provided to counteract these web cracks.

21. A few typical 'stress' plots are presented as results of the finite element analysis in Appendix A representing the overall behaviour of the structure. In general the limit state design principle is followed in accordance with CP110 recommendations. The criteria is to restrict the development of principal tensile stress within the concrete section in the form of web shear cracks. The design is, therefore, carried out by providing web reinforcement for this diagonal tension. The total shear resistance is made by both the concrete and reinforcement with the assumption that the concrete section remains uncracked in flexure.

22. In spite of the horizontal web reinforcement which provides the effective shear reinforcement, an orthogonal mesh reinforcement is obligatory and a minimum 0.4% vertical reinforcement is recommended. This amount of vertical reinforcement is normally sufficient to cater for the tensile stresses due to bending at the two ends of the shear wall. Shear walls have inherently large moment capacity and the most likely mode of failure is web shear cracking. It is prudent, therefore, to limit this type of crack by overdesigning the section in shear either by reducing the upper limit on allowable shear stress on concrete or by providing reinforcement to counteract the complete shear without any assistance from concrete.

Salient features of aseismic design

23. There are several factors which control the effectiveness of a seismic restraint structure. Simple and symmetrical layout is the most effective form of restraint; too many non-symmetrical arrangement of load bearing members cause instability and large lateral movements which in cyclic loading conditions may cause severe cracks. Having selected a symmetrical arrangement of structural members, the next step is to provide sufficient stiffness requirements in accordance with sub-soil properties. A slender structure should normally be avoided since it produces large lateral movements. Shear walls are often provided to withstand the effect of seismic forces. In addition to the potential strength, which shear walls possess, they offer considerable lateral stiffness and thus protect a building against lateral displacement or sway during an earthquake. However, shear walls must be designed for considerations of flexural strength, shear strength and energy absorption capacity or ductility.

24. Meticulous reinforcement detailing is also a prime consideration in an earthquake resistant structure. Some recommendations are listed in paragraph 28.

SOME CONSIDERATIONS FOR SHEAR WALL DESIGN

Geometry or shape

25. The behaviour of a shear wall is dictated to a large extent by its aspect ratio. It is this height to depth ratio of wall, which controls the mode of failure in a cantilever shear wall. Two types of cracks normally propagate in a shear wall, 'flexure-shear' cracks or 'web-shear' cracks. When the aspect ratio is approximately between 1 and 2, 'web-shear' cracks normally predominate in a shear wall as shown in Fig. 6. Reactor building box walls almost conform to this category. In this type of wall, moment capacity is usually large and hence to be designed for sections uncracked in flexure in the form of web-shear reinforcement.

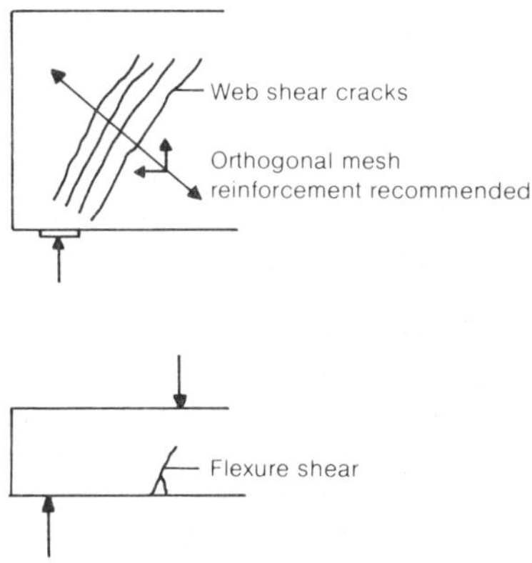

Fig. 6 Web shear cracks common for walls with aspect ratio between 1 to 2

26. The box shear walls of the reactor building essentially form the two flanges and web walls of the overall structure. The flanged walls do not change the overall behaviour of the web walls to a large extent though the flange action increases the moment capacity particularly for tall cantilever shear walls. The effect is obviously more pronounced at the ends of the box where shear action terminates and the bending action predominates. The effect of a flanged wall on the distribution of vertical bending stresses is illustrated in Fig. 7. The stresses are concentrated mostly at ends of walls both for flanged and web walls, relieving almost $^2/_3$ lengths of wall nominally stressed. The effect is almost negligible on web walls for this considerable length. The variation in vertical stress distribution is due to the action of shear lag phenomenon from the very wide flange lengths. As aspect ratio of shear walls approaches one, this effect could be more and more pronounced. However this is an important consideration when wide lengths of flanged walls are attached to web shear walls where the question arises as to what width of flange or web plate would be considered effective. In very tall shear

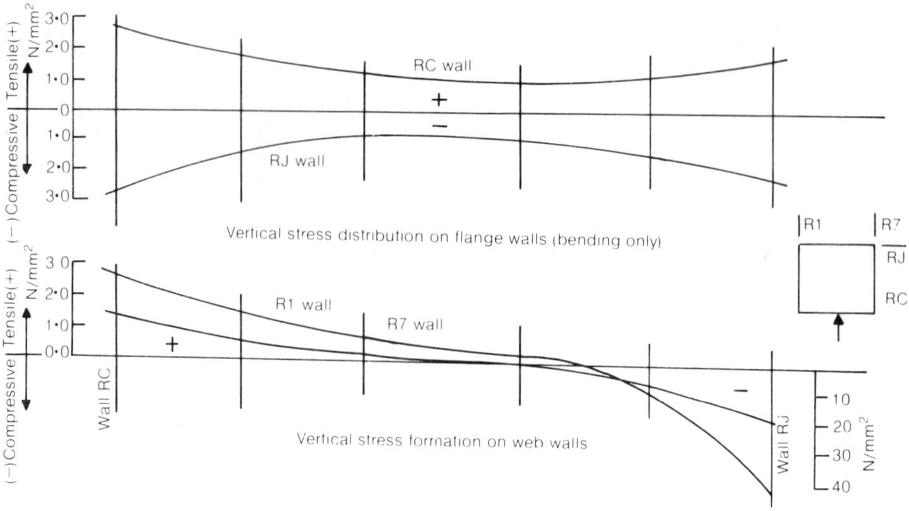

Fig. 7 Effect of shear lag on vertical stress distribution of box walls (bending only)

walls, however, the flange reaction is felt almost uniformly right throughout the web plate which is primarily due to the predominance of bending in such a member. Fig. 8 illustrates this uniform nature of stress distribution. The shear lag effect can be either determined by a numerical solution like finite element method or by a mathematical solution (ref. 2).

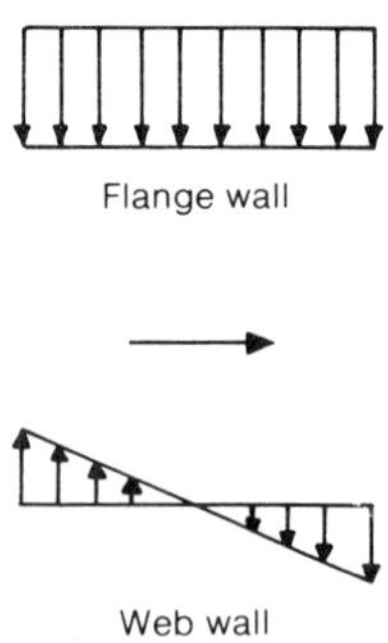

Fig. 8 Vertical stress distribution on tall wall

Horizontal construction joints

27. To minimise earthquake damage to shear walls at horizontal construction joints due to sliding movements, a rough concrete interface combined with vertical reinforcements should be provided in order to provide the required clamping force. The following equation (ref. 1) may be used to work out the vertical reinforcement required at a horizontal joint. An allowance has been made in this equation for the beneficial

effect of gravity loading. An alternative approach is to adopt 'shear-friction' theory as illustrated in CP110.

$$p_v = \frac{1.25 v_u - (N_u/A_c)}{f_y} \not< 0.0025 \qquad (1)$$

where N_u = ultimate axial load

Ac = gross concrete area

Av = Area of vertical web reinforcement

$p_v = \frac{A_v}{A_c}$

<u>Reinforcement detailing</u>

28. The objective of a careful detailing in a reinforced concrete structure during an earthquake is to render sufficient ductility and strength to the structure. Following recommendations are made to achieve this criteria.

(a) use of transverse or shear reinforcement to make the strength in shear greater than the ultimate strength in flexure

(b) limitations on the amount of tensile reinforcement or the use of compressive reinforcement, to increase energy-absorbing capacity

(c) use of confinement by hoops or spirals at critical sections of stress concentration, such as flange-web intersection of walls, ends of web walls as shown in Fig. 9.

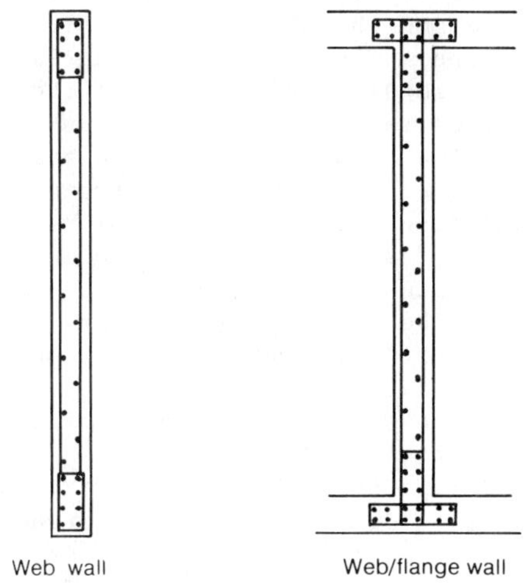

Fig. 9 Confinement of reinforcement at ends or intersections of walls

(d) special attention to details such as anchorage, continuity and bar splicing.

Ductility considerations

29. Ductility is a structural property describing the behaviour under large deformations. It applies to any structural entity of a building. Ductility implies high deformations without appreciable loss of strength and it also implies inelastic deformation or energy dissipation. These two factors are essential for earthquake resistant structures, since they must survive high deformations with no loss of strength and dissipate as quickly as possible the high input of energy. The ductility factor 'μ' has already been defined in paragraph 13. The 'deformation' mentioned therein can be expressed in terms of deflection or rotation as shown in Fig. 10.

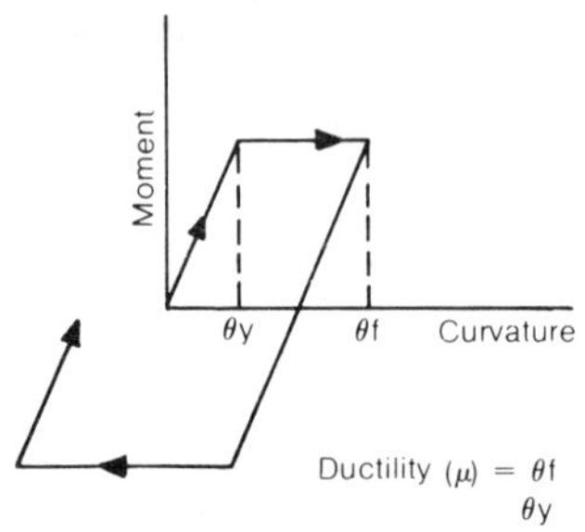

Fig. 10 Idealised elastoplastic cycle

30. An interesting hypothesis can be drawn from the elasto-plastic behaviour of the material against a cyclic loading which can exceed beyond the elastic limit of structure without causing total collapse of structure or material. The limit to this dynamic cyclic loading is dictated by the amount of inherent ductility or rotational capacity in the structure. Fig. 11 shows a typical dynamic impulse on a structure gradually stretched beyond the elastic limit of structure. This inter-action between the structure response (moment-rotation characteristics) and the cyclic loading is demonstrated in Fig. 12 where the structure or material goes through a complete hysteris loop under the influence of a dynamic load. It also shows the peak intensity of load corresponding to the maximum possible structure rotation. This optimum point of design could be significantly above the elastic limit point depending on the plasticity of the material.

31. In order to define this ultimate point to the dynamic load (equivalent elastic loading) imposed on a structure, a simple method of combining the two different responses is illustrated in Fig. 12 in terms of deflection. The shaded quadilateral abcd represents the response due to structure and the unshaded triangle aef represents the equivalent elastic loading where point b corresponds to the elastic limit of structure and point e to the maximum limit of cyclic load. It can be seen from simple geometry that ductility $\mu = \frac{af}{ag} = \frac{ef}{cf}$. In other words, a more ductile structure will allow a large impulse load,

above the limit of elasticity, travel throughout the structure without causing any undue distress. This increment factor ef/cf could be anything between 2-6 depending upon the ductility of material.

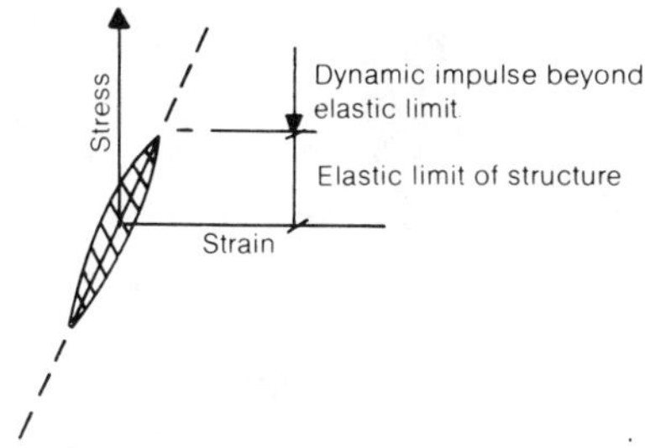

Fig. 11 Dynamic impulse on stress/strain limits of structure or material

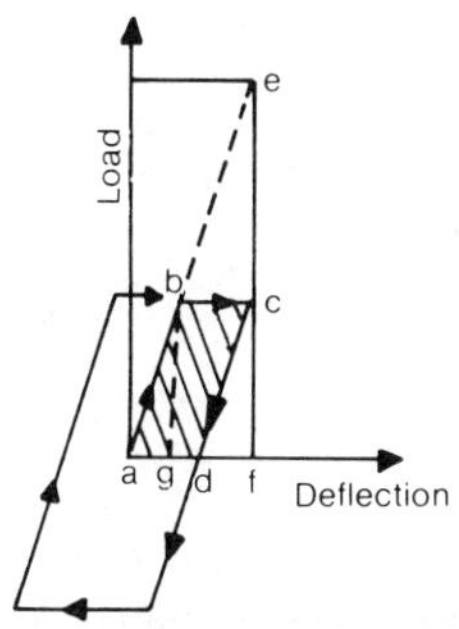

Fig. 12 Interaction between structure response and cyclic loading

32. A portal frame example is illustrated here to show the effect of incremental collapse on the ductility of a structure. Fig. 13 shows the sequence of formations of plastic hinges at 5, 4, 3 and 1 with a collapse load of Wc = 3MP/L with a corresponding deflection of 0.34 MpL^2/EI. The detail calculations of bending moment and corresponding deflections are not presented here. It can be seen that the ratio of Wc to Wy is 3/2.424 or 1.24 with a corresponding deflection ratio of 0.34/0.18 or 1.88. The horizontal deflection Δ is calculated at the top of each vertical member and expressed non-dimensionally as $\Delta\ EI/MpL^2$ and plotted against the value of either load expressed as WL/MP. The ductility factor of 1.88 in this particular example will ensure that the structure has a capacity of absorbing approximately twice the intensity of loading beyond its elastic limit.

33. The principles, outlined in above example, of incremental loads being applied in different stages to the same frame structure until it reaches failure, can be applied to shear walls as well. Detail calculations for such a wall is not presented here. However, it is usually the base of a vertical cantilever wall which suffers the first point of 'yield'. Any

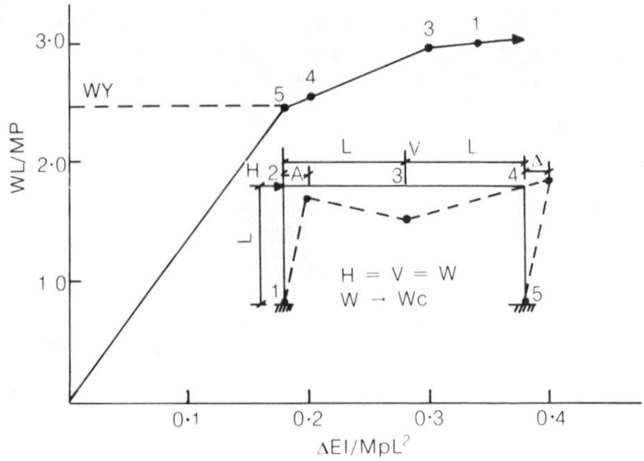

Fig. 13 Portal frame showing progressive collapse

further ductility beyond this point will be very much influenced by the degree of freedom available in the structure like ends of members or frames, interconnections, points of contraflesure etc. representing the potential areas of hinge formations. It is ideal, therefore, for a seismic restraint structure to have several degrees of freedom in the structure. The flanged shear walls of the reactor building is of course a multidegree freedom structure and can be designed and reinforced suitably to have a desirable ductility.

CONCLUSIONS

34. The paper summarises the essential design features of a seismic restraint structure based on strength and ductility requirements. It outlines the basic ingredients for a seismically effective structure. As is set out in the paper, the design is very much flexible and it is entirely up to the designer to decide upon a right form of structure from the very beginning. Suggestions have been made in this paper as to choosing the right amount of strength and ductility in a structure. The proportions have to be right to provide the best form of seismic restraint. However, before the engineer gets himself involved in the intricacies of a finite element analysis followed by a rigorous check on strength and energy absorption capacity of a structure, one important and prudent consideration should be the choice of a right material combined with just sufficient stiffness in the structure. These two factors, ie material and structure stiffness alone control the overall behaviour of a structure to a large extent.

35. Behind the underlying assumptions made in this paper as to providing on optimum safe structure to withstand seismic forces, is a fundamental requirement to provide adequate safety factors. Safety factors in a conventional construction project may influence design but is not a dominant aspect and normally do not pose any major problem. But in an earthquake design for a nuclear power station, safety is a consideration to be borne in mind at every stage.

APPENDIX A - Results of the Finite Element Analysis (Two typical plots are presented only)

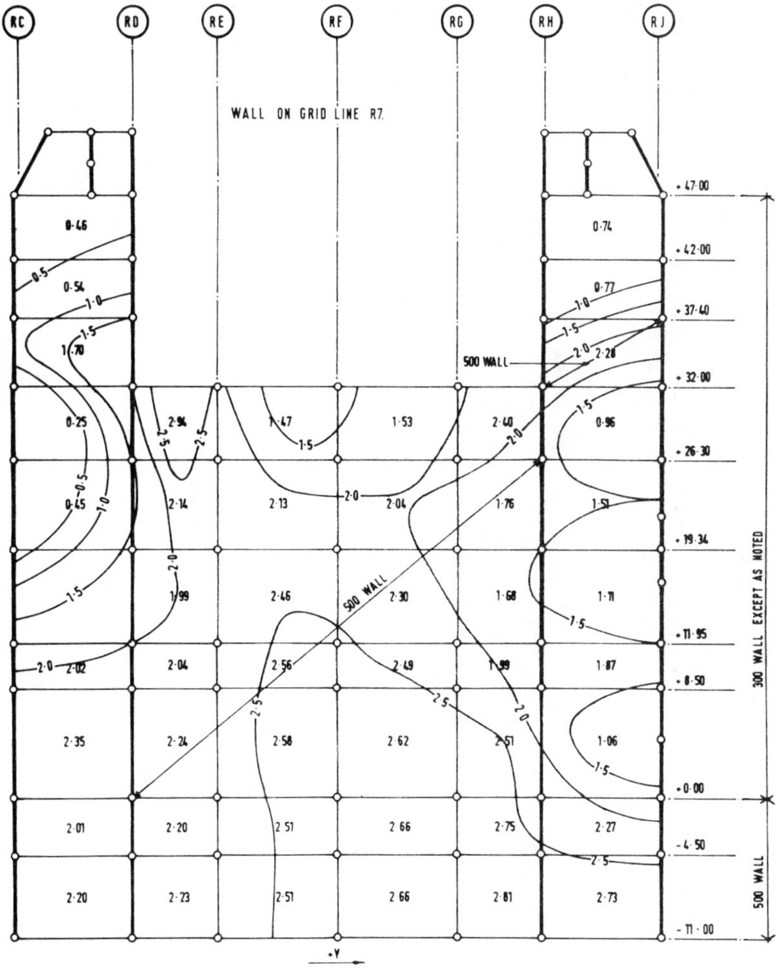

HORIZONTAL SHEAR STRESS CONTOURS
+Y SEISMIC LOADING

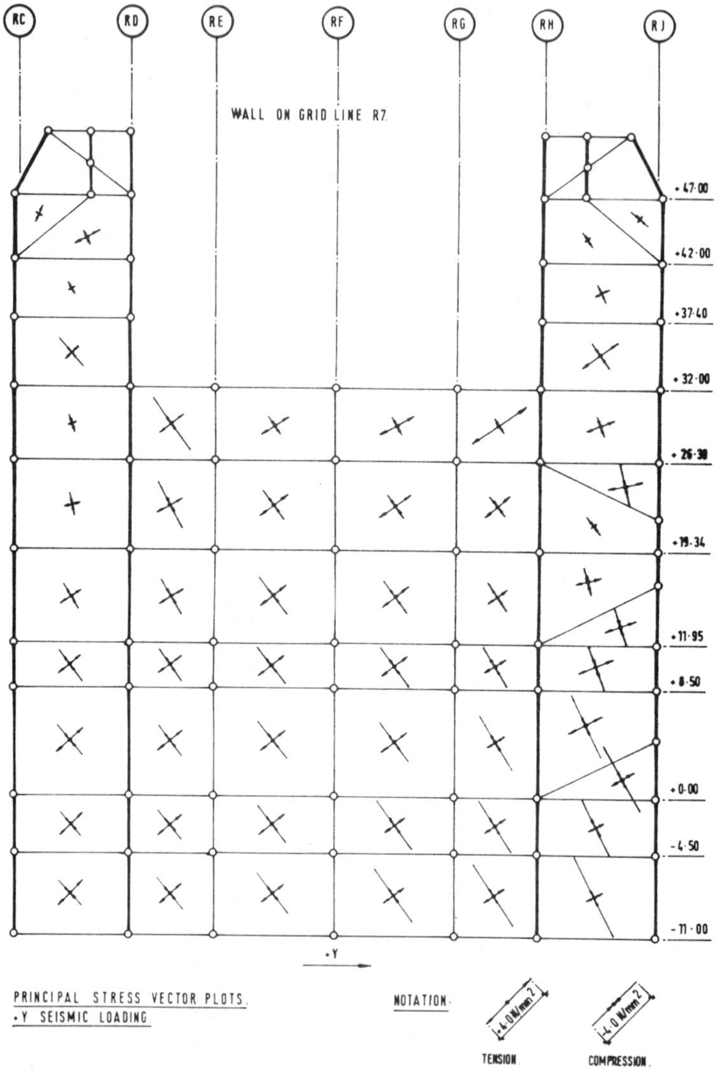

REFERENCES

1. A.C.I. Building Code requirements for reinforced concrete (ACI 318-71), 1971.

2. Gupta A.K. Modelling of shear wall buildings'Nuclear Engineering and Design'79 (1984) 69-80.

25. Seismic analysis of the Upper and Lower Glendevon dams

C. A. TAYLOR, BSc, PhD, A. A. DUMANOGLU, BSc, PhD, J. TREVELYAN, BSc, PhD, J. R. MAGUIRE, BSc, PhD, MICE, R. T. SEVERN, DSc, FEng, University of Bristol, and A. C. ALLEN, BSc, FICE, FStructE, FIWES, FGS, Allen, Gordon and Company, Consulting Engineers

SYNOPSIS. The responses of two Scottish dams to earthquake ground motions appropriate to the UK were investigated using dynamic finite element models. The results support the hypothesis that well constructed dams have sufficient inherent strength to withstand earthquakes of the magnitude likely to occur in the UK, even if no special provision for earthquake loading is made in the design.

INTRODUCTION

1. The Upper and Lower Glendevon dams are located on the River Devon in the Ochil Hills, north of Alloa in Central Scotland. This area has a long history of seismic activity, particularly around Comrie and Menstrie. Since 1978, there has been a swarm of small earthquakes of Richter magnitudes up to 3.5, having epicentres within a few kilometres of the dams. This activity, although of a low level, prompted a precautionary assessment of the behaviour of the two dams under likely earthquake loading. A series of dynamic reponse analyses, using representative earthquake input motions, and forced vibration tests of the prototype dams formed the basis of the study. The vibration tests were carried out by the Building Research Establishment. This paper concentrates on the results of the response analyses.

2. The 45m high concrete gravity Upper dam lies immediately at the head of the Lower Glendevon Reservoir. It has a crest length of 360m and its largest transverse cross-section is 35m wide at the base. The dam was constructed in 1955. The Lower dam is of earthfill construction with a vertical clay core. It is 35m high, has a crest length of 165m and is located in a narrow, V-shaped valley. It has slopes of 1 in 3 on the upstream face and 1 in 2.5 on the downstream face. The fill materials are sandy clays, with the shell material having a slightly higher proportion of fines than the core. Construction of the Lower dam was completed in 1922. The dams lie upstream of the Castlehill concrete arch dam and a substantial built-up area.

3. The scope of the response analyses was restricted to a general assessment of the susceptibility of the two dams to

damage by earthquake. Measurements of the dynamic material properties of the concrete and fill materials were not considered appropriate at this stage since a series of preliminary analyses of the type undertaken would be necessary before a detailed materials testing programme could be formulated. Reliance was placed on published data for similar materials. Some information on the static properties was available and this enabled appropriate dynamic property data to be chosen. Input motions were based on spectral curves and time histories which have been suggested as being relevant to the UK and the Glendevon area.

LOCAL SEISMICITY IN THE GLENDEVON AREA

4. Of primary interest in the study were near-field earthquakes occurring close to the dams. The Menstrie area has experienced many small tremors and these have tended to occur in swarms. About 320 shocks occurred during a particularly active period between 1839 and 1848. The current swarm, which began in 1978 following a quiescent period of over ten years, has averaged over three shocks greater than magnitude 3.0 every six months. These shocks have occurred in an area of 10km radius centred between the Upper and Lower dams and about 4km to the north. Several of the epicentres lie on a straight line between the dams. Focal depths of 2km to 15km have been located for the present swarm.

5. Although the tremors have had small magnitudes, the shallow focal depths and close proximity to the dams may result in quite high peak accelerations, perhaps 0.3g. Recently, strong motion instruments have been installed on and around the dams, but no ground or dam response data are currently available.

DESIGN EARTHQUAKES

6. In a review of UK seismicity, Burton (ref. 1) suggested that, in the absence of any UK strong motion data, the ground acceleration measured at Temblor, California during the Parkfield earthquake of 27th, June, 1966 could be taken as being representative of the type of motion likely to occur in the UK. This motion (Fig. 1) was the result of a magnitude 5.6 earthquake and had a peak horizontal acceleration of about 0.3g. It has been estimated that the measurement point was between 23km and 60km away from the epicentre. The Temblor record therefore has several features which are consistent with conditions in the Glendevon area, i.e. near-field, moderate magnitude and appropriate peak acceleration.

7. A set of response spectra (Fig. 2) for UK earthquakes has been proposed in a report commissioned by the Central Electricity Generating Board and British Nuclear Fuels Ltd. (ref. 2). These curves cover a range of ground conditions and damping values and are normalised to a peak ground acceleration of 1.0g. The report recommends that the vertical motion spectra be taken as two-thirds the horizontal motion spectra over the entire frequency range.

8. Both the Temblor time history and the UK response spectra were used in the response analyses of the two dams. Peak accelerations of 0.3g horizontally and 0.2g vertically were assumed.

ANALYSIS OF THE UPPER GLENDEVON DAM

9. The primary objective of the analysis of the Upper dam was to determine the magnitude of the dynamic stresses which might be induced in the dam as a result of possible earthquake shaking. From this information it would be possible to make a preliminary assessment of the susceptibility of the dam to damage by earthquake. A more detailed study could then be undertaken if the preliminary results indicated that this was necessary.

Method of analysis

10. To maintain a fairly simple approach and to be consistent with the available information on material properties, the dynamic stresses were computed by the response spectrum method using the finite element program SAPIV (ref. 3). In this method, the dam is assumed to vibrate elastically in a number of modes. The maximum responses due to all the modes are averaged to produce an estimate of the maximum overall response of the dam. The input motion is defined by a response spectrum (e.g. Fig. 2).

11. Each point on a response spectrum curve represents the maximum response, in terms of acceleration, velocity or displacement, of a simple oscillator having a known natural frequency and viscous damping factor. For a multi-degree of freedom system, the maximum absolute displacement vector in the x-direction due to a particular mode can be calculated from

$$\underline{U}_{i,x} = \frac{-\underline{\emptyset}_i^T \; M_x \; S_a^x(\omega_i, \xi_i) \; \underline{\emptyset}_i}{\omega_i^2} \qquad (1)$$

where $\underline{U}_{i,x}$ is the displacement vector in the x-direction for the ith mode, $\emptyset_i$, is the mode shape vector for that mode, M_x is the x-component of the mass matrix of the system. S_a^x is the spectral acceleration specified for the x-direction at a natural frequency, ω_i, for a damping factor, ξ_i . Similar expressions to (1) may be written for the y- and z-components of the mode. The displacement vectors, $\underline{U}_i$, for a number of modes are usually calculated and then combined by the square root of the sum of the squares technique to obtain an estimate of the overall maximum absolute displacements, from which the maximum absolute stresses may be computed.

12. A static analysis was performed in addition to the response spectrum analysis to determine the pre-earthquake stress distribution in the dam. The static and dynamic

stresses were then combined to find the maximum tensile and compressive stresses that might occur in each finite element.

Finite element model

13. The largest transverse cross-section of the dam was represented by a 2-D plane strain finite element model consisting of 57, 4- and 3-noded isoparametric element and 73 nodes. The foundation rock was modelled to a depth of 27.5m. Both the foundation rock and the mass concrete of the dam are known to be of moderate quality and, therefore, a suitably low elastic modulus of $10000N/mm^2$ was assumed for both materials. The forced vibration tests of the prototype dam undertaken by the BRE indicated high damping values of the order of 10% critical damping. This value of damping was therefore assumed in the analysis.

14. Hydrodynamic effects were accounted for by the Westergaard "added mass" technique (ref. 4). Westergaard showed that the hydrodynamic pressures acting on the upstream face of a dam are opposite in phase to the ground motion and can be therefore be represented as equivalent inertial forces. The Westergaard method assumes the dam to be rigid and leads to a parabolically increasing distribution of hydrodynamic pressures, and hence added mass, with depth. It has been shown (ref. 5) that this method tends to underestimate the hydrodynamic pressures in the upper third of a dam and overestimate the pressures elsewhere. However, for simplicity, the Westergaard method was adopted in the present analysis in preference to more sophisticated and expensive methods of dynamic fluid-structure coupling.

Results of the analysis

15. The pre-earthquake static principal stresses in the dam and foundation are shown in Fig. 3. The stresses are in N/mm^2 with a double bar denoting tension. As is typical of gravity dams, the stresses are low, the maximum compressive stress of $1.5N/mm^2$ occurring at the toe of the dam and the maximum tensile stress of $-0.28N/mm^2$ occurring at the heel. The static stresses do not include creep effects.

16. The dynamic displacement and stress responses were estimated from the first three modes which the finite element anaylsis predicted at 3.56Hz, 7.61Hz and 8.65Hz respectively. The overall maximum absolute displacement response, equal to the square root of the sum of the squares of the individual modal responses, is shown in Fig. 4. These displacements represent the envelope of maximum displacements and may be positive or negative. The maximum upstream-downstream displacement of 27mm and the maximum vertical displacement of 9mm occur at the crest.

17. The response spectrum method gives the overall maximum absolute direct stresses (σ_{xx}, σ_{yy}) and shear stress (τ_{xy}) acting at any particular point. Like the displacements, these stresses are equal to the square root of the sum of the squares of the individual responses and each

component may be positive or negative. It is not, therefore, appropriate to present the dynamic stresses in the form of principal stresses. To assess the significance of the dynamic stresses, they were combined with the corresponding static stresses to find the largest compressive and tensile stresses which might occur at the centroid of each finite element. As each component of the dynamic stresses can be either positive or negative, these stresses can be added to the static stresses in six different combinations. Each combination was evaluated to obtain the maximum tensile and compressive stresses which are shown in Fig. 5. The stresses are in N/mm^2 with tension being negative. The orientation of the stress values is purely for clarity and has no physical significance. The maximum compressive stress of $2.85N/mm^2$ occurs at the toe of the dam while the maximum tensile stress of $-2.07N/mm^2$ occurs at the heel. As is to be expected, the upstream and downstream faces of the dam suffer the largest stress variations.

18. There is surprisingly little information on the dynamic strength of mass concrete. Limited experimental work in the USA and Japan (ref. 6) has indicated that the failure strength of mass concrete when loaded at rates associated with earthquake loading may be 30% higher than the failure strength obtained in normal, slow loading rate laboratory tests. Many seismic design codes for buildings allow a 30% increase in the normal permissible stresses to accommodate earthquake loads. Typical static permissible stresses for mass concrete in dams are $7.0N/mm^2$ in compression and $1.5N/mm^2$ in tension (ref. 7), giving corresponding dynamic permissible stresses of $9.0N/mm^2$ and $2.0N/mm^2$ respectively, if a 30% increase is allowed. Examination of Fig. 5 shows that the maximum compressive stresses in the dam due to static and earthquake loading combined lie well within the permissible stress, while the maximum tensile stress of $-2.07N/mm^2$ marginally exceeds the corresponding permissible stress.

19. The stress results indicate that the main portion of the dam will not be overstressed as a result of any likely earthquake in the Glendevon area. Further work is in progress to assess the limitations of the analysis and its accuracy. This work includes 3-D dynamic response analyses which should give a better estimate of the dam's behaviour.

ANALYSIS OF THE LOWER GLENDEVON DAM

20. Several reviews of the actual behaviour of embankment dams during earthquakes (refs 8,9,10) have indicated that well constructed embankments, particularly those composed of cohesive fill materials, can be expected to perform satisfactorily in earthquakes of up to magnitude 6.0, having peak accelerations up to 0.3g, without special provision for earthquake loading having been made in the design. As the Lower Glendevon embankment is composed of sandy clay fills and the seismicity of the area is low, it is reasonable to assume that the dam will have sufficient inherent strength to

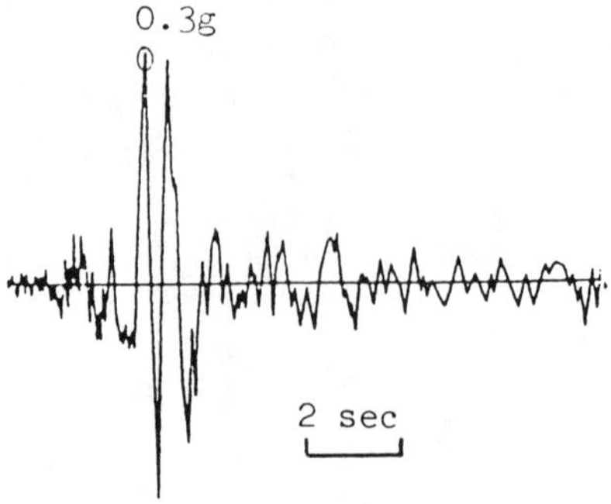

Fig.1 Temblor ground acceleration

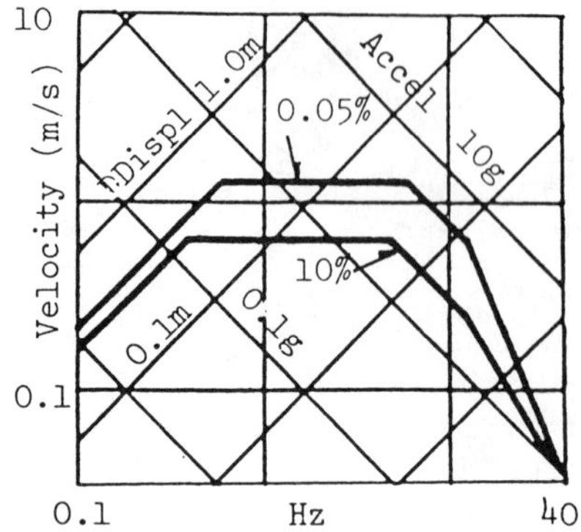

Fig.2 UK spectral curves

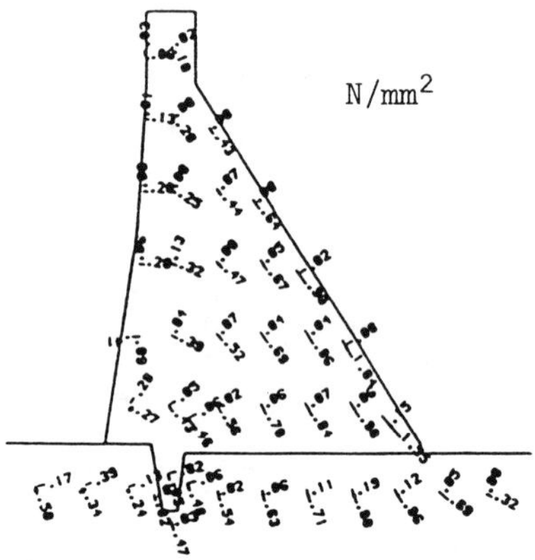

Fig.3 Upper dam, static stresses

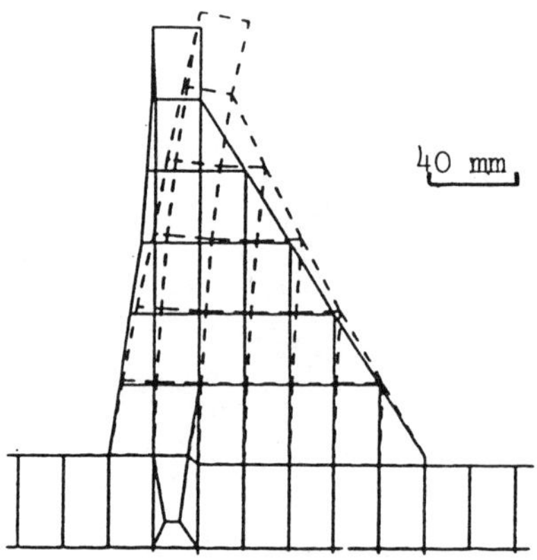

Fig.4 Upper dam, dynamic displacements

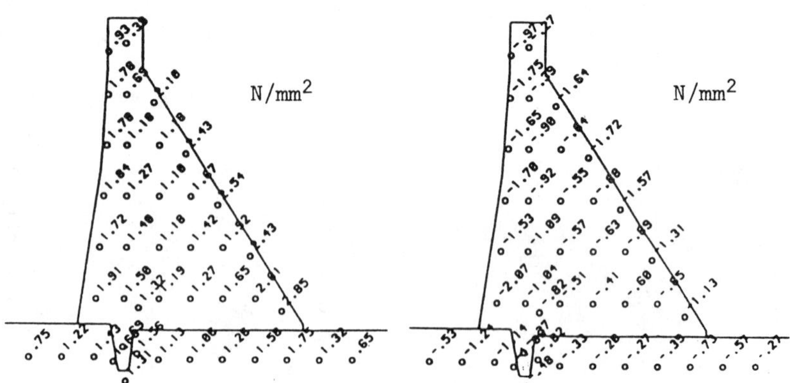

(a) Compression (b) Tension

Fig.5 Upper dam, dynamic stresses

withstand any earthquake to which it is likely to be subjected. To confirm this view, however, the response of the dam to the Temblor ground motion was investigated to assess, in broad terms, its dynamic strength.

Method of analysis

21. Generally, the seismic response of an embankment will be non-linear due to the non-linear stress-strain behaviour of the fill materials. This problem is the subject of considerable research, but, as yet, no fully reliable method of analysis has been developed. Various simplified methods based on slope stability considerations have been proposed (refs 11,12,13), while more sophisticated methods utilising non-linear constitutive models of the fill materials are under development (ref. 14).

22. One method which attempts to account for the non-linear behaviour of fill materials in an approximate manner and can give an estimate of the mobilised dynamic strength throughout the embankment is that proposed by Seed (ref. 15). The method has found quite wide application but it is recognised to have several major limitations. The method may be summarised as follows:

(i) Determine the static stress distribution in the dam prior to the earthquake.
(ii) Determine the shear stress time history of elements in the dams due to the earthquake.
(iii) Test representative soil samples under uniform cyclic loading to determine the dynamic strength properties of the soil. The dynamic strength is defined as the number of cycles at a given shear stress amplitude to cause liquefaction or a given amount of permanent strain.
(iv) Convert the irregular shear stress history determined in (ii) to an equivalent number of uniform cycles and compare with the measured dynamic strength.
(v) Using engineering judgement, estimate the probability of failure of the embankment from the distribution of mobilised dynamic strength.

23. The most important limitations of the method are associated with steps (ii) and (iii). The determination of the dynamic shear stress histories for step (ii) requires a dynamic response analysis of the dam. Two finite element programs, QUAD4 (ref. 16) and LUSH (ref. 17), have been developed for this purpose by Seed and his co-workers. QUAD4, which was used in this study, performs a time integration analysis, while LUSH performs the response calculations in the frequency domain using a Fast Fourier Transform algorithm. Both programs model the strain dependent, non-linear properties of soils by using equivalent, linear elastic shear moduli and damping factors. In QUAD4, these equivalent moduli are determined iteratively by computing the elastic response of the dam to the whole earthquake using a starting set of

shear moduli and damping factors. The material properties for each finite element are then adjusted to be compatible with the maximum shear strain occurring in the element and the whole analysis repeated. The analysis continues iteratively until the changes in shear moduli and damping factors between successive analyses reach a specified tolerance. The final shear moduli and damping factors are taken to be the equivalent, linear elastic material properties and the computed stress histories based on these parameters are used in step (iv).

24. Apart from the use of equivalent, linear elastic properties, this method of response analysis suffers from being a total stress analysis and, therefore, neglects any effects due to the generation of pore pressures in the fill during shaking. For cohesive fills, however, pore pressure generation is not a significant problem and the use of QUAD4 in the analysis of the Lower dam was, therefore, considered reasonable.

25. Other limitations of the Seed method relate to the laboratory testing methods used in step (iii). These are usually cyclic triaxial tests which are known to yield unreliable results near failure due to non-uniformities developing in the test samples. Despite all these limitations, the Seed method can give a useful indication of the performance of a dam under earthquake loading and for this reason was adopted in the present study.

Material properties

26. The shear strength, stress-strain moduli and damping characteristics of soils are all stress and strain dependent. In general, the shear strength and stress-strain moduli increase with increasing confining pressure, while the shear moduli decrease with increasing shear strain. Soil damping is relatively insensitive to confining pressure but increases significantly with increasing shear strain. In the analysis of any soil structure it is therefore important to select strength and moduli values which are compatible with the states of stress and strain in the structure. All dynamic material properties used in the analysis of the Lower dam were taken from published data since an expensive soils testing programme was not considered warranted at this stage.

27. QUAD4 computes the strain dependent shear modulus for each finite element by applying a reduction factor, which is itself dependent on the shear strain in the element, to the low strain initial shear modulus specified at the start of the analysis. The reduction factor is defined by the curve shown in Fig. 6 which was derived by Seed and Idriss (ref. 18) from a large number of cyclic tests of clay materials. At the end of each iteration, QUAD4 uses the maximum shear strain computed in each finite element to define a new reduction factor. This factor is then applied to the low strain initial shear modulus for the element to obtain the new modulus which will be used in the next iteration. The low strain initial

shear modulus, G, for each element was taken from the curve shown in Fig. 7, also taken from ref. 18, which relates the modulus to the undrained shear strength of the soil, S_u, and the shear strain, γ. The undrained shear strength was estimated using the Mohr-Coulomb criterion, assuming a cohesion and angle of friction determined from static triaxial tests of the fill materials. An initial shear strain of 0.001% was assumed throughout the dam in order to determine the initial shear modulus.

28. Strain dependent damping ratios for the fills were based on a curve proposed in ref. 18 and shown in Fig. 6. This is an average curve derived from cyclic tests of many cohesive soils. At the end of each iteration, QUAD4 updates the damping factor for each element in accordance with this curve and the maximum shear strain computed for that element.

29. Suitable cyclic strength data appropriate to the Glendevon fill materials were taken from ref. 19 which describes the seismic analysis of the Guri dam in Venezuela. This dam, which is a 120m high embankment and was analysed by the Seed method, contains an impervious fill which is broadly similar to the Glendevon fills. The cyclic strength of this fill was characterized by the maximum amplitude of the applied uniform cyclic shear stress which caused a permanent axial strain of 5% after 3 cycles in an undrained cyclic triaxial test. When defined in this manner, the cyclic strength of a soil is dependent on the initial mean confining pressure in the test and the initial consolidation conditions. Anisotropic consolidation (i.e. when the maximum principal stress is greater than the minimum principal stress) has a marked effect on the cyclic strength of a soil, generally leading to an increase in the strength. Since anisotropic conditions exist in most parts of an embankment prior to an earthquake, it is important to take account of this effect in a seismic analysis. The results of the cyclic triaxial tests on the Guri fill are shown in Fig. 8. This shows the variation of the cyclic shear stress, τ_c, causing 5% axial strain in 3 cycles with the initial confining pressure. Curves for initial consolidation ratios, K_c of 1.0 (isotropic) and 2.0 (anisotropic) are shown. Strengths for other values of K_c may be found by interpolation.

Finite element model

30. The largest transverse cross-section of the dam was modelled as a 2-D plane strain structure. The mesh (Fig. 9) consisted of 261 nodes and 274, 4- and 3-noded, isoparametric elements. The foundation rock was modelled to a depth of 35m.

Static analysis

31. Step (i) of the Seed method requires a static analysis to determine the stress distribution in the dam prior to the earthquake. This information is necessary to select appropriate initial shear moduli for the dynamic analysis and

to select the cyclic strength of the fill materials which is compatible with the initial anisotropic stress conditions in the dam.

32. The static stress distribution was determined from a simple, elastic finite element analysis using the program SAPIV. Only dead weight loads were considered. These loads were applied on the assumption that the dam was constructed in a single lift. Such an analysis predicts unrealistic displacements since an embankment is constructed in a large number of lifts. However, the stress distributions predicted by a single lift analysis do not differ significantly from those predicted by a multi-lift analysis. Since only the static stresses, not displacements, were required for the dynamic analysis, a single lift analysis was adequate.

Dynamic analysis

33. The response of the dam was computed using program QUAD4. The Temblor ground acceleration record (Fig. 1), scaled to a peak acceleration of 0.3g was applied horizontally, with the same record scaled to a peak acceleration of 0.2g being applied vertically. The final shear stress history for each finite element in the body of the dam was saved for subsequent processing.

Calculation of dynamic strength mobilisation

34. The purpose of the dynamic analysis of the Lower dam was to assess the proportion of the inherent dynamic strength of the embankment which would be mobilised if it was shaken by the Temblor ground motion. For each finite element, a strength mobilisation factor, F, was calculated as follows

$$F = \frac{\tau_d}{\tau_c} \qquad (2)$$

τ_c is the maximum amplitude of the uniform cyclic shear stress required to cause 5% permanent axial strain in 3 cycles in a cyclic triaxial test and is defined by the cyclic strength curves of Fig. 8. τ_d is the maximum amplitude of the uniform cyclic shear stress which, when applied for 3 cycles, is equivalent to the whole shear stress history for the element computed in the dynamic analysis. Thus, if the value of F is greater than or equal to 1.0, the dynamic strength of the soil element is fully mobilised. τ_d was calculated from the irregular shear stress history using a weighting procedure proposed by Seed et al (ref. 20). This procedure has been evaluated experimentally (ref. 21) and was found to be satisfactory. A detailed description will not be repeated here.

Results of the analysis

35. Fig. 10 shows a typical shear stress history computed

by QUAD4 for an element in the downstream shell adjacent to the base of the core. By the Seed weighting procedure, this irregular history was found to be equivalent to three uniform cycles at 62kN/m^2 peak amplitude, which is 70% of the maximum shear stress achieved in the stress history. The element has a consolidation ratio, K_c of 1.73 and an initial mean confining pressure of 440kN/m^2 giving a cyclic failure stress, τ_d, of 92kN/m^2 from Fig. 8. The element therefore has a dynamic strength mobilisation factor, F, of 0.67, i.e. two-thirds of the dynamic strength is mobilised.

36. The distribution of dynamic strength mobilisation factors throughout the dam is shown in Fig. 11. The highest factor of 0.77 occurs at the bottom of the upstream edge of the core. The majority of mobilisation factors are less than 0.5, indicating that approximately half of the overall dynamic strength of the embankment is mobilised.

37. These results appear to support the hypothesis that well constructed embankments have a significant degree of inherent dynamic strength.

GENERAL CONCLUSIONS

38. The results of the analyses of the Upper and Lower Glendevon dams tend to support the hypothesis that well constructed dams can be expected to perform satisfactorily in earthquakes up to magnitude 6.0 without special allowance for earthquake loading having been made in the design. As a magnitude 6.0 earthquake is an extremely rare event for the UK, a detailed seismic analysis will not normally be a necessary part of the design of dams to be built in this country. However, the possible effects of earthquakes should not be completely neglected. Further studies of the Glendevon dams are in progress to verify this view.

ACKNOWLEDGEMENTS

39. The authors gratefully acknowledge the permission of Fife Regional Council to publish the results of the analyses of the Upper and Lower Glendevon dams.

REFERENCES

1. BURTON P.W. Assessment of seismic hazard in the UK. Instrumentation for ground vibration and earthquakes. Institution of Civil Engineers, London, 1978, 35-48.
2. PRINCIPIA MECHANICA LTD. Seismic ground motions for UK design. A report commissioned by British Nuclear Fuels Ltd and the Central Electricity Generating Board, April, 1981.
3. BATHE K-J et al. SAPIV - A structural analysis program for static and dynamic response of linear systems. EERC Report no. 73-11, University of California, Berkeley, 1974.
4. WESTERGAARD H.M. Water pressures on dams during earthquakes. Proceedings, ASCE, 1931, November, 1303-1318.
5. ALTINISIK D. and SEVERN R.T. Natural frequencies and response characteristics of gravity dams. Dams and earthquake, p.205, Thomas Telford Ltd, London, 1981.

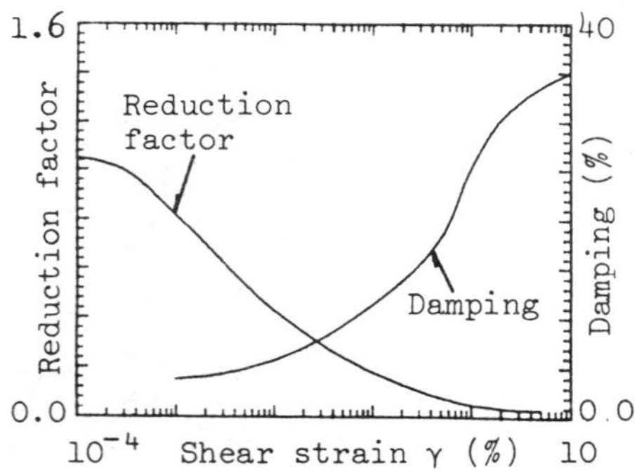

Fig.6 Shear modulus reduction factor and damping for clays

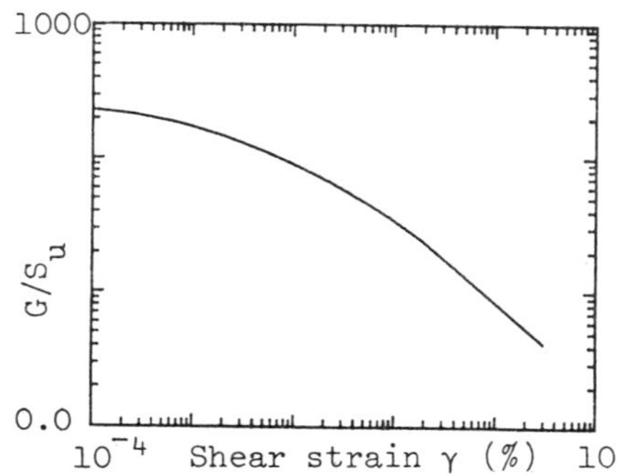

Fig.7 Initial shear modulus for clays

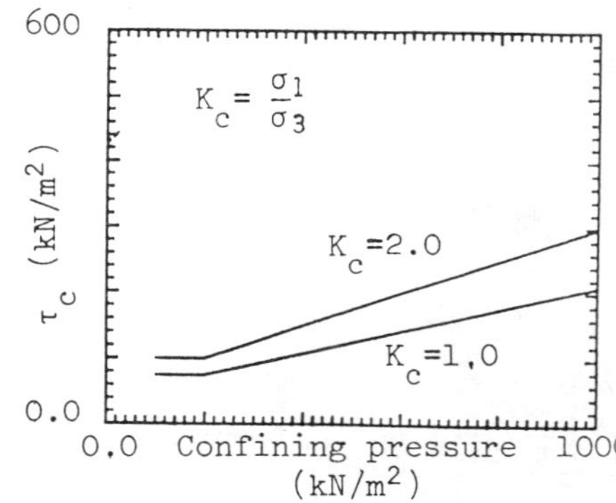

Fig.8 Guri fill, cyclic strength

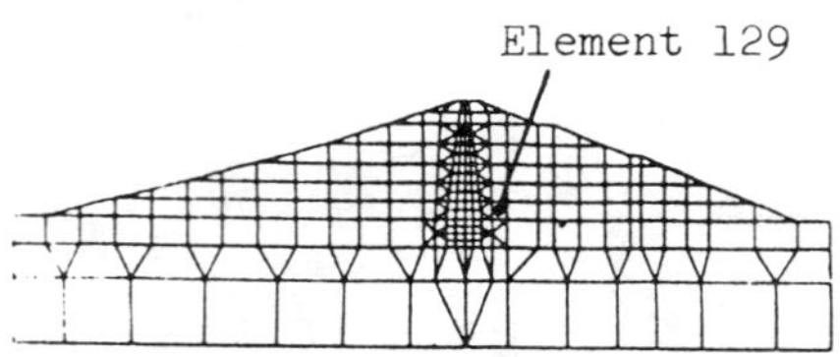

Fig.9 Lower dam, F.E. mesh

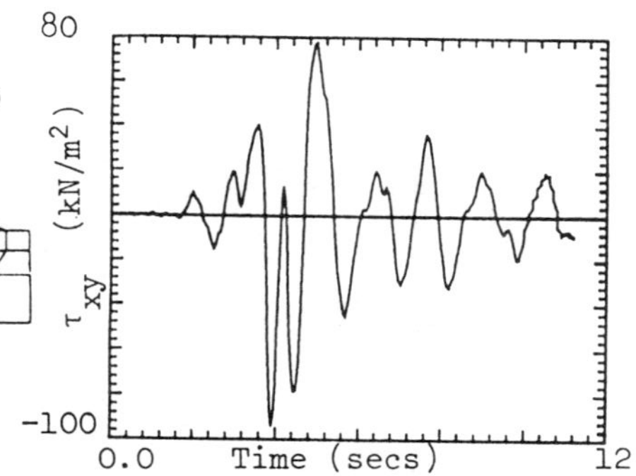

Fig.10 Shear stress history element 129

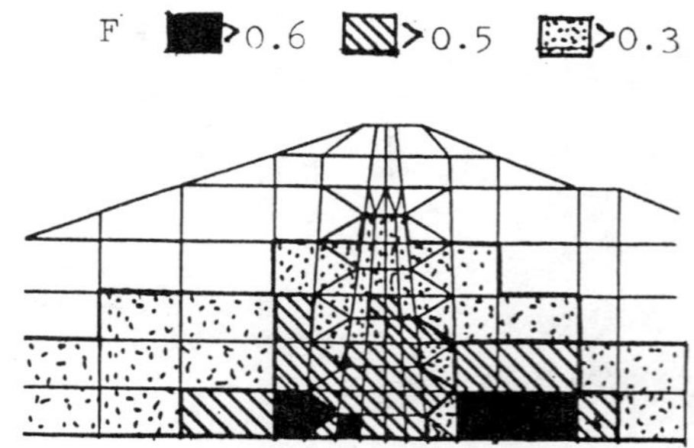

Fig.11 Lower dam, dynamic strength mobilisation

6. RAPHAEL J.M. The nature of mass concrete in dams. Douglas McHenry International Symposium on Concrete and Concrete Structures, p.133, publication SP-55. American Concrete Institute, Detroit, 1978.
7. THOMAS H.H. The engineering of large dams, p. 300. John Wiley & Sons, London, 1976.
8. AMBRASEYS N.N. On the seismic behaviour of earth dams. Proceedings, 2nd World Conference on Earthquake Engineering, Tokyo, 1960, 331-356.
9. SEED H.B. et al. Performance of earth dams during earthquakes. Journal of the Geotechnical Engineering Division, ASCE, 1978, vol. 104, no. GT7, July, 967-994.
10. SEED H.B. Lessons from the performance of earth dams during earthquakes. Dams and earthquake, p.97, Thomas Telford Ltd, London, 1981.
11. NEWMARK N.M. Effects of earthquakes on dams and embankments. Geotechnique, 1965, vol. 15, no. 2, June, p. 139.
12. SARMA S.K. Seismic stability of earth dams and embankments. Geotechnique, 1975, vol. 25, no. 4, December, p.743.
13. MAKDISI F.I. and SEED H.B. Simplified procedure for estimating dam and embankment earthquake-induced deformations. Journal of the Geotechnical Engineering Division, ASCE, 1978, vol. 104, no. GT7, July, p.849.
14. PANDE G.N. and ZIENKIEWICZ O.C. (eds) Soil mechanics - Transient and cyclic loads. John Wiley & Sons, London, 1982.
15. SEED H.B. et al. Analysis of the slides in the San Fernando dams during the earthquake of February 9, 1971. EERC Report no. 73-2, University of California, Berkeley, 1972.
16. IDRISS I.M. et al. QUAD4 - A computer program for evaluating the seismic response of soil structures by variable damping finite element procedures. EERC Report no. 73-16, University of California, Berkeley, 1978.
17. LYSMER J. et al. LUSH - A computer program for complex response analysis of soil-structure systems. EERC Report no. 74-14, University of California, Berkeley, 1974.
18. SEED H.B. and IDRISS I.M. Soil moduli and damping factors for dynamic response analyses. EERC Report no. 70-10, University of California, Berkeley, 1970.
19. CARRERA E. et al. Investigations of the Guri embankment dams under seismic loading. Proceedings, 30th Congress, ICOLD, New Delhi, 1979, Q51-R4.
20. SEED H.B. et al. Representation of irregular stress time histories by equivalent uniform stress series in liquefaction analyses. EERC Report no. 75-29, University of California, Berkeley, 1975.
21. ANNAKI M. and LEE K.L. Equivalent uniform cycle concept for soil dynamics. Journal of the Geotechnical Engineering Division, ASCE, 1979, vol. 103, no. GT6, June, 549-564.

Discussion

DR C. A. Taylor, Introductory remarks

The analysis of the Upper and Lower Glendevon dams was undertaken at Bristol University for Allen, Gordon and Co, Consulting Engineers, acting on behalf of Fife Regional Council. The Authors wish to thank the Council for the permission to present the results of the analysis and also thank Mr D. Hamilton, supervising engineer for the Lower Dam, for his assistance.

MR MENU

In his paper Dr Taylor used only one time history; a peak acceleration scaled strong motion recorded at Parkfield Temblor during the earthquake of 27 June 1966. This motion possesses specific features such as duration, energy level, energy flux density, frequency content, phase distribution density and so on at least as important as peak acceleration or peak velocity.

It is good practice to use several strong motion records to cover a range of those characteristic for analysis.

From the various stress histories that Dr Taylor obtained, what made him consider it unnecessary to envisage a more complete analysis with a few more input time histories and still be confident of his conclusions?

DR TAYLOR

The authors agree with Mr Menu that it is preferable to use several different earthquakes in a seismic analysis, especially when non-linear effects are of interest. However, the cost effectiveness of such an approach should be considered for each structure, since non-linear time history analyses and the associated materials testing programmes are expensive. In the case of the Lower Glendevon embankment, the seismicity of the area is low and , based on the good performance of similar clay fill dams in much stronger earthquakes than are likely to occur in the region, it was considered that the embankment should have adequate inherent strength. The analysis was a precautionary measure to help assess the scale of the problem; it was not an exhaustive study.

26. Probability of seismically induced liquefaction in British sector of North Sea

W. D. LIAM FINN, BSc, MASc, PhD, University of British Columbia, Vancouver, and GAIL M. ATKINSON, BSc, MEngSc, Earth Physics Branch, Ottawa

SYNOPSIS. The probability of seismically induced liquefaction in the British Sector of the North Sea at a depth of 10 ft is estimated to be less than 1/700 per annum if the cohesionless soils have cone penetration resistances greater than those equivalent to a standard penetration resistance of 10 blows/ft. The probabilities are estimated using the computer program PROLIQ which computes the joint probability of occurrence of critical magnitudes and acceleration amplitudes capable of generating liquefaction for given in-situ conditions.

INTRODUCTION

1. Preliminary estimates of the probability of seismically induced liquefaction in the British Sector of the North Sea are determined using procedures developed by Atkinson, Finn and Charlwood (ref. 1). The method is incorporated in the computer program PROLIQ (Atkinson, Finn and Charlwood, ref. 2) available from the University of British Columbia. The program combines Seed's simplified procedure for assessing liquefaction potential (ref. 3) with the Cornell (ref. 4) method of assessing seismic hazard, by modifying the latter's procedure to include the joint probability of occurrence of magnitude and acceleration amplitude. The rationale for this approach is that critical ground motion levels which result in liquefaction are explicit functions of duration or implicitly of magnitude. An attractive feature of the method for offshore applications is that the resistance of the seafloor sands and silts to liquefaction may be characterised by cone penetration data which are usually obtained in offshore site investigations.

2. Seed's method of liquefaction assessment is well-known and widely used and will be briefly referred to here. The essential features of PROLIQ will be presented; more details may be found in Atkinson, Finn and Charlwood (ref. 5).

3. In Seed's method, the state of the soil during an earthquake is characterised by the modified penetration resistance N_1 and the cyclic stress ratio CSR. The modified penetration resistance N_1 is derived from standard penetration test data N, modified to a value compatible with an overburden pressure of 1 tsf in the manner described by Seed and Idriss (ref. 6).

4. The CSR is defined as

$$\mathrm{CSR} = 0.65\,\frac{\sigma_v}{\sigma'_{vo}} \cdot A \cdot r_d \qquad (1)$$

where σ_v = total overburden pressure; σ'_{vo} = initial effective overburden pressure; A = maximum peak ground acceleration as a fraction of acceleration due to gravity; and r_d = a reduction factor to account for soil flexibility (ref. 3).

5. The probable performance of a site during an earthquake is inferred from field data on how cohesionless soils with a similar state pair (N_1, CSR) fared in the past. For an earthquake of given magnitude, a lower bound line can be drawn dividing state pairs of sites which liquefied from those which did not. Such a line defines critical values of cyclic stress ratio CSR_{crit} for various values of N_1. Liquefaction is, therefore, considered to occur when the CSR in the soil exceeds CSR_{crit} for the given N_1. CSR_{crit} lines determined by Seed and Idriss (ref. 6) are shown in Fig. 1.

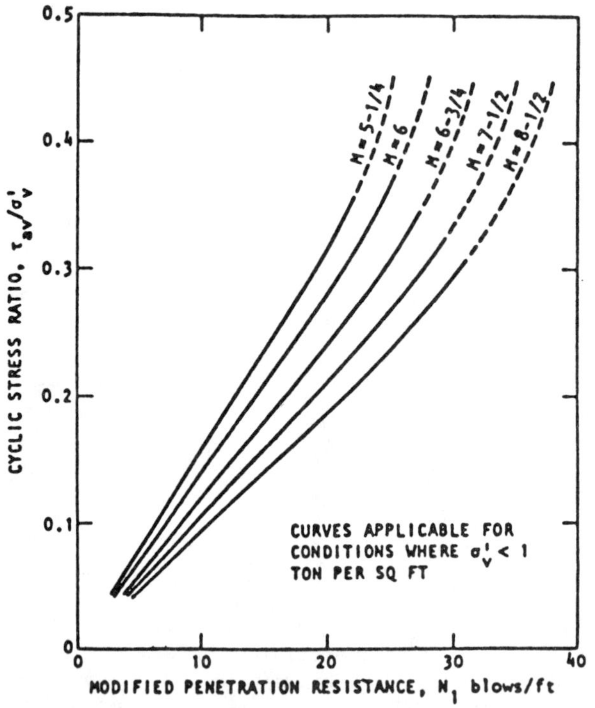

Fig. 1. Chart for evaluation of liquefaction potential for sands for different magnitude earthquakes (ref. 6). Curves define lower bound CSR values required to cause liquefaction.

6. Offshore penetration data will be in the form of cone penetration resistances q_c. These resistances can be converted to equivalent standard penetration resistances N by an

appropriate conversion factor. The conversion factor proposed by Robertson, Campanella and Wightman (ref. 7) which depends on mean-grain size is shown in Fig. 2.

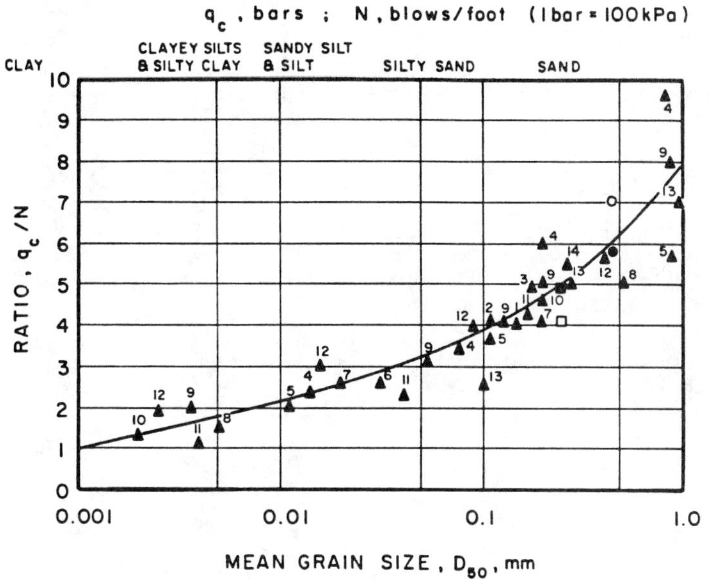

Fig. 2. SPT-CPT correlations.

THEORETICAL DEVELOPMENT

7. Consider a site located within a large homogeneous zone of earthquake activity. The seismicity of a homogeneous zone can be characterised by its discrete magnitude recurrence relation:

$$N(M) = N_o \exp(-\beta M) \qquad M \leq M_x \qquad (2)$$
$$N(M) = 0 \qquad M > M_x$$

where (N(M) = number of earthquakes of magnitude M, per annum, occurring in the zone; M_x = maximum magnitude; and N_o, β = zone recurrence constants determined by fit of the observed seismicity data to Eqn. (2). In the discrete form of the recurrence relation adopted here, M is sorted into 1/2 unit windows, such that M = 5, for example, refers to $4.75 \leq M < 5.25$.

8. For any earthquake occurring in the zone, the attenuation of acceleration from the source to the site is of the form:

$$A = C_1 \exp(C_2 M) \{R + C_4 \exp(C_5 M)\}^{C_3} \qquad (3a)$$

or

$$A = C_1 \exp(C_2 M) R^{C_3} \exp(C_4 R) \qquad (3b)$$

where R = the distance from the source to the site (e.g., hypocentral, epicentral or closest distance to fault rupture), and C_1 through C_5 = the regional attenuation coefficients. The functional form of Eqn. (3a) is that used by Campbell (ref. 8) in his recent attenuation relations for California. Note that if C_4 and/or C_5 are set equal to zero, other common and simpler functional forms are obtained. The functional form of Eqn. (3b) is that used by Joyner and Boore (ref. 9).

9. The curves of Fig. 1, showing the CSR required to cause liquefaction as a function of N_1 and earthquake magnitude, were defined as lower bound values. To obtain the expected or average probability of liquefaction, it is more appropriate to define equivalent average CSR_{crit} values. It will also be convenient to have a curve corresponding to each 1/2 magnitude unit, since the magnitude recurrence relation is discretized into 1/2 magnitude unit classes. Therefore, using Seed and Idriss' (ref. 6) raw data, the curves of Fig. 1 were redefined to obtain mean CSR_{crit} values, for each 1/2 magnitude unit, as shown in Fig. 3. The mean was defined by the method of least squares of the misclassified points, as described by

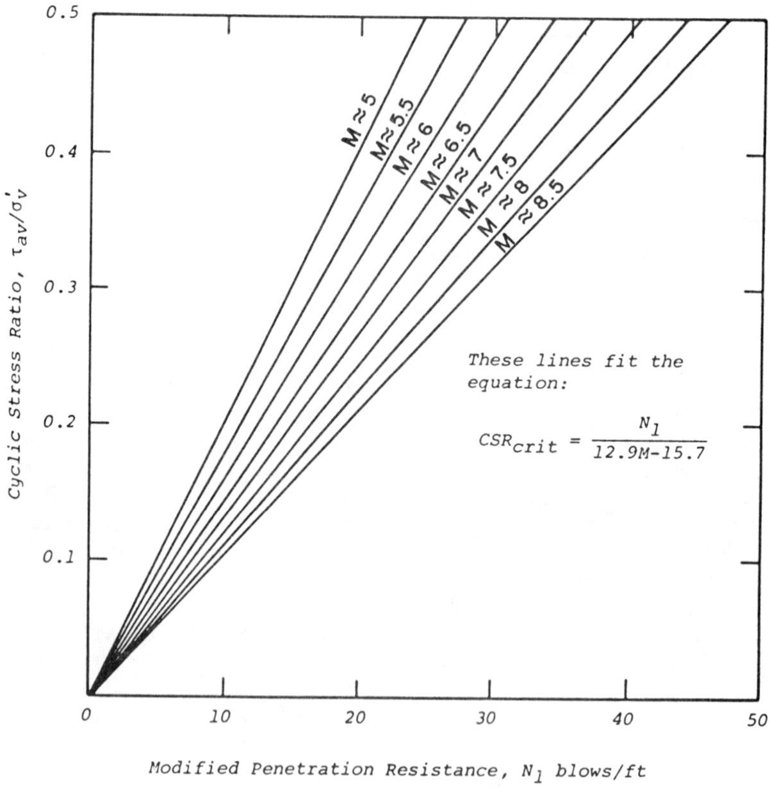

Fig. 3. Mean liquefaction criteria for sands for different magnitude earthquakes.

Yegian and Whitman (ref. 10). Figure 4 compares the mean and Seed lower bound curves for M7.5 earthquakes with available data.

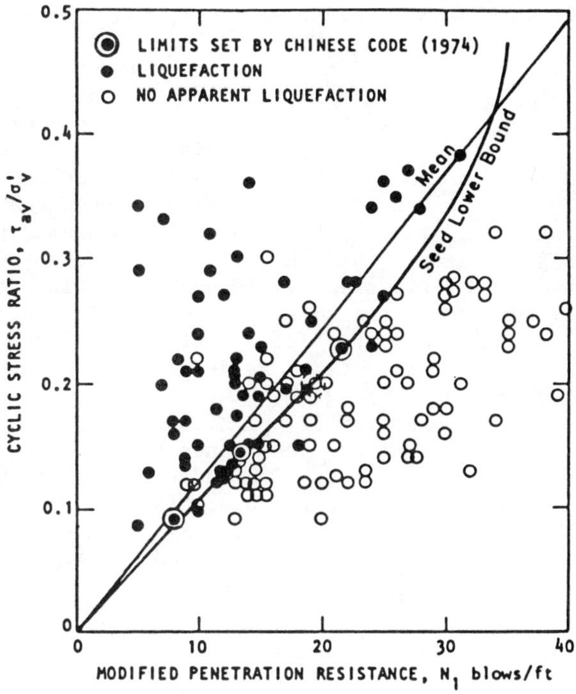

Fig. 4. Field behaviour of sands for M7.5 earthquakes (ref. 6).

10. Since earthquakes of magnitude less than 5 are not generally of sufficient duration to cause liquefaction, M5 is the lowest magnitude which contributes to the probability of liquefaction. The largest magnitude lies somewhere between 8.5 and 9 (global maximum). This leaves 8 (1/2 unit) significant magnitude classes. Consider a particular magnitude class, M_i. For this magnitude value, $(CSR_{crit})_i$ for a given modified penetration resistance (N_1) can be calculated from the relation shown in Fig. 3:

$$(CSR_{crit})_i = N_1/(12.9M_i - 15.7) \qquad (4)$$

Substitution of $(CSR_{crit})_i$ into Eqn. (1) allows its solution for the critical acceleration required to cause liquefaction, $(A_{crit})_i$. It should be noted that for some sites, detailed dynamic laboratory test data may be available to more closely define the $(A_{crit})_i$ values. The program has been designed to accommodate such cases through direct input of $(A_{crit})_i$ values. The probability of liquefaction, due to earthquakes of magnitude M_i, is the probability of exceedence of $(A_{crit})_i$ at the site, due to all earthquakes of magnitude M_i.

11. $(A_{crit})_i$ will be exceeded if an earthquake of magnitude M_i occurs sufficiently close to the site. This maximum distance, within which epicentres must occur, shall be referred to as $(R_{crit})_i$. Its value is obtained by solving the attenuation relation (Eqn. (3)) for R, given an M_i and $(A_{crit})_i$ pair.

12. It should be noted that if the specified attenuation relation is given in terms of closest distance to fault rupture, rather than epicentral or hypocentral distance, it is necessary to make geometric corrections to the R values calculated from Eqn. (3). This is because the calculation of occurrence rates for events initiating liquefaction implicitly assumes that R_{crit} encloses the area within which epicentres must occur to cause liquefaction. If a closest distance to fault attenuation relation has been used, then some earthquakes with epicentres outside $(R_{crit})_i$ may initiate site liquefaction. The program can account for this distinction, as described by Atkinson, Finn and Charlwood (ref. 2).

13. To find the probability of the occurrence of M_i within $(R_{crit})_i$, we use the magnitude recurrence relation (Eqn. (2)), to calculate the per annum rate of occurrence of earthquakes of magnitude M_i within the entire zone, $N(M_i)$. The rate of liquefaction due to events of magnitude M_i is given by:

$$(\lambda_{LIQN})_i = \frac{(S_{crit})_i}{S_{zone}} N(M_i) \qquad (5)$$

where $(S_{crit})_i$ = the area within the zone that is enclosed by $(R_{crit})_i$, and S_{zone} = the area of the entire seismic zone. In the simple case of the large homogeneous zone, the enclosed area is circular and $(S_{crit})_i = \pi(R_{crit})_i^2$.

14. The probability of no occurrences of soil liquefaction due to earthquakes of magnitude M_i, assuming a Poisson process, is:

$$(p_o)_i = \exp(-\lambda_{LIQN})_i \qquad (6)$$

and it follows that the probability of liquefaction (one or more occurrences) for M_i is $1-(p_o)_i$.

15. For each of the eight discrete values of M_i, a different $(p_o)_i$ is calculated. For any M_i in exceedence of M_x for the zone, $(p_o)_i = 1$. The probability of no soil liquefaction for earthquakes of all magnitudes occurring within the zone is:

$$(p_o)_{zone} = \prod_{i=1}^{8} (p_o)_i \qquad (7)$$

where Π denotes multiplication. The probability of liquefaction due to all earthquakes in the zone is $1-(p_o)_{zone}$.

16. The method described above can readily be used for the general case of a site which is influenced by several seismic source zones, each characterised by a different magnitude

recurrence relation. In this case, each area $(S_{crit})_i$ enclosed by $(R_{crit})_i$ would generally be non-circular; up to 8 $(S_{crit})_i$ values (one for each M_i) would be computed for each seismic zone, based on the zone's geometry. For each seismic source zone the probability of no soil liquefaction due to all earthquakes, $(p_o)_{zone}$, would be computed by Eqn. (7). The probability of no liquefaction due to all earthquakes of all zones is then:

$$P(\text{NO LIQN}) = \prod_{\text{all zones}} (p_o)_{zone} \quad (8)$$

and the overall probability of liquefaction is 1-P(NO LIQN).

17. The Venture gas field off Nova Scotia is an example of a site subject to seismic excitation from the several sources shown in Fig. 5. The probability of liquefaction in this area has been estimated by Finn (ref. 12) using PROLIQ.

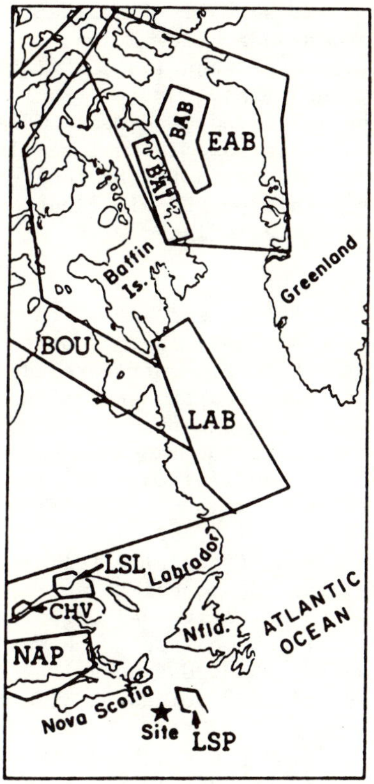

Fig. 5. Location of example application site in relation to seismic source zones (ref. 11). The background zone has been removed since its upper magnitude bound of 5.0 implies it cannot produce initiating events for liquefaction.

18. The development described above treats the attenuation of acceleration as a deterministic process, i.e., for a given earthquake magnitude and distance the ground acceleration is precisely defined. However, it is well-known that there is considerable scatter of observed accelerations about the hypothetical values predicted from any attenuation relation; this will cause the realised acceleration A to differ from the value predicted by the attenuation relation $\bar{A}$ by some factor k, where $k(\equiv A/\bar{A})$ is a lognormally distributed random variable with mean k = 1; its standard deviation ($\sigma_{\ln k}$) is dependent on many factors, but generally is within the range 0.4 to 0.6.

19. The uncertainty in the attenuation relation implies that the $(R_{crit})_i$ values are also subject to uncertainty; some epicentres outside $(R_{crit})_i$ would actually cause liquefaction at the site, while others within the radius would not. If we make the simplifying assumption that the variability in A can be equivalently expressed in terms of variability in R, it follows that R_{crit} is lognormally distributed with mean $j \equiv (R_{crit}/\bar{R}_{crit}) = 1$ and standard deviation $\sigma_{\ln j} \simeq -1/C_3 \, \sigma_{\ln k}$, where C_3 is the distance attenuation coefficient (always of negative sign) from Eqn. (3a).

20. Because of the R-squared dependence of area, the scatter of R_{crit} will cause an increase in realised liquefaction probability. This is accounted for by integrating (numerically) the contribution to area from the distribution of $(R_{crit})_i$, to obtain the expected area associated with each $(R_{crit})_i$.

SEISMICITY OF THE NORTH SEA

21. A major study of the seismicity of the North Sea was reported by Husebye et al (ref. 13), as part of an assessment of earthquake risk for Norwegian offshore installations. The historical records of large earthquakes offshore showed that they occurred over a wide geographical area (Fig. 6). The largest earthquakes were in the range M = 6.0-6.5 but magnitudes up to M = 7.5 were considered possible. The locations of major earthquakes in relation to significant geological features are shown in Fig. 7. As a result of the paucity of data on large earthquakes, especially reliable instrumental data, no attempt was made to associate the earthquakes with geologic features or to delineate seismogenic source zones. Husebye et al (ref. 13) assumed the North Sea and adjacent land masses to be a large homogeneous seismic zone. Later studies of earthquake risk by Selnes (1981) are based on the same assumption.

22. The present study adopts the assumption of the homogeneous seismic zone. The risk anywhere in the interior of a large homogeneous seismic zone far enough from the boundaries to be beyond the influence of seismicity outside the boundaries is constant. Therefore, the probability of liquefaction over most of the British Sector of the North Sea may be estimated by selecting a single site in the interior of the seismic zone and assigning to that site a wide range in soil properties representative of the range over the British Sector.

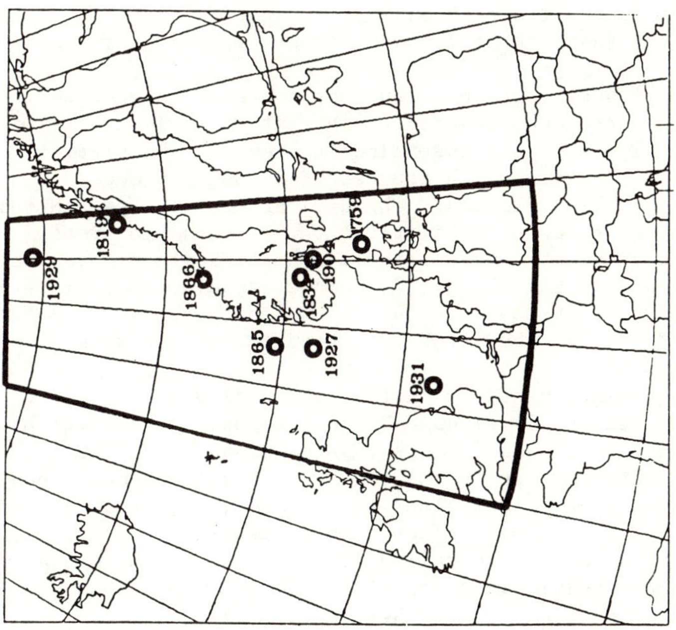

Fig. 6. Large earthquakes in North Sea and environs (ref. 13).

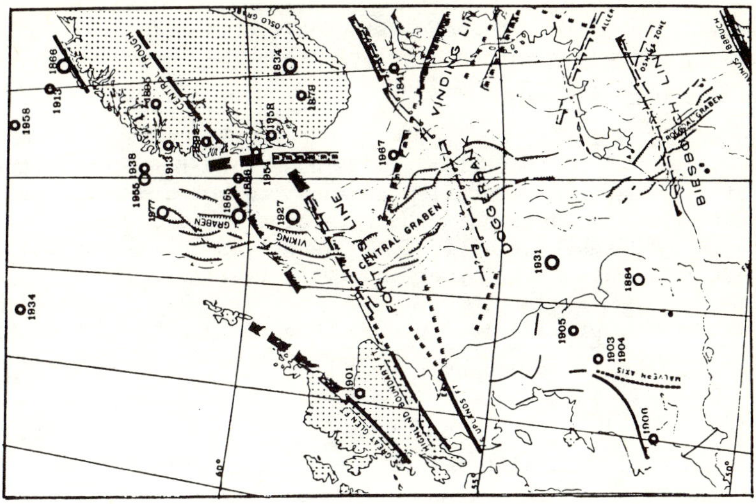

Fig. 7. Relation of large earthquakes to geological features in North Sea (ref. 13).

23. For the present study, a site is located at 3°E-53°N in the seismic zone bounded by longitudes 5°W and 15°E and latitudes 50°N-60°N.

24. Using the same data base as Husebye et al (ref. 13) a discrete magnitude-recurrence relationship was derived

$$N = 38{,}900 \exp(-1.91M) \tag{9}$$

for an area A = 1,454,000 km^2.

25. Husebye et al (ref. 13) and Selnes (ref. 14) discuss the lack of a reliable acceleration attenuation relationship for the North Sea and environs. The rate of amplitude decay, presented by Husebye et al (ref. 13) supports the use of the attenuation relation of Joyner and Boore (ref. 9), so

$$a/g = 0.095 \exp(0.573M)R^{-1} \exp(-0.00587R) \tag{10}$$

in which R in km is given by

$$R = (d^2 + 7.3^2)^{1/2} \tag{11}$$

in which d = closest distance to the surface projection of fault rupture.

PROBABILITY OF LIQUEFACTION

26. The probability of liquefaction at the chosen site was computed at a depth of 10 ft over a range of N values from N = 5 to N = 30. The saturated unit weight of the soil was taken as 125 pcf. The maximum earthquake magnitude for the zone was assumed to be M = 7.5 following Husebye et al (ref. 13).

27. The results of the PROLIQ analysis are summarised in Fig. 8 which shows the probability of liquefaction at a depth of 10 ft as a function of standard penetration values N at the site.

28. Table 1 shows the contributions to the liquefaction rate λ from each earthquake magnitude class, the critical acceleration a_{crit} necessary to cause liquefaction and the critical distance R_{crit} within which the earthquake should occur in order to cause liquefaction. The data are for the case N = 10 blows. For very low probabilities, the liquefaction rate is approximately equal to the probability of liquefaction. The liquefaction rate for the M = 7.5 class is adjusted by interpolation to reflect the fact that although the class is defined by $7.25 \leq M < 7.75$, the largest earthquake for the seismic zone is M = 7.5.

29. It is clear, based on current seismicity estimates, that the risk of seismically induced liquefaction in the British Sector of the North Sea is less than 1/700 per annum at a depth of 10 ft anywhere in the sector if the in-situ CPT bearing pressures q_c are equivalent to a standard penetration resistance of N = 10 or greater. These estimates, however, may change radically if detailed tectonic and seismological studies should result in a departure from the homogeneous seismogenic zone model of the North Sea.

30. Even a redistribution of current seismicity to more

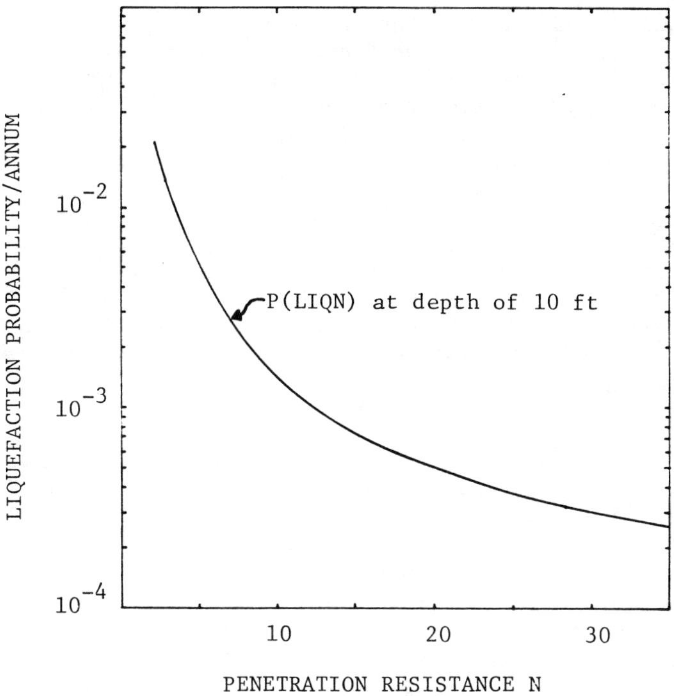

Fig. 8. Probability of liquefaction anywhere in the British Sector of the North Sea at a depth of 10 ft.

TABLE 1. Contributions to liquefaction rate by each magnitude class. N = 10

Magnitude Class	A_{crit} (g)	d_{crit} (km)	Liquefaction Rate per annum $x10^{-3}$
5.0	0.26	0.0	—
5.5	0.23	6.4	0.137
6.0	0.20	13.7	0.242
6.5	0.18	25.2	0.316
7.0	0.17	46.0	0.405
7.5	0.16	87.5	0.280

restricted zones would result in sharp increases in the probability of liquefaction at sites in these zones. Similarly, the limitation of large earthquakes to certain geological features would also radically alter the probability of liquefaction at adjacent sites.

31. If the assumption that the North Sea and environs is a homogeneous seismic zone were dropped, the probability of liquefaction in the British Sector of the North Sea would be established by computing the probabilities for a number of sites distributed over the entire sector and then drawing contours of equal probability. A set of contours would be drawn for each depth of interest.

ACKNOWLEDGEMENTS

32. This research was supported by PRAI Grant No. 67-8304 of the Natural Sciences and Engineering Research Council, Canada to the first author. The use of McGuire's (1976) algorithms for source area calculations greatly reduced the required programming effort.

REFERENCES

1. ATKINSON G.M., FINN W.D.L. and CHARLWOOD R.G. Simple computation of liquefaction probability for seismic hazard applications. Earthquake Spectra, vol. 1, no. 1, pp. 107-124, Earthquake Engineering Research Institute, Berkeley, California, 1984.
2. ATKINSON G.M., FINN W.D.L. and CHARLWOOD R.G. PROLIQ - a computer program for probability of liquefaction. Research Report, University of British Columbia, Vancouver, in press, 1984.
3. SEED H.B. and IDRISS I.M. Simplified procedure for evaluating soil liquefaction potential. Jour. of the Soil Mech. and Found. Div., ASCE, Vol. 97, SM9, pp. 1249-1273, 1971.
4. CORNELL C.A. Engineering seismic risk analysis. Bull. Seism. Soc. Am., 58, pp. 1583-1606, 1968.
5. ATKINSON G.M., FINN W.D.L. and CHARLWOOD R.G. A simplified method for calculating the probatility of seismically induced soil liquefaction. Research Report, University of British Columbia, Vancouver, in press, 1984.
6. SEED H.B. and IDRISS I.M. Ground motions and soil liquefaction during earthquakes. Engineering Monographs on Earthquake Criteria, Structural Design and Strong Motion Records, 5, Earthquake Engineering Research Institute, Berkeley, Calif., 1983.
7. ROBERTSON P.K., CAMPANELLA R.G. and WIGHTMAN A. SPT-CPT correlations. Soil Mechanics Series Report No. 62, Dept. of Civil Engineering, University of British Columbia, 1982.
8. CAMPBELL K.W. Near-source attenuation of peak horizontal acceleration. Bull. Seism. Soc. Am., 71, pp. 2039-2070, 1981.
9. JOYNER W.B. and BOORE D.M. Peak horizontal acceleration and velcoity from strong motion records including records from the 1979 Imperial Valley, California earthquake. Bull. Seism. Soc. Am., 71, pp. 2011-2038, 1981.
10. YEGIAN M.K. and WHITMAN R.V. Risk analysis for ground failure by liquefaction. Jour. of the Geotech. Eng. Div., ASCE, Vol. 104, GT7, pp. 921-938, 1978.

11. BASHAM P.W., WEICHERT D.H., ANGLIN F.M. and BERRY M.J. New probabilistic strong seismic ground motion maps of Canada - a compilation of earthquake source zones, methods and results. Earth Physics Branch, Open-file 82-33, 1982.
12. FINN W.D. Liam. Probability of liquefaction at the Venture site. Report to Nolan-Ertec Ltd., Halifax, Nova Scotia, 1983.
13. HYSEBYE E.S., RINGDAL F., FYEN J. and SANDVIN. Earthquake risk for Norwegian offshore installations. Norsar, Kjeller, Norway, Contribution 263, 107 pp., 1979.
14. SELNES P.B. Earthquake risk on the Norwegian Continental Shelf. Publication No. 134, Norwegian Geotechnical Institute, Oslo, 7 pp., 1981.

OPEN DISCUSSION

Presentations and discussion

DR R. S. STEEDMAN, Cambridge University

The papers by Smith, Booth _et al._ and Taylor _et al._ all describe the application of numerical analyses to dynamic soil structure problems. Without validation by comparison with centrifuge model data, however, the solutions given by such numerical analyses can only be of limited value.

In a centrifuge model under high acceleration homologous points in model and prototype experience identical stress levels. Under subsequent dynamic shaking the model deforms as a prototype, exhibiting pore pressure generation, loss of bearing strength etc. Detailed instrumentation monitors acceleration and pore pressure response leading to a more fundamental understanding of damage mechanisms.

Liquefaction is a phenomenon associated with soil at low effective stress levels under high hydraulic gradients and is characterized by features such as piping, boiling and fracture.

Such events, which may severely and permanently damage an associated structure, cannot yet be modelled by numerical methods alone since continuum mechanics ceases to be applicable.

A dynamic facility has been in operation at the Cambridge Geotechnical Centrifuge for over 3 years. Tests have studied the earthquake behaviour of embankments, retaining walls, offshore islands, coastal dykes and their foundations, and embedded structures. Outside the UK centrifuge data is increasingly being accepted as the data for validation of numerical analyses. In Europe four large centrifuges are being designed and built. In the USA two large centrifuges will shortly be commissioned. In Japan six centrifuges are currently in operation and will shortly have a dynamic capability.

The British Earthquake Engineering community must respond or they will lose the edge they currently possess.

MR J. A. CROWDER, Civil Engineering Department, Imperial College, London

I hope to show why an earthquake strong motion data base is relevant to the study of earthquake risk and how it can help designers in the problem of seismic qualification for hazardous facilities such as dams and nuclear plant.

The major problem we suffer from in the UK is the lack of any locally generated strong motion records. Mr Alderson overcomes this by saying that they would not have been recorded at the site in question, and even if they had they could not be expected to repeat themselves. He proposes the use of synthetic and modified signals instead. However, the importance of the local geology and site conditions and the variations that they produce are emphasized.

Mr Irving explained the CEGB approach of using a peak acceleration of 0.25 g from the BGS seismicity data. This led, in 1979, to the selection of the Parkfield 5 and Temblor records of the 1966 Parkfield earthquake, in the absence of any other more suitable records at the time. Both these records were scaled to match the desired criteria of 0.25 g and 26 cm/s peak values. Because soil behaves in a non-linear manner, and because soil conditions will not be the same in Britain, this can only be considered a step towards the use of real data.

Another approach in establishing the safety of small, essential plant under seismic loading is to use a shake table and examine the response spectrum of the plant. The required response spectrum is derived by the designer and provided to the shake table staff, who then select a set of frequencies to define the response spectrum and from them derive a synthetic time history (with discrete frequencies and no consideration of phase differences). This means that there are an infinite number of synthetic time histories which can be used to test for each desired response spectrum.

Although discrete frequencies can be acceptable, and for a synthetic signal are virtually unavoidable, the phase spectrum of the time history is important.Consider a single time history shaking a structure with an elasto-plastic response. If the time history is used in the normal manner, the final deformation of the structure can be calculated. If, however, the history is reversed, then a different final deformation will be observed. The change required to do this is simply a pulse sequence pattern modification which is closely related to the phase, since the frequency content and peak values remain the same.

At Imperial College, we feel that a better approach to the problem of assessment of seismic safety is to use real time histories since they contain both amplitude and phase information, although research is being carried out into simulation of time histories involving phase spectra. For this reason the Earthquake Strong Motion data base was set up. We currently have about 600 3-component records, collected from around the world, which are being processed and catalogued. Although critics will say that a selected time history will not, and in fact cannot occur at a specified location some time in the future, we can at least point to the fact that it has already been recorded in similar conditions and so is more representative than either the response spectrum approach or the use of modified time histories selected on peak criteria only. It is also possible to use time histories both in modelling studies and on a shake table such as the one at Imperial College.

The time histories are catalogued under about 40 different

parameters. These range from simple identification of the record such as earthquake name, continent, date and station name, through a description of the earthquake (focal depth, magnitude, stress drop, fault area and type) to a description of the actual record such as local soil conditions, location of instrument, intensity at station, frequency content and peak values observed on the record. The distance from the instrument to the fault is also recorded.

Thus the catalogue may be searched for any predetermined parameters and a record selected to use in design or research.

MR J. M. H. MENU

One of the conclusions reached during the work carried out at Imperial College on interpretation of macroseismic maps of small and moderate earthquakes was the determination of realistic focal depths.

For such shocks, a point source model for macroseismic analysis is certainly reasonable. The three following equations are the starting point of the model:

$$I = C + b \log (E) \quad (1a)$$

$$E = E_o R^{-n} e^{-aR} \quad (1b)$$

$$M_S = \alpha + \beta \log (E_o) \quad (1c)$$

where $R = (D^2 + H^2)^{1/2}$ (1d)

D is the epicentral distance, H is focal depth (actually depth of some energy centre but both are similar for small shocks), M_S is surface wave magnitude, E_o is seismic energy at the point source, E is seismic energy density at point of observation and I is seismic intensity at the point of observation; α and β are the constant coefficients of the Gutemberg energy relationship; n is the coefficient of geometric spreading and a is average near-field absorption coefficient.

They express respectively the proportionality of intensity with seismic energy density at a given site (1a), the decay of this energy density with distance as the sum of geometric and dispersive term and an inelastic attenuation term (1b) and the Gutemberg formula relating magnitude and seismic energy released at the source.

From these equations, the following macroseismic field equations can be deduced:

$$I = I_o - bn [\log (R/H) + aH\log(e)/n (R/H - 1)] \quad (2a)$$

$$M_S = 1/p [Io - q + bn (\log (H) + aH\log(e)/n)] \quad (2b)$$

where I_o is epicentral intensity; and p and q can be expressed in terms of α, β, b and C.

Equation (2a) is seen to be a generalization of Kovesligethy's formulae where a variable bn replaces the assumption bn = 3, or of Blake's formulae (bn = 3, a = 0). It can be used to determine I_0, bn, aH/n and H as soon as at least 4 isoseismals are available and their average radii computed. A non-linear least square fitting on $I-I_0$ based on its Taylor's expansion was used with careful choice of initial estimates to ensure convergence.

When M_S is also known, equation (2b) allows to reduce the minimum number of isoseismals needed, although it was preferred not to use this possibility.

The best estimates for bn varied typically from 1.5 to 4.5 in accordance with the average value of 3 chosen by Kovesligethy. Values of aH/n ranged from 0.002 to 0.2 showing that strong inelastic attenuation very near the source can be detected. This was in particular possible for very shallow earthquakes and explosions. In paper 4 (Ambraseys and Jackson) expressions for b, n, and a are given.

These parameters were found to be depth dependent, showing the major effect of H on the various factors affecting intensities.

Another interesting aspect revealed by the study was that the obtained focal depths varied, sometimes considerably, from values obtained by different methods (teleseismic). Table 1 shows some of the extreme cases.

Table 1. Discrepancies in focal depth determinations

	Teleseismic	Macroseismic
France : 17 Feb 1927	45 Km (Damman 1927)	20 Km
Denmark: 31 Oct 1980	65 Km (Lehmann 1956)	15 Km
Belgium: 11 June 1938	50 Km (Rothe 1941)	12 Km
Germany: 11 July 1949	37 Km (Berg 1950)	5 Km
France : 2 Jan 1959	40 Km (Rothe 1967)	14 Km

Focal depths obtained from macroseismic map analysis are well within the earth's crust, (which is not always the case for teleseismic determination), and hence are more likely to be exact. It must be said, however, that the teleseismic estimates are from the early days of seismic measurements and depth determination. Despite the roughness of macroseismic data and the limitation of the model, encouraging results are worth noting.

Discussion

Dr Skipp put a question to Mr Menu on his presentation in connection with comparisons of focal depths derived from macroseismic and teleseismic data, cautioning that many reported focal depths based on early European teleseismic data are of dubious accuracy. Dr Browitt was not surprised that comparisons were not good and commented that more objective data on focal depth is required. Dr Aspinall considered there is evidence to support the teleseismic data and advised engineers to leave such questions to seismologists.

Dr Kunar supported the plea made in Dr Steedman's presentation for the use of available experimental facilities to validate non linear soil/structure interaction analysis. However, he doubted whether nuclear facilities, in which a certain amount of non linearity is included in the soil structure interaction analysis, would be built on saturated sands, where such effects are significant, and considered this contributes to the lack of interest by the nuclear industry in the facilities and methods described by Professor Finn and Dr Steedman. Mr E.D. Booth noted that in the offshore industry, there is no choice of site and Professor Finn, while agreeing with Dr Kunar that such facilities should only be used where required, noted that in the offshore industry non linearities can be very pronounced and can cause problems. Dr Steedman considered that an understanding of failure mechanisms is an essential element of any design process and this is only possible by comparison with physical models. Mr Mallard queried the boundary conditions available in the centrifuge and Dr Steedman confirmed that the model is contained in a rigid box with a rigid base. However, travelling wave effects can be modelled.

Dr Skipp foresaw difficulties in incorporating physical modelling into the design process of a typical project and considered that only well funded, long term projects could contemplate the time scales involved in physical testing. Physical testing is not consistent with commercial time scales.

Mr Parker noted that both analytical and physical models depend upon data on soil properties, and that when using analytical models sensitivity studies could be carried out to overcome uncertainties in material properties. Perhaps physical testing should be used to validate particular types of analysis, rather than specific analysis. Mr Parker did not see how the facility at Cambridge could help in his work.

Dr Steedman emphasized the importance of physical modelling by reference to the decoupling shown by Professor Finn's model. Dr Kunar noted that Mr C. Smith's paper predicted decoupling by analytical methods and Professor Finn completed the discussion on physical testing by reiterating the need to predict behaviour qualitatively.

Dr Skipp questioned Professor Finn on the effect of slope on his results. Professor Finn confirmed his results are based on level ground and commented that reasonable slopes would tend to retard pored water pressure.

Panel discussion

Panel D. E. KEY (Chairman), C. W. A. BROWITT, J. IRVING, R. R. KUNAR, D. J. MALLARD, D. PAPASTAMATIOU and B. O. SKIPP
Reporter J. H. MILLS

British seismicity

The Chairman began the discussion by introducing the question of the level of British seismicity. Dr Woo and Dr Muir Wood responded to comments by Professor Finn that recent work suggested that a new siesmic model is required for the North Sea, by drawing attention to their work carried out for NORSAR, which led to NORSAR abandoning their catalogue of 1978 in favour of the catalogue recently produced by Principia Mechanica Ltd. Professor Finn asked whether the area is still considered homogeneous and Dr Muir Wood replied that, following work by Ambraseys and Melville, the seismic hazard in the Central Graben is considered higher than elsewhere. Professor Finn asked if a report was available showing sub-zones of the polygonal type used in Canadian specifications. Dr Muir Wood considered such zoning to have no geological or tectonic basis, preferring site specific studies based on source data.

Dr Skipp continued the discussion on the level of British seismicity by reference to comments in previous sessions by commenting that Professor Ambraseys' suggested upper limit of magnitude of 5.5 would cause no problems to engineered structures. Dr Skipp wondered if overmuch has been made of seismic activity in Britain. Mr Irving considered Professor Ambraseys' remarks significant, but noted that an upper limit of 5.5 on magnitude does not solve all the engineer's seismic problems, as the major contribution to seismic risk is from magnitude 4.5 to 5.5 earthquakes close to the site, and such events could still be significant for certain sites. A part of the problem is the assumption that a large fault can generate a large magnitude earthquake, a problem compounded in Dr Skipp's opinion by the sometimes unwarranted association of earthquakes with the largest faults known locally. Mr Mallard suggested there is some evidence in the British historical record of large magnitude events, in particular the St. Andrews event in 811. Dr Aspinall questioned the appropriateness of Professor Ambraseys' remarks for hazardous installations. Drs Skipp and Papastamatiou suggested broadening the scope of the discussion to include historical and geological evidence from the whole of NW Europe. In a written reply to the remarks by Drs Aspinall, Skipp and Papastamatiou, Professor Ambraseys and Dr Jackson said that the authors state quite clearly what they think in their paper and that it had not been read carefully.

Dr Muir Wood was not certain how the figures of 5.5 had been derived. He knew of three earthquakes in Norway greater than magnitude 6, and considered it unwise to assume an upper limit on magnitude, quoting inter alia, geological evidence of a magnitude 8 event in North Sweden approximately 9000 years BP and events on the coastal margin of Norway of magnitude 7. These should be compared with a maximum magnitude of 6.5 for the whole of Europe north of the Alps based on examination of historical evidence over a period of 700 years. In response to a question from Dr Browitt, Dr Muir Wood explained the evidence to support the conclusion that the event in North Sweden was a single event of magnitude 8 and was not several magnitude 6 events.

Dr Kunar considered the conference and discussion reflected the ups and downs of the last two years; yesterday, according to Professor Ambraseys, large earthquakes do not occur in Britain; today it is considered unwise to assume an upper limit on magnitude, with magnitudes of 7 and 8 being quoted! This frustrates engineers who have to design structures for the British earthquake. Dr Kunar understood that the response spectra derived for design in the UK are based on an upper limit on magnitude of 6. Mr Mallard stated that the design spectra are based upon earthquakes which contribute most to the risk. Dr Woo explained the basis of the spectra and in reply to Dr Kunar's assertion that a deterministic upper limit on magnitude had been used, considered that the design spectra would not change significantly if larger earthquakes were properly weighted into the study.

Dr Papastamatiou pointed out that the magnitude 7 and 8 events were specific to a site and must be related to particular geological structures, but Dr Muir Wood stated that the events to which he was referring are thought to be caused by post glacial rebound. As such they are not typical of the present seismic environment in the area, North Sweden today being an area of low seismicity. Mr E.D. Booth reminded the conference that Professor Ambraseys' comments were specific to the UK and asked Mr Irving if magnitude 5.5 events contribute most to the seismic risk. Mr Irving said no, and referred to studies carried out in connection with Sizewell 'B' concerning the contribution to risk of different magnitude earthquakes. The major contribution to risk is from magnitude 4Ms to 5 to 5.5 M_s events. Mr E.D. Booth noted that attenuation formulae tend to give high zero period accelerations for small events and asked if acceleration is an important measure of possible damage; in particular could a magnitude 5.5 event cause damage to the proposed structures at Sizewell? Mr Irving agreed with the comments on attenuation formulae and sero period acceleration, and confirmed this is the reason the CEGB are interested in damage factors as a measure of the damage potential of earthquakes rather than peak acceleration. Mr Irving would not consider a magnitude 5.5 earthquake to be worthy of consideration unless it occurred within 10 km of the site.

Dr Musson agreed with Dr Aspinall's earlier criticism of Professor Ambraseys' comments, believing Professor Ambraseys should have included the qualifications that 'there is no evidence' in his remarks. Professor Price sought the CEGB's views on Professor Ambraseys' comment that the largest British event is trivial, and Dr Steedman raised the question of cost. In reply to the latter, Mr Alderson considered that the nuclear industry could not ignore the risk of a release from a nuclear power station due to the seismic hazard. Professor Heidebrecht supported the need to establish the damage potential of small, near-field earthquakes, quoting a magnitude 5 event near New Brunswick which had a peak acceleration of approximately 0.3 g. However, the response spectra of the motion contained only a single spike at about 23 Hz and did not damage nearby log cabins or dislodge crockery. The peak acceleration should be discarded as a measure of damage. This contribution completed the discussion of the level of British seismicity, Professor Ambraseys' comments the day before seemingly being the main item of discussion.

Further discussion of papers

With reference to paper 19 Mr E.D. Booth asked Mr Irving to comment on the fact that the structural fragilities shown were that higher than pipework and equipment fragilities. Mr Irving confirmed that equipment fragilities, which were based on data from Diablo Canyon Study in 1978-79, had a higher scatter and lower mean than structural fragilites.

With respect to the Paper by Dr Kunar, Professor Heidebrecht asked whether a system analysis had been carried out to establish the relationships between qualified and non qualified plant. Dr Kunar replied that for existing plants such relationships had been established in an ad-hoc manner on a walk-through of the plant; no systematic review had been made. Mr C. Smith confirmed that a systematic approach to this problem had been adopted for Sizewell 'B' and Mr Thomson commented on qualification of plant for Heysham 2 and Torness. In connection with the latter Mr Thomson questioned the zero period acceleration of floor response spectra used in the qualification and was assured by Mr Parker, who had derived the spectra, that the accelerations are realistic in terms of free-field ground accelerations and a true reflection of the response of the structure and the site. However, Dr Aspinall stated that as the floor response spectrum is the peak response, such spectra contain no information on phasing or duration and as such should be classed as a device of the devil! Dr Kunar rose in defence of response spectrum methods for linear calculations.

General discussion

Mr C. Smith questioned Dr Muir Wood on the value of mathematical analysis in the seismotectonic field and asked how, in the absence of such analysis, can the geologist's judgement be validated. Dr Muir Wood agreed mathematical analysis is possible and may help, but more data is required. Dr Aspinall commented on the funding required to carry out such analysis and considered the geologist's judgement should be treated in the same way as engineering judgement. Dr Skipp stated that all geological interpretation is not judgemental but considered that there has been a tendency to rush into mathematical analysis. There is no substitute for the accumulation and critical appraisal of data. Dr Papastamatiou considered Mr C. Smith's suggestion to be positive but the difference in accuracy between structural models and ground models must be appreciated. Professor Price concluded this discussion with the comment that the structural engineer has the advantage of knowing what his structure looks like. The geologist does not have thi detail available.

Dr Kunar drew attention to validation of analytical techniques by comparisons with the full scale tests and structures which have been subjected to earthquakes, noting that databases of the performance of plant and structures during earthquakes are being constructed in the USA and elsewhere. Professor Finn advised that dynamic tests on an AGW facility are being planned and Mr C. Smith drew attention to low level testing of this facility reported in SMIRT 1981. The value of the latter is reduced to lack of soils data.

Dr Muir Wood made the comment that the scientific study of seismicity began in this country, and that there is a body of expertise in this country which can match anything available in the USA or elsewhere. Engineers should be encouraged to take advantage of this expertise.